EUROPEAN COATINGS LIBRARY

Gerhard Jonschker

Sol-Gel-Technology in Praxis

Gerhard Jonschker: Sol-Gel-Technology in Praxis

ISBN: 978-3-86630-886-2

Preface

This is not a conventional textbook about colloidal chemistry or the sol-gel process. Many excellent textbooks already have been written about these subjects [1, 2, 5, 23, 60, 137]. This book aims to be a guide for the practical worker and uses an unusual approach. It focuses on practical examples mainly from the patent literature, accompanied by interpretations and explanations. The theoretical basis for these explanations is given in a preface to each example at a minimum level.

When I was a beginner in sol-gel chemistry, now almost 22 years ago, I often felt kind of lost because in publications, textbooks, or the patent literature the authors did not reveal the rationale behind their decisions to use exactly this solvent, that catalyst etc.

I always wished for a book for the praxis-orientated beginner which explains all these questions. Now, since I never came across such a book, you hold in your hands my answer to this need.

The interpretations and explanations stem from my own lab experience and are not official statements of the inventors or applicants. The explanations and examples cannot and will not be complete and are not meant to be so. They focus on some practical aspects and try to explain the theoretical background with the help of the examples. Where necessary, all examples have been reworked and adapted for better readability.

Such a book always is a balancing act. A practical guide should be a pragmatic shortcut to a basic understanding and getting quick results. The necessary simplifications should be tolerated by the more experienced reader. Wherever it was possible and appropriate, further literature is recommended for a deeper understanding of the subject.

I hope you can feel the joy and excitement I felt during my lab work and during the writing of this book. Even more so, I hope that you feel inspired and can enjoy your own experiences with sol-gel- and nanotechnology.

Your feedback and comments are highly welcome and can be directed to the mail address sol-gel@arcor.de.

Gerhard Jonschker

Heppenheim, Germany, January 2014

Gerhard Jonschker: Sol-Gel-Technology in Praxis

ISBN: 978-3-86630-886-2

Important: Intellectual property rights:

For sure the special feature of this book are the many examples and their interpretation. They are meant to teach the theoretical background and should help to transfer knowledge quickly to practical applications.

The sources of these examples in most cases are patents. The author and publisher would like to emphasize, that by printing the examples and by explaining several aspects of them by no way the rights of the patent owners are restricted. Therefore, since some of these patents are still in place and binding, the buyer or reader of the book has no right to use these patents without permission of the owners.

Please respect the intellectual property and contact prior to a commercialization of your product the owner of patents you could possibly infringe.

The author and publisher do not accept any liability which might result from not respecting the legal frameworks.

EC
European Coatings
LIVE
www.european-coatings.com/live
Take two!
Enjoy more information
EC
live
EC
rnal
Enjoy lead articles in a new way. Live on the Internet.
European Coatings LIVE – the new editorial online-supplement to the JOURNAL! Each month, the author of the lead article in the current issue will expand on it in more depth and provide a great deal more context.
Live! European Coatings JOURNAL + LIVE = twice the enjoyment.
Registration is free – Listen in and air your views!
EC European Coatings JOURNAL
More Information at:
www.european-coatings.com/live
VINCENTZ

EC European Coatings
Tech Shelf
Grab the App!
Your mobile source for coatings technology information
Here's how!
Scan the code below with your iPad or iPhone, or simply enter the URL address into your browser. (For subscribers: please enter your subscription number in the 'settings' menu of the app.)
www.european-coatings.com/techshelf
Do you prefer reading a digital edition of European Coatings Journal? Or are you searching for eBooks on coatings technology? Now you can have both – thanks to the European Coatings Tech Shelf!
Keep up with all the expert knowledge you need – anytime, anywhere, on your mobile device
» Already a subscriber to European Coatings Journal? Then enjoy free access to all current issues of the eJournal as well as to the editorial archives.
» See the complete range of European Coatings eBooks at a glance – and browse inside to find exactly the information you need.
Sponsored by
BYK
Additives & Instruments
Vincentz Network · Tel. +49 511 9910-025 · info@european-coatings.com · www.european-coatings.com
VINCENTZ

Contents

Gerhard Jonschker: Sol-Gel-Technology in Praxis

ISBN: 978-3-86630-886-2

1 What is sol-gel and nanotechnology?

1.1 Definitions

According to *W. Ostwald*, colloids are a field which is only for practical reasons separated within the larger field of dispersed systems [7].

Definitions are important because they allow that a topic can be discussed without misunderstandings. When talking about sol-gel and nanotechnology this seems to be especially important, because nanotechnology is no scientific field on its own, but rather a conglomerate of almost all scientific disciplines, ranging from chemistry and physics over materials science to e.g. medicine.

The BMBF (the German Ministry of education and research) defines the term nanotechnology as follows: *"Nanotechnology describes the production, analysis and application of structures, molecular materials, inner interfaces and surfaces exhibiting at least one critical dimension or production tolerance smaller than 100 nm. The nanoscale nature of the components of the system alone is responsible for the new functions and properties to improve existing or develop new products"* [3].

The second part of this definition is very important, because it separates nanotechnology from many products which contain nanoparticles by nature or by accident, or those products where the nanostructures are not relevant for the value at use. Nanotechnology thus incorporates the knowledge about structure-property relationships and the deliberate production of particles or structure in the nanoscale.

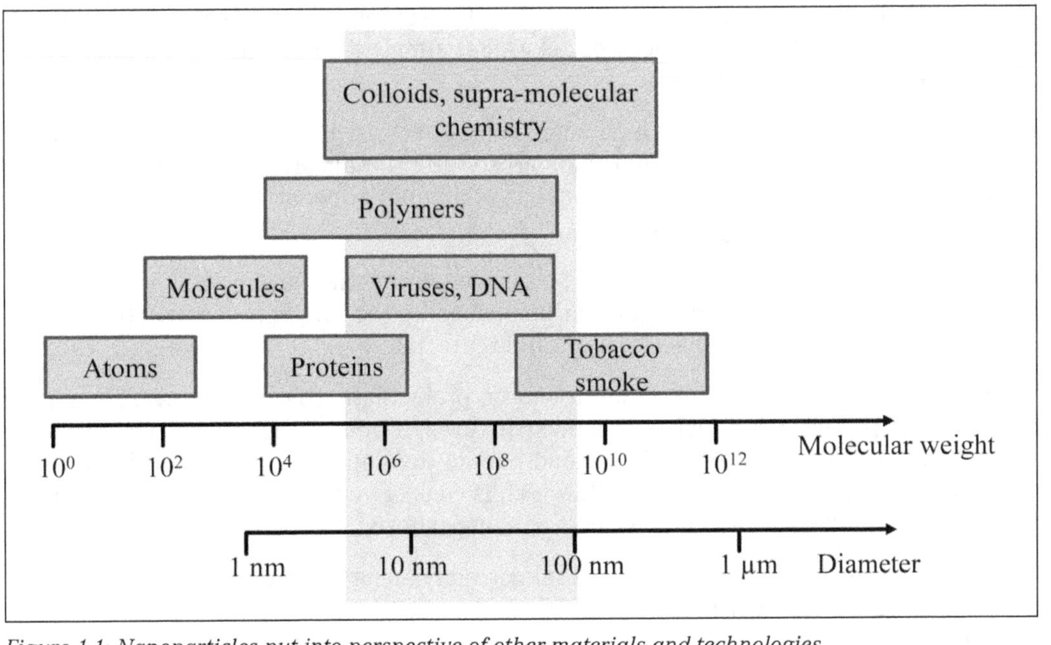

Figure 1.1: Nanoparticles put into perspective of other materials and technologies

Gerhard Jonschker: Sol-Gel-Technology in Praxis

ISBN: 978-3-86630-886-2

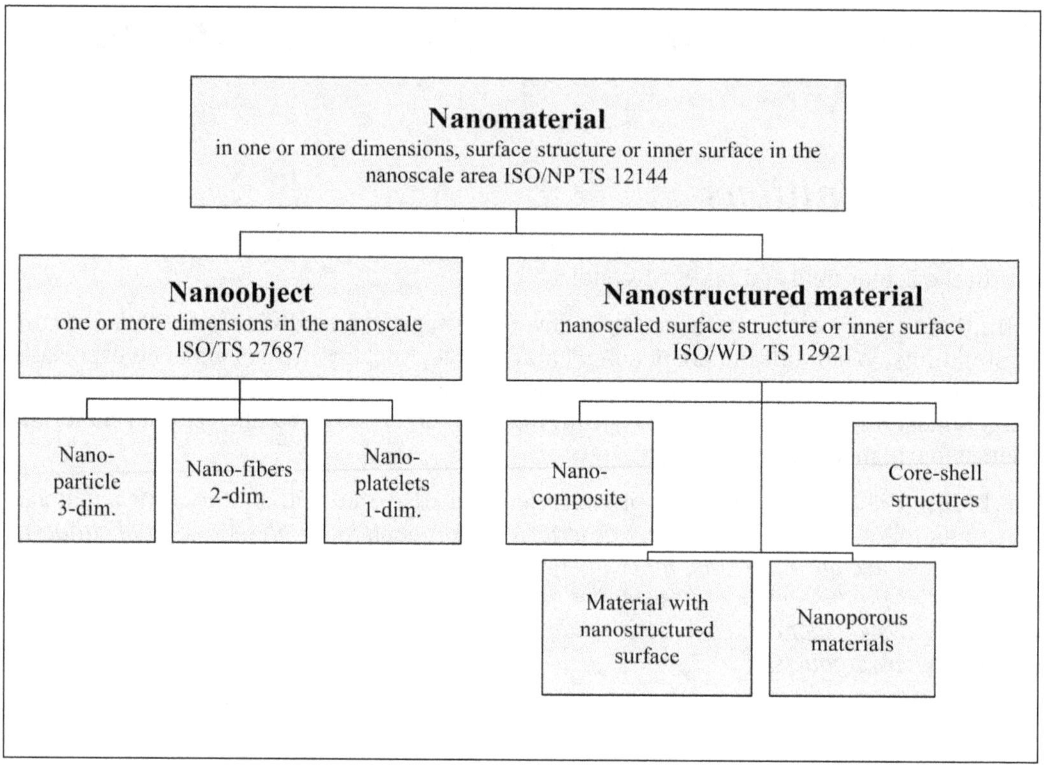

Figure 1.2: Important definitions of nanomaterials

Now it becomes clear, that nanotechnology cannot be a defined market or a precisely defined industry. Nanotechnology rather is an artificially designed scientific discipline, which covers different materials and processes, as well as simply a description of a dimension in material- and process development.

The sol-gel process is one aspect of nanotechnology. It describes the preparation of nano-scaled objects like e.g. particles and their processing to nanostructured materials. One process step thereby includes the solidification of a liquid nanoparticle dispersion (sol) by interparticulate forces to a gel.

Figure 1.1 can help to put nanoparticles into perspective of other materials or technologies. The gray shading illustrates the relevant dimensions for the sol-gel process and nanotechnology. Some further definitions are shown in Figure 1.2.

With regard to a consistent definition, it can be problematic that almost every material contains nanoparticles or consists of nanostructures, if you just look close enough with suitable microscopes. Therefore, the second part of the nanotechnology definition refers to the special, deliberate property profile which is caused by the nanostructures (compare "Health, safety and environmental aspects of nanoparticles", Chapter 7).

In order to be able to talk effectively about sol-gel- and nanotechnology, some further definitions are necessary. The german norms "ISO" help us out with useful descriptions. According to ISO TC 24/SC 4, TC 146, TC 209:

Particle

A very small piece of material with defined physical limits.

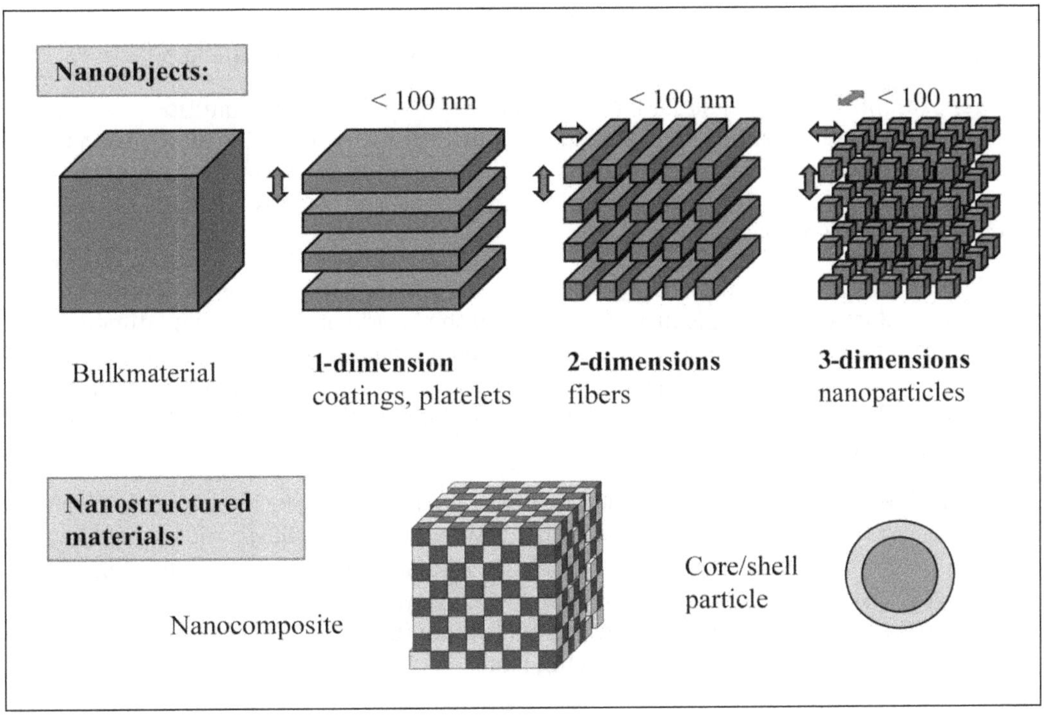

Figure 1.3: Illustration of different forms of nanomaterials

Aggregate

A collection of particles with strong interparticulate bonds (covalent or even sintered necks), whose resulting surface area is significantly smaller than the calculated sum of the surface areas of its components.

Agglomerate

A collection of weekly bonded particles or aggregates, whose surface area is comparable to the sum of the calculated surface areas of its components.

The term "monodisperse", in contrary to "polydisperse" describes a group of particles which are homogeneous with regard to their size and shape. Officially a special limit does not exist, but ±10 % variation seems to be a suitable value. Monodispersity in most cases is a desired property of the dispersion, because it allows a more precise control of the properties of the material which is to be synthesized. This is important, especially for the preparation of transparent nanocomposites or coatings to avoid haziness caused by larger particles or agglomerates.

1.2 Historical and actual facts

"It's a pity that even now, colloidal chemistry has become an almost unmanageable field of science" (W. Ostwald 1927).

Nano- and sol-gel technology are not new and since centuries are used to prepare materials with extraordinary properties. Gold-ruby glass for example is one of the oldest nanomaterials. Described already by the sumerians in 700 BC, German chemist Kunkel rediscovered this ancient technology [4]. Nano gold particles are responsible for the beautiful red color.

Their agglomeration is prevented by addition of SnO_2 to the liquid glass melt. In 1900 Zsigmondy described monodisperse gold particles as the basis for this technology [5].

Another example are carbon nanotubes, which can be found e.g. in the multiply folded steel blades of Japanese master forgers. In this case nanotechnology was used without knowing the structure-property relationships. Today's discussion tends to use the term nanotechnology only in these cases, where knowingly and deliberately nanostructures are used to develop materials with new properties.

In 1947 for the first time Taniguchi proposed the term nanotechnology for these applications [6]. Decades before him, Ostwald already described the "world of the neglected dimensions" in a lecture series around the year 1910. He focused on materials ranging in size from 1 to 100 nm known as colloids, materials, which are not held back by conventional filters. Still today, his book is a fascinating source of inspiration, describing some early commercial applications of nanotechnology [7].

A decisive breakthrough for today's popularity of chemical nanotechnology however was besides others the work of Schmidt et al who dealt with the synthesis and applications of organically modified inorganic materials [8, 87]. These materials, which are often referred to as "ormocers", "ormosils", or "ceramers", open up a new world full of possibilities for materials science by combining organic polymers with inorganic building blocks. Some of them are described in more detail in the following pages.

1.2.1 Nanotechnology as part of our daily life

We cannot avoid coming into contact with nanotechnology, because nanotechnology is a universal building principle of nature, even more so, it is a basis for our existence. It makes sense to look at the deliberate and coincidental points of contact with nanotechnology in our lives to better understand the breadth and the nature of nanotechnology.

A typical day with nanotechnology

We start the day with a refreshing shower in the nanocoated easy-to-clean shower cabin, put on our silver nanoparticle impregnated socks and impregnated stain-proof tie. On our way to work, we drive with fuel-saver tires, reinforced with nanoscaled carbon black and silica.

During the drive with our new car with a scratch-proof clear coat due to silica nanoparticles we produce exhaust gases loaded with nanoscaled carbon black, even though they had been purified by our noble metal nanoparticles containing catalyst. At work, we use the newest CPU made by "26 nm technology" and print our documents with laser printers using nanostructured toner particles.

After work we relax with a tennis match, of course with the carbon nanotubes reinforced new racket. Finally at home at candlelight we read the novel "prey" from Michael Crichton [9] and shiver at the thought of wild swarms of nanoorganisms turning the world into a gray goo. During reading, we inhale the nanoparticle soot produced by the candle flame. While brushing our teeth at the evening of this busy day with a toothpaste containing nano-apatite which remineralizes our teeth, we look in the mirror and see the most perfect nanomachine of all - ourselves, because most processes in our body run on the nanoscale [10].

The different reports and estimations of the current and future market potential of nanotechnology were collected in fall 2011 by the German BMBF (Ministry of education and research). They outbid each other in euphoria and differed by billions of euro - why [11-13]?

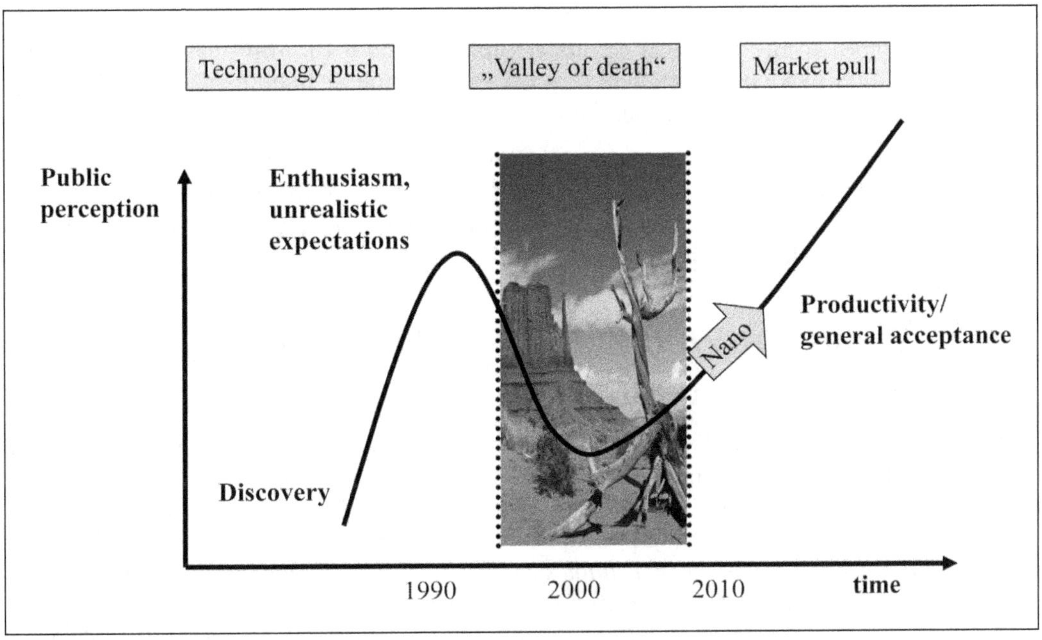

Figure 1.4: Changes of "breakthrough"-technologies in the public perception over time
Source (picture): Sumba, www.piqs.de

Nanotechnology is, as already discussed before, no market of its own, therefore it is difficult to draw a line. Today's computer processors are without doubt nanotechnology, proudly Intel advertises its 22 nm architecture.

What turnover can now be accredited to nanotechnology? That of the processor, the motherboard, or that of the whole computer? How this question is answered will significantly affect the market potential of nanotechnology which is published.

Of the whole computer, only a tiny fraction is nanotechnology, however a decisive one. This opens up the question how to generate turnover and profit with nanotechnology. This is occasionly called the "nano-trap" [14].

A practical example can be an easy-to-clean coating of shower cabins. Calculated from the typical raw materials, a 100 nm thick coating, applied at 0.1 g/m^2 causes € 2 - € 5 material costs when an area of 10 m^2 is coated. An end customer however pays a premium of up to 250 € per shower cabin, thereby generating a value of roughly € 25,000 per kilogram nano-coating which is kept by the producer or seller of the shower cabin.

For a material producer it should be difficult to get a significant piece of that cake if the technology and the application are not thoroughly protected by patents or other hurdles. At the end, the major part of the generated value stays with the producer or seller, in spite of the fact that the R&D of that nanotechnology coating might have been a risky venture, burning a lot of money. These and further problems in realizing the full monetary potential of nanotechnology innovations led to the discrepancy between the high expectations and the so far realized commercial success.

Meanwhile, a more realistic approach to the market potential of nanotechnology is favored. Not nanotechnology by itself, but rather the achievable and affordable customer advantage is in the focus.

This situation is nothing specific. Not only nanotechnology, but almost every new technology breakthrough in the beginning causes euphoria about the limitless opportunities. With time, market participants realize that not the technology itself can be sold, but only economically sound problem solutions for a paying customer.

A consequence is, that the high expectations cannot be fulfilled and disappointment arouses. Sooner or later the new technology gets absorbed and integrated into current product development cycles and a market pull for the new problem solutions develops. It is this stage of development, nanotechnology is in today.

1.3 Nanotechnology in the development of paints and coatings

Nanotechnology has already been used in the development of paints and coatings since a long time. Many pigments exhibit dimensions in the nanoscale and water-based coating resins consist of nanosized polymer droplets. So it is not surprising that the German organization of the coatings industry (VdL) describes the future perspective of nanotechnology like this: *"Nanotechnology is a key technology of the future and gains more and more importance also in the fields of paints and coatings. The improvement of conventional coatings and the realization of new functions with the help of nanotechnology will increase in the following years"* [353].

Before nanomaterials can be used in the development of coatings, they have to be synthesized first. The problems and possibilities of the manifold synthesis processes are discussed in the next Chapter 2.

2 Synthesis of nanomaterials

"The colloidal state is a universal state of matter" Wolfgang Ostwald wrote in 1922. Nanotechnology describes only a dimension, therefore in principle almost all materials can be brought into a nanoscaled state. During the synthesis of nanomaterials, three critical success factors are important and shown in Figure 2.1.

The synthesis of nanomaterials with defined specifications in many cases is difficult, because structure-property relationships are not known and must be determined in large-scale series experiments. After their synthesis, agglomeration of nanoparticles has to be prevented in order to make use of their specific properties. The third important point is the adaption of nanomaterials to their environment. Nanoparticles usually are not synthesized directly in the medium they are used in at a later stage of the process and the transfer requires a suitable surface modification.

In the following chapters, step by step the critical success factors for the synthesis of nanoparticle dispersions are discussed, starting with the synthesis. Two basic routes are known to synthesize nanomaterials. Either something big is broken down into smaller pieces or something very small is grown in a controlled way to yield nanoparticles. These approaches are known as top-down and bottom-up synthesis pathways.

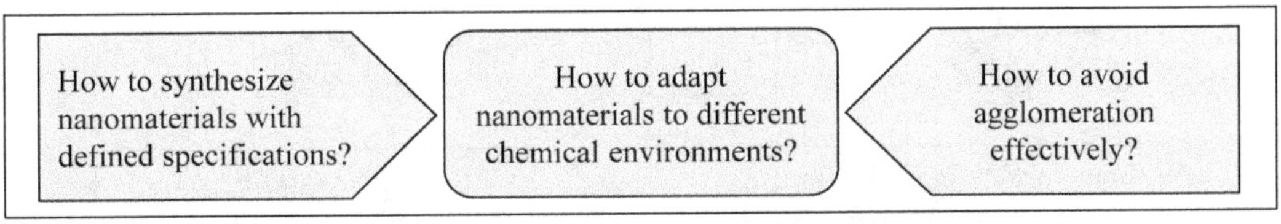

Figure 2.1: Critical parameters of success during the synthesis of nanomaterials [15]

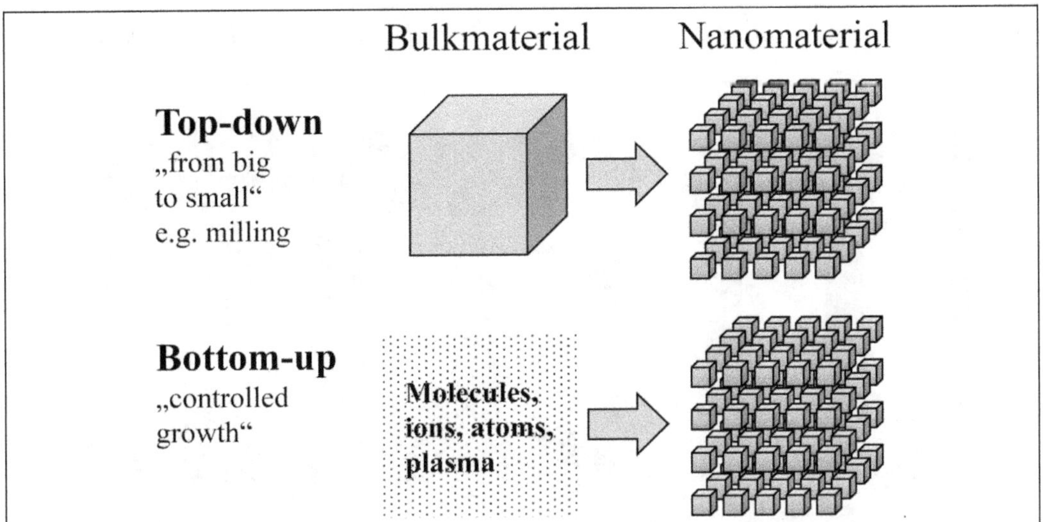

Figure 2.2: Top-down und bottom-up approach for the synthesis of nanomaterials

Gerhard Jonschker: Sol-Gel-Technology in Praxis

ISBN: 978-3-86630-886-2

2.1 Top-down processes

During a top-down synthesis, bigger structures are broken down to smaller structures by energy-intensive processes like milling. In order to protect the particles from agglomeration, the generated new surfaces have to be coated with significant amounts of surface modifiers. For 20 nm sized particles, up to 15 weight% are necessary. Smaller particles due to their larger surface can need even more.

During milling, more and more energy is put into the system in the form of freshly generated surfaces. At some point in time, the tendency of the system to reduce its energy content by agglomeration becomes so dominant, that longer milling does not lead to a further decrease of particle size

A disadvantage of top-down processes is the high energy consumption. For dispersing an agglomerated, coarse powder into a nano-dispersion, ball mills of some kilowatts per hour are necessary, which need up to 10 hours to break down the bigger particles. During this process, a contamination of the nanoparticles by the mill has to be taken into account.

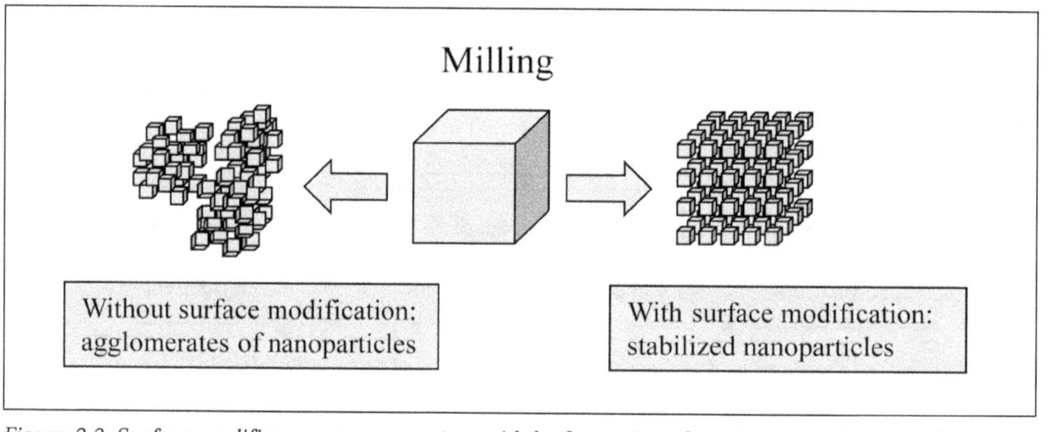

Figure 2.3: Surface modifiers are necessary to avoid the formation of agglomerates during milling

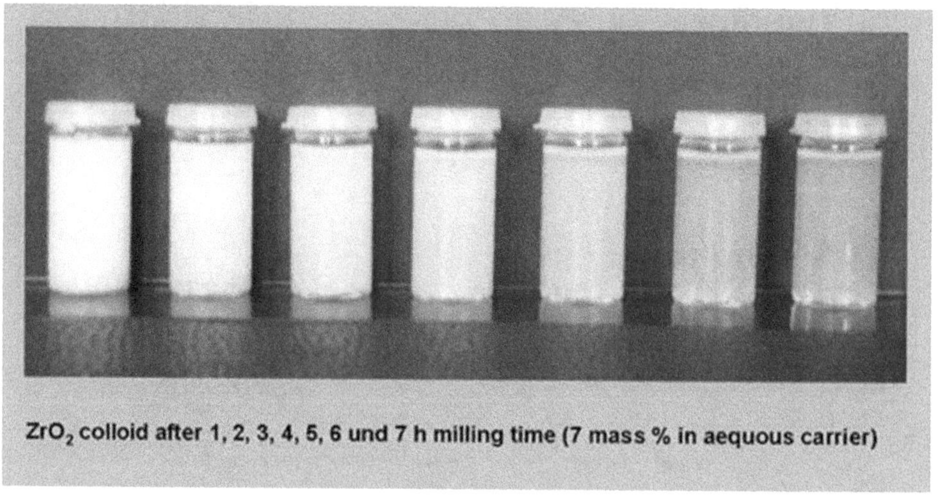

Figure 2.4: Transparency of zirconium dioxide dispersions as a function of milling time in a high energy mill

Source: Bühler AG

➤ ***Example 1: Preparation of a ZrO_2 (zirconium dioxide) nanoparticle dispersion by milling*** [16]

1,000 ml of distilled water, 400 g of zirconium dioxide (BET surface 150 ± 10 m^2/g) and 60 g of 3,6,9-trioxadecanic acid are placed in a reaction vessel and mixed for 30 minutes while stirring. The obtained mixture is milled in an agitating ball mill for 4 hours (Drais Perl Mill PML-H/V, 1,700 g milling balls, zirconium silicate, ball diameter 0.3 to 0.4 mm, continuous operation in circular mode). The colloid obtained in this way contains particles with an average particle diameter of d_{50} = 0.0118 µm (UPA).

Avoiding agglomeration

During milling, the decisive point to avoid agglomeration is to occupy the newly generated surfaces quickly with a surface modifier. Trioxadecanic acid is such a modifier with an exceptionally strong affinity to metal oxide surfaces, which also is used to control the wet chemical synthesis of other oxides, like e.g. ITO.

15 to 20 weight% surface modifier, in relation to the weight of the nanoparticles is a common value. This large amount of organic contamination is carried through all the following process steps and usually has to be removed at some point. Also, a contamination by the material of the mill should be considered.

➤ ***Example 2: Preparation of an ITO indium tin oxide nanoparticle dispersion*** [17, 18]

140 g indium (III) chloride (0.63 mol, anhydrous), 18 g tin (IV) chloride · 5 H_2O and 5.6 g caprolactam were added to 1,400 ml water and stirred. After a clear solution had been formed, this was heated to 50 °C. Once this temperature had been reached, 105 ml ammonium hydroxide solution (25 %) were added dropwise under vigorous agitation. The produced suspension was stirred for a further 24 hours at a temperature of 50 °C. A further 280 ml ammonium hydroxide solution were then added to the mixture for full precipitation.

A white deposit consisting of indium hydroxide was formed, which was centrifuged off (30 min at 4000 rpm). The powder was dried in a vacuum drying oven at 190 °C until a slight yellowing of the powder could be determined. The dried powder was finely ground in a mortar, spread out in crystallising trays and placed in a forming gas oven. The oven was evacuated, then flooded with nitrogen. The oven was heated at a heating rate of 250 °C/hour to 250 °C with a nitrogen flow of 200 l/h. This temperature was maintained for 60 minutes under a forming gas atmosphere at a gas flow of 300 l/h. The oven then cooled to room temperature under a nitrogen atmosphere (duration approx. 5 hours). This resulted in dark blue ITO powder.

A mortar mill is charged with 25 g of a mixture of 50 % by weight ethylene glycol, 50 % by weight diethylene glycol monobutylether, and 5.6 g of 3,6,9-trioxadecanic acid. 75 g of ITO powder are added slowly, and milling is carried out for 2 hours. This results in a dark blue suspension of high viscosity which is homogenized on a roller bed for approximately 20 minutes. The resulting suspension is redispersed in ethanol by introducing 43 g of the suspension thus obtained into 57 g of ethanol and stirring.

Separation of the ethanol produces ITO powders which are redispersible in ethanol to a particle size of less than 20 nm. The primary particle size is 10

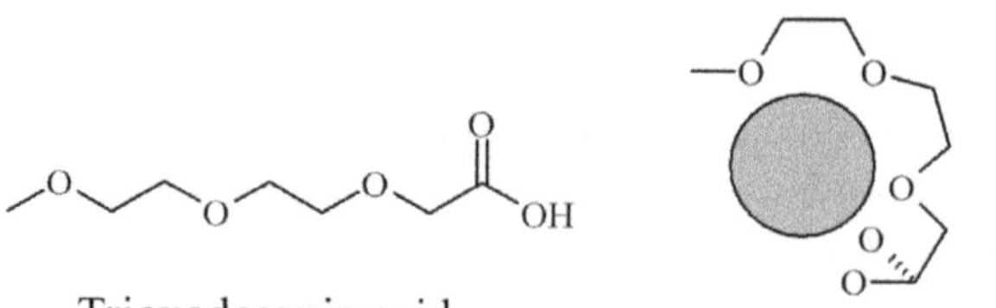

Trioxadecanic acid

Figure 2.5: Trioxadecanic acid is a surface modifier with the capability of establishing chelate bridges to the surface

to 11 nm, the specific surface area 70 m^2/g. Isoelectric point: 7.2. The tin content is usually 8 mol-%. From these powders it is possible to apply sol-gel layers which, at a film thickness of 400 nm with a baking temperature of 550 °C, are able to realize a transmission > 90 % and a surface resistance of 160 Ω/square on glass.

ITO indium tin oxide nanoparticle dispersion

When synthesizing nanoparticles via controlled wet chemical methods, it is important to avoid the formation of irreversible aggregates. One possibility is to use surface modifiers right from the beginning during the precipitation step. Thereby the newly generated surfaces are immediately covered by the modifier and agglomeration as well as aggregation is reduced to a great extent.

The use of caprolactam in this example is such a case. The co-precipitation of the ITO yields agglomerated nanoparticles which can be redispersed to primary particle size, because their active surface had been covered and growth had been stopped by the surface modifier. A further effect is that the size of the primary particles can be controlled by the amount of the surface modifier if the adsorption on the surface is strong enough (compare Figure 3.18, page 49) [63].

If thermally treated nanoparticle agglomerates have to be dispersed, strong shear forces are necessary. In this example the thermal treatment was used to foster crystallization and thus a three roll mill is necessary to break down the aggregates. As pointed out before, the newly generated surfaces have to be protected again with a surface modifier. Trioxadecanic acid is a very effective modifier which coordinates with its carbonic acid function and the ether oxygen atoms to the surface of the nanoparticles.

The first step in the nanoparticle synthesis of the example is the solubilizing, diluting and heating of indium and tin chloride. Under these conditions, already hydrolysis and formation of chloride-stabilized seed particles takes place (compare Table 2.2, page 25). It is not known if this step is decisive for the product quality, however it is important to understand, that even the smallest detail like the dilution of the salt solution can be very important for the final product.

2.2 Bottom-up processes

Bottom-up processes use atoms, ions or molecules to build up the nanoparticles via physical or chemical processes. A rough segmentation can be done via the phase in which the reaction takes place, so gaseous, fluid and solid state processes are distinguished. Table 2.1 shows examples of several known bottom-up processes.

Table 2.1: Physical and chemical methods for nanoparticle synthesis

	Physical processes	Chemical reactive processes
Gas phase processes	PVD LASER-evaporation Plasma processes Metal wire explosion	Flame pyrolysis CVD Hot wall reactors
Liquid phase processes	Dissolving and precipitation in solvents Dissolving in supercritical gases followed by relaxation	Controlled precipitation Sol-gel processes Electrochemical processes
Solid phase processes	Phase separation in glasses and metals during cooling (e.g. gold ruby glass)	Reactive milling [19]

In a first step, the components/educts of the to-be-synthesized nanoparticle are evaporated or solubilized. Supersaturation leads to seed particle formation, which then aggregate to bigger structures (compare Figure 2.8, page 25).

At this point in time it is decided whether the nanoparticles can be protected by surface modifiers against agglomeration and aggregation. Once built, bigger clusters are very hard to separate again in further processing steps.

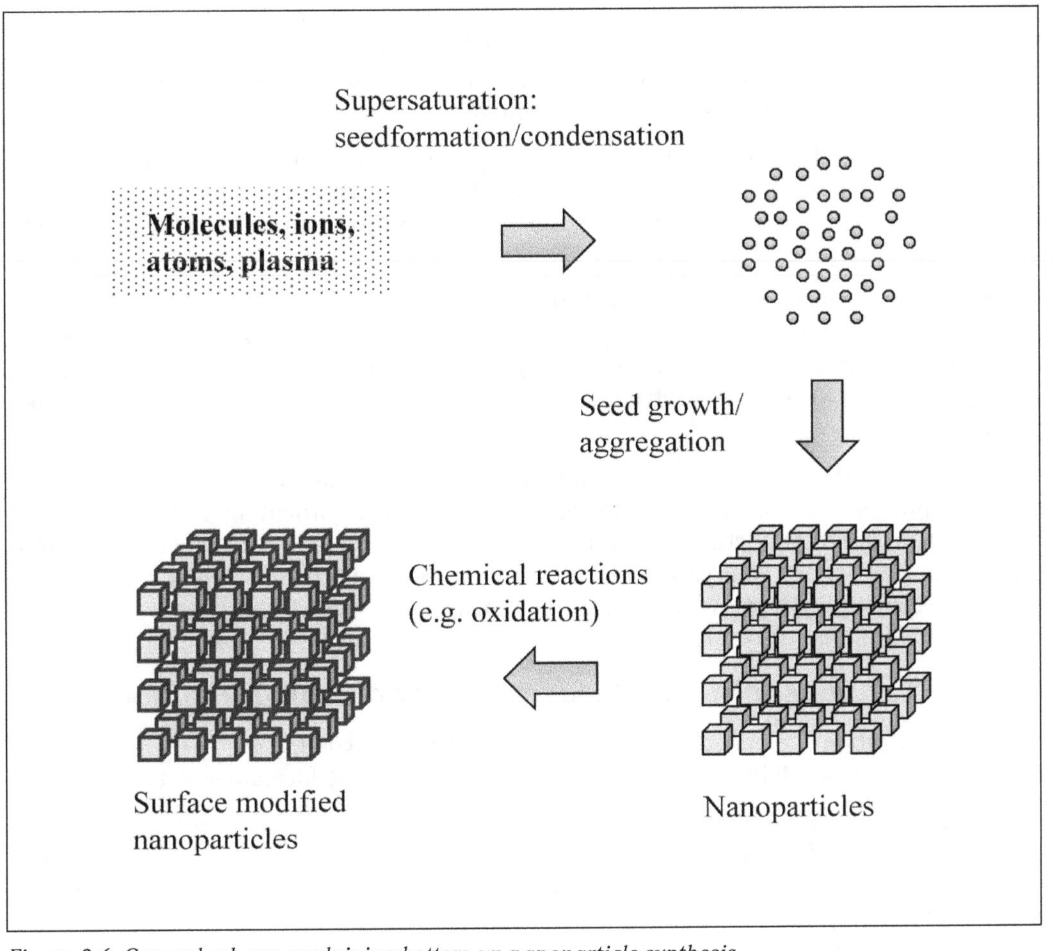

Figure 2.6: General scheme explaining bottom-up nanoparticle synthesis

2.2.1 Nanoparticles synthesized by gas phase processes

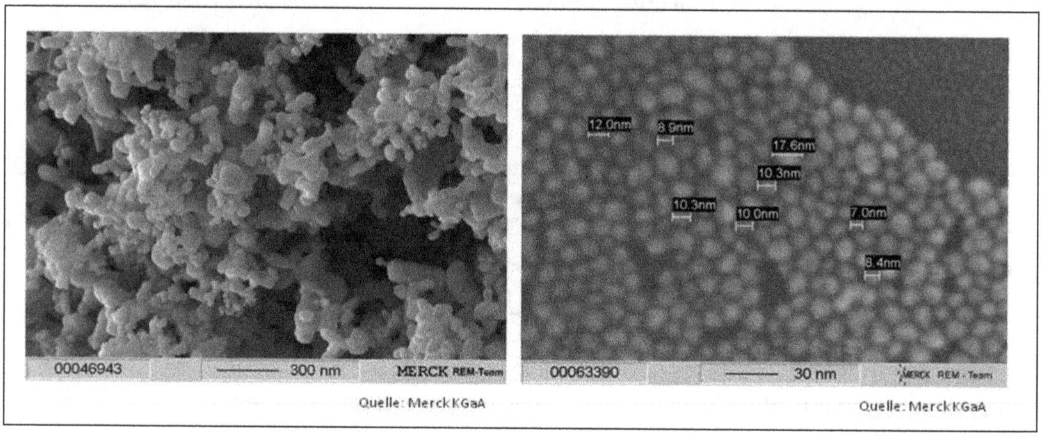

Figure 2.7: Electronmicroscopy pictures of nanoscaled zinc oxide, synthesized via gas phase reaction (left) and by wet chemistry (right) *Source: Merck KGaA*

Gas phase processes usually run at very high temperatures. Therefore organic surface modifiers cannot be introduced at the process step when seeds and particles grow. Only after having cooled down below the decomposition temperature of the organic component, the surfaces can be modified. At this point in time however, most aggregates and agglomerates are already generated. Big efforts therefore are necessary to break down these bigger particles down to primary particle size in formulations.

One of the most well-known examples of the gas phase process for nanoparticle production is the flame hydrolysis of $SiCl_4$ to pyrogenic silica, known as "Aerosil" or "Wacker HDK". The highly aggregated structure of the SiO_2 is the basis for its application as thickener in paints and coatings (compare Figure 3.4, page 38).

In wet chemical processes, organic surface modifiers can be used right from the start of the reaction and thereby are able to suppress agglomerates and aggregates from the beginning. Smaller and monomodal particles can thus be produced more easily. This is evident when particles produced by wet phase and gas phase processes are compared directly like in Figure 2.7.

The particles in the left picture show aggregates, whereas the particles in the right picture (wet process) are much more homogeneous in size and without aggregates. It should be noted however, that particles stemming from the gas phase synthesis were treated at higher temperature and thus can show higher mechanical and chemical stability compared to particles stemming from wet chemistry methods.

2.2.2 Nanoparticles synthesized by wet chemical methods

If the concentration of a material in a solvent is increased beyond its solubility limits, seed formation takes place, followed by growth, aggregation and the formation of a precipitate.

One of the first lessons in practical chemistry: If solutions of $BaCl_2$ and Na_2SO_4 are mixed, a white precipitate of $BaSO_4$ appears. What once was feared in the practical training now is one of the goals of nanotechnology: the formation of a colloidal precipitate, which cannot be held back by conventional filter paper. The following part of this book focuses on the aspects which should be considered to succeed in this goal.

Table 2.2: Possibilities to synthesize nanoparticles by wet chemical synthesis [20]

Superordinate principle	Execution	Example
Y is displaced by X	Controlled precipitation of a hardly soluble compound MeX	$BaSO_4$ from $BaCl_2$ and Na_2SO_4
	Hydrolysis by heating or diluting a concentrated salt solution	TiO_2 Synthesis from $TiOCl_2$ or $TiOSO_4$ [20]
	Hydrolysis of metal-organic groups in the sol-gel process	SiO_2 "Stöber" synthesis (compare Example 3)
Y is removed or decomposed, X takes the free place	Chemical decomposition of Y by a redox-reaction or electrolysis	Electrochemical synthesis of ZrO_2 starting from $ZrOCl_2$ with Ir-electrodes
	Thermal decomposition of Y	Cr_2O_3 from Cr-acetate (compare Example 19)
	Ion exchange of Y against X	Synthesis of colloidal silica sol (compare Example 5)

Many different processes to produce nanoparticles are known, but they all can be traced back to some basic concepts. When a nanoparticle of the composition MeX is to be synthesized from an educt MeY, Y has to be replaced by X. This can be done by different methods (Table 2.2).

In a wet chemical synthesis of nanoparticles, first a supersaturated solution is prepared from which seeds are growing and in a following step aggregate to bigger clusters. In contrast to processes in the gas phase, the reactive species are not generated by evaporation but by a chemical reaction. The crucial point which determines the decisive difference in homogeneity and dispersability between gas phase and wet chemical processes is the possibility to introduce soluble surface modifiers and thereby controlling the seed formation and growth.

LaMer [22] influenced the common understanding of these processes with his model. According to his model, seed formation happens in a homogeneously supersaturated solution.

The rising concentration of the hardly soluble product in the reaction mixture first exceeds the saturation limit and enters the supersaturated area. Seed formation is kinetically hindered until the concentration reaches the critical seed formation concentration. Starting from this point, the speed of seed formation becomes measurable [21, 22].

The seed formation and growth lowers the concentration of solubilized material until at some point in time it is again lower than the critical seed formation concentration. The time frame between exceeding and falling under this critical seed formation concentration is called seed formation phase.

After that, in the supersaturated area, seed growth takes place until the super saturation approaches the equilibrium concentration (compare Figure 2.11, page 27, Figure 2.8) [5].

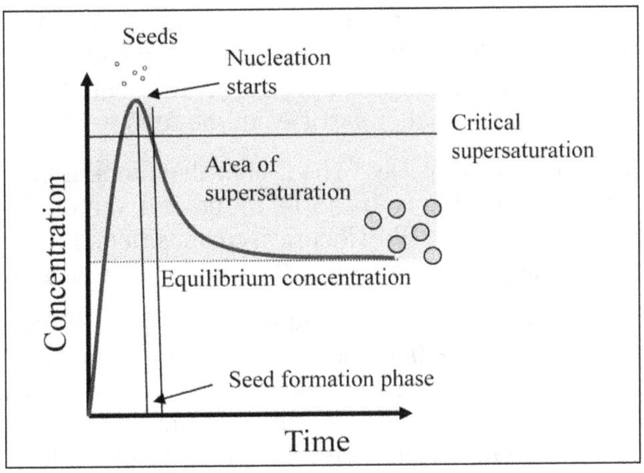

Figure 2.8: LaMer model of seed formation in homogenous supersaturated solutions [22]

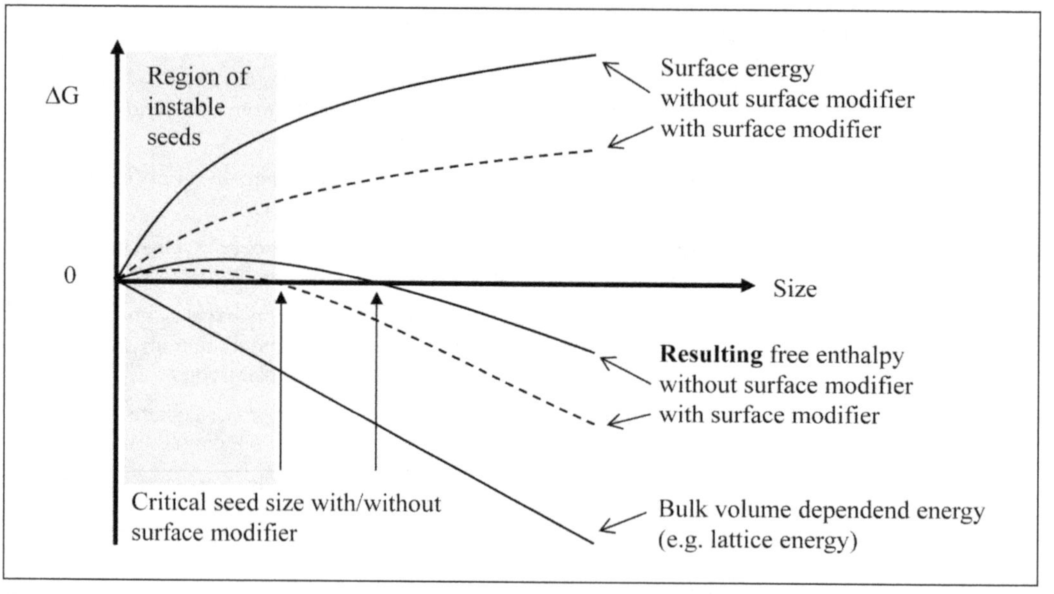

Figure 2.9: Correlation between seed stability and surface tension. Surface modifiers decrease the region of instable seeds

During the seed formation phase, formation and dissolution of seeds are concurring processes driven by the surface energy and the lattice energy of the seed.

If the energy gain by the lattice formation is higher than the energy which is necessary for generating the new surface, the seed will grow. If not, the seed is instable and will dissolve again.

Lattice energy is a material constant and can hardly be changed. Surface energy however can be influenced by additives, solvents and surface modifiers. If lowered, more stable seeds are generated earlier in the process, leading as a consequence to smaller particles in the final product (Figure 2.9).

Finally, the concentration plays a decisive role. At concentrations below the critical seed formation concentration, dissolution of the seeds is favored, whereas at concentrations above, growing can lead to stable seeds more easily [23].

By increasing the temperature, the critical seed formation concentration is reached faster and thus a greater number of seeds are generated compared to lower temperatures. The available material for seed growth is distributed between the larger number of surviving seeds, causing smaller particles in the final product.

A general rule thus says: The higher the temperature, the smaller the particles. A practical application of this rule can be found in Example 3 on page 29, the Stöber synthesis of SiO_2 nanoparticles. With identical composition, the final particle size can be varied in a wide range of approximately 15 to 800 nm just by changing the temperature [24, 25, 67].

The structure of the resulting nanoparticle is influenced to a great extent by the colloidal stability of the seeds in the reaction media (compare "Stabilization of nanoparticles against agglomeration", page 40). Stable seeds grow by deposition of soluble material on the seed, whereas instable seeds aggregate to bigger clusters. The formation of colloidal silica sol by ion exchange (compare Example 5, page 31) is carried out under conditions of stable seeds, whereas the Stöber process runs under destabilizing conditions due to the high-temperature and high ion strength of the reaction medium.

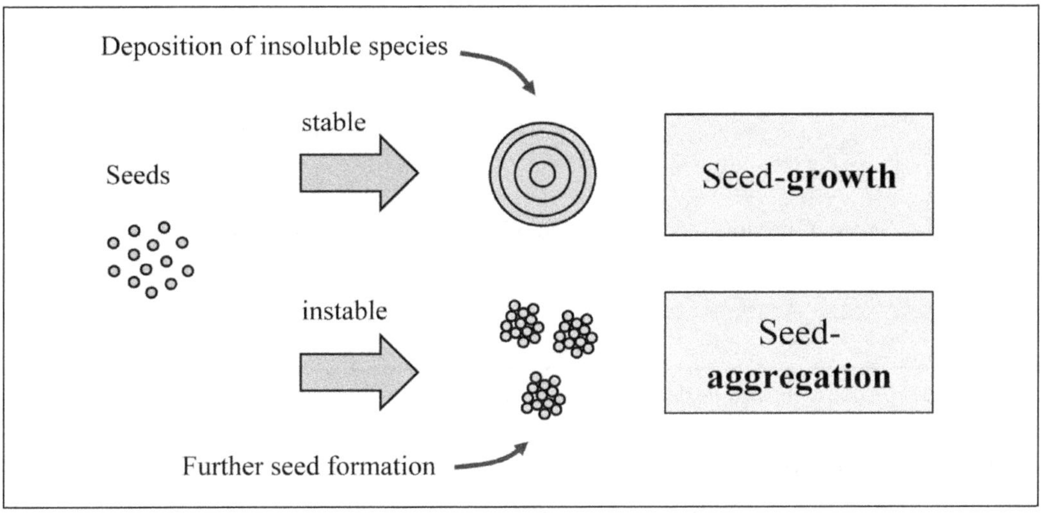

Figure 2.10: The further development of the seeds and thus the stability of the resulting particle is dependent on the colloidal stability of the seeds in the reaction medium (compare "Electrostatical stabilization", page 41)

Influence on the seed formation rate

As a result, particle morphology and the resulting final material properties like density or porosity differ. For the synthesis of nanoparticulate dispersions and in order to avoid precipitation, the seed formation has to stop at a point, when the particles are still in the nanoscale region. This can be achieved by keeping the seed formation as short as possible, which translates into a very high seed formation rate. By keeping the seed formation short, the formation of monomodal dispersions, which means dispersions with a narrow particle size distribution is favored [26].

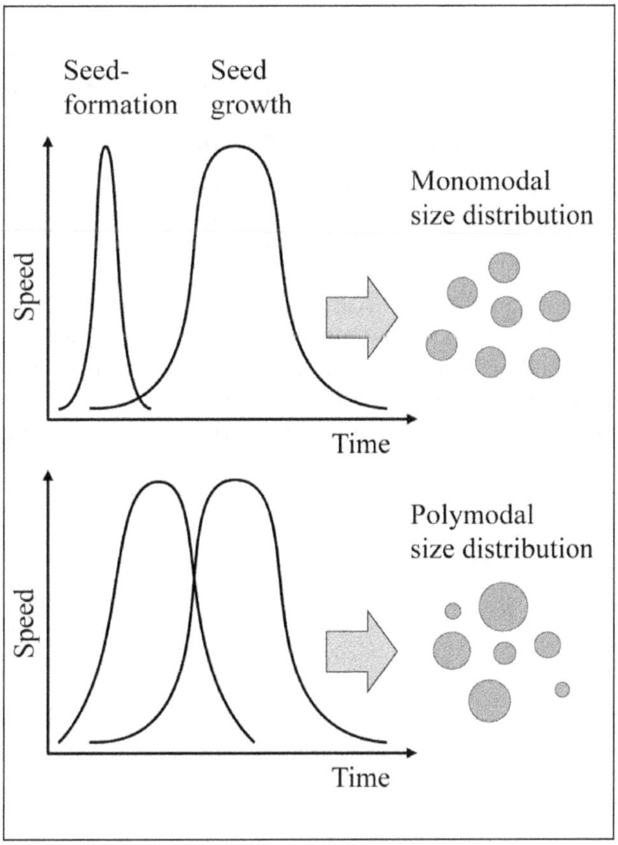

Figure 2.11: The separation of seed formation and seed growth is important for the synthesis of monodisperse nanoparticles

In literature the separation of seed formation and seed growth is a widely accepted concept for the preparation of stable nanoparticle dispersions [5]. The nucleation period should be as short as possible and during the following phase of seed growth the concentration should never exceed again the critical supersaturation threshold (compare Example 3, page 29).

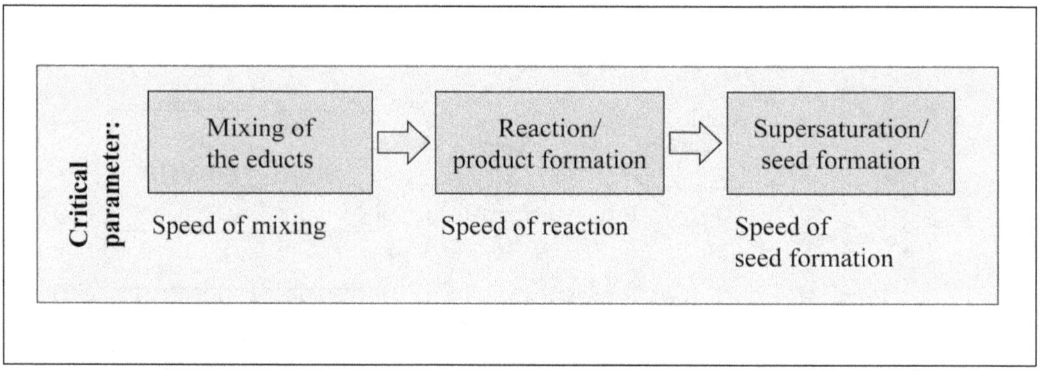

Figure 2.12: Critical parameters during seed formation.

2.2.2.1 Ostwald ripening

If in a dispersion particle sizes differ to a great extent, Ostwald ripening can be observed. The smallest particles dissolve and the solubilized material deposits on the bigger particles. The driving force for Ostwald ripening is the correlation of the solubility with the curvature radius of the particles. Analytical chemistry uses Ostwald ripening to improve the retention of fine precipitates in filtering processes (e.g. barium sulfate) [27].

A very short nucleation phase needs a fast, complete and homogeneous mixing of the educts as a prerequisite. Unfortunately this is not often the case, because reaction speed exceeds mixing speed in most cases. So the reaction is completed before the educts have been mixed thoroughly, resulting in inhomogeneous seed formation and growth.

Critical parameters of seed formation

If, like it is the case for ionic reactions, the reaction speed is very high, Figure 2.13 shows two possibilities of particle size distribution development, depending on the speed of seed formation and mixing speed.

If monomodal particle dispersions are the objective of the synthesis, mixing speed should always be higher than the speed of seed formation. Conventional mixing devices cannot always deliver such a high mixing speed, therefore processes have been developed, which ensure a thorough homogenization in short time. The "micro jet high gravity reactor" [28] or the "controlled double jet precipitation" [29] are examples of such devices.

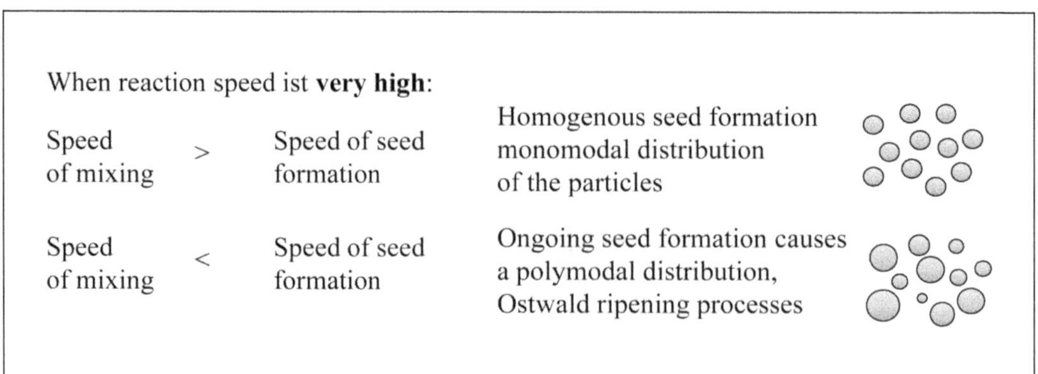

Figure 2.13: Dependence of the particle size distribution on the speed of mixing and the speed of seed formation

Synthesis of titanium dioxide or copper oxide with the "high gravity reactor" have been described [30, 31]. Since such special reactors cannot be used in every case, nanoparticle synthesis in conventional stirring vessels are conducted at low solid content to avoid uncontrolled agglomeration and precipitation.

Even particles like dust which were brought in by accident can act as seeds for the formation of nanoparticles (compare "Interference pigments", page 76). Therefore during synthesis and handling of colloids, cleanliness is important. Literally each dust particle is a potential seed.

Special emphasis should be put on impurities and byproducts which act as electrolytes. A precipitation of $BaSO_4$ from $BaCl_2$ and Na_2SO_4, besides the product also generates NaCl which destabilizes the synthesized colloid (compare "Electrostatical stabilization", page 41). If a simple peptisation by removal of the salt (e.g. by washing with deionized water) is not feasible, alternative sources for Ba^{2+} and SO_4^{2-} have to be taken into consideration.

2.2.2.2 Hydrolysis

The basics of the hydrolysis of aqueous salt solutions are known to every chemist. If solutions of $ZnCl_2$ or $FeCl_3$ are diluted and/or heated, after some time a precipitate of the corresponding hydroxide or oxide is generated. OH- ions have displaced the chloride ions and the hydroxide precipitates.

Commercially, this principle is used for the synthesis of titanium dioxide pigments. Details can be found in the reviews of Matijevic [32, 33].

The hydrolysis of metal and silicon alkoxides is a central reaction step in the sol-gel process. The basics of these reactions are discussed in "Coating resins made by nanotechnology – The sol-gel process" starting in Chapter 5, page 91.

➤ *Example 3: Preparation of a SiO_2 dispersion via the Stöber process* [34]

A hydrolysis mixture is prepared which consists of 13.5 g (0.75 mol) of water, 64.4 g (1.4 mol) of ethanol and 6.4 g (0.38 mol) of ammonia. To this hydrolysis mixture, thermostatted at 40 °C, 4.2 g (0.02 mol) of tetraethoxysilane, likewise thermostatted, is added in one batch with thorough stirring. A sol of primary particles having mean particle diameters of 58 nm with a standard deviation of 5 % is obtained.

To the sol of primary particles thus obtained, 650 g (3.1 mol) of tetraethoxysilane and 5.9 l of hydrolysis mixture are added dropwise while stirring over a period of 5 days. Spherical SiO_2 particles having mean particle diameters of 3.1 µm with a standard deviation of 1.3 % are obtained.

Factors influencing the seed stability

During the synthesis of colloidal silica sols (compare Example 5, page 31) a solution of metastable silicic acid Si(OH)4 is prepared via sodium ion exchange, starting from water glass solutions. The Stöber process however is performed under alkaline conditions using ammonia as a catalyst and organoalkoxysilanes as a silica source (compare: "Hydrolysis and condensation", page 94).

Under these conditions, hydrolysis, but especially condensation reactions take place at a very high speed and within seconds a large number of SiO_2 seeds are generated which aggregate to larger particles. By variation of the temperature, the particle size can be increased or decreased in a controllable manner. The higher the reaction temperature, the smaller the particles will be.

The pH during the synthesis of colloidal silica sol by ion exchange is acidic and it is only the final product which is stabilized by adding defined amounts of alkaline ions to adjust a high pH. But not only the difference in pH is crucial. Whereas the ion exchange process generates almost pure Si(OH)4, during the Stöber process large amounts of ions [NH^4+, OH-] are present. This implies consequences for the stability of the seeds.

In both processes, in a first step seeds are generated, which either can grow or aggregate to bigger particles. The ion exchange favors seed growth, whereas the Stöber process leads to seed aggregation. Thus the structures of the resulting SiO_2 particles are different.

In a variation of the Stöber process, seeds which have been prepared in a separate reaction are introduced to direct the particle growth. Patchwork-like the seed grows by aggregation of the newly created seeds. In the ion exchange process, new seed formation is suppressed, because freshly generated Si(OH)4 deposits on and reacts with the particles which are already present.

2.2.2.3 Controlled precipitation and peptisation

The first step when using peptisation for the preparation of nanoparticle dispersions is the precipitation of a hardly soluble product. Typically, the precipitate consists of agglomerates of nanoparticles which cannot remain dispersed in the reaction media. The reason may be a high ion strength caused by salt formation and/or a low zeta potential of the product at the pH of the reaction medium.

As long as the precipitate is surrounded by a mother liquid with a high salt concentration, the nanoparticles cannot be redispersed again. During peptisation, the precipitate is washed until the salt concentration is significantly reduced.

In some cases at this point the precipitate redisperses itself spontaneously, but more often the zeta potential has to be adjusted by a shift of pH or the adsorption of a surface modifier (compare “Electrostatical stabilization”, page 41). Shear forces, like intensive stirring support the dispersion step.

➤ *Example 4: Preparation of a colloidal ZrO_2 dispersion by peptisation [35]*

To a zirconium oxychloride aqueous solution (1 l, concentration 0.5 M) are added 335 ml of an ammonia solution (concentration 3 M) under intense stirring by means of a homogenizer-type mixer (“Ultraturax type”). Thereafter the precipitate is recovered by filtration onto a sintered glass (porosity #4) or by centrifugation. The thus obtained wet cake is purified by alternate redispersion and filtration processes until the pH value of the washing water has become stable around pH 7. As observed by transmission electron microscopy of the thus purified zirconium hydroxide precipitate, an amorphous product is provided.

The purified zirconium hydroxide cake is redispersed into a given amount of water under vigorous stirring, then under sonication. The thus obtained “milk” is introduced into a PTFE beaker and placed in an autoclave to carry out the hydrothermal treatment under continuous stirring. At the end of this step, a slightly blue-colored opalescent zirconia aqueous colloidal solution is obtained. To significantly improve the sol stability, it is necessary to reduce the ionic charge through a dialysis step. After sonication a stable sol containing about 7 % ZrO_2 is obtained.

Stabilizing a colloidal dispersion

In this example a fresh precipitate is dispersed by removal of the salt load and by a following electrostatical stabilization.

During the precipitation of ZrO_2 from $ZrOCl_2$ with NH_3, NH_4Cl is generated. The high ion strength decreases the range of the repulsive electric potential of the particle surface. This is the reason, why the dispersion is unstable and agglomerates precipitate. If the disturbing ions are removed before stable interparticulate bonds are generated, the agglomerates can be separated and redispersed again (compare Figure 3.11, page 42).

However stabile dispersions can only be obtained, if the zeta potential of the particle surface is sufficiently high (higher than 20 mV). If crystalline particles should be obtained, usually a temperature treatment is necessary. The lack of suitable high boiling solvents makes the usage of an autoclave mandatory.

2.2.2.4 Ion exchange

According to the survey in Table 2.2, nanoparticles can also be synthesized by the removal of stabilizing an- or cations. The resulting hydroxides condense to oxide nanoparticles if suitable reaction conditions are given. The following example shows how such a process might look like.

➤ ***Example 5: Preparation of a colloidal SiO_2 dispersion in water*** [36, 37]

An aqueous alkali silicate solution having a water content of 47 % and a ratio of SiO_2 to Na_2O of 2.4 was diluted with demineralized water to a water content of 97 %. 100 parts of this diluted solution were passed at a rate of 20 parts per hour through a column packed with an acidic ion exchanger and subsequently was supplied to a distillation receiver in which the incoming deionized silicate solution was held at boiling temperature and the water distilling off was removed from the solution. After the end of the introduc tion, the silica sol formed was concentrated by further heating to 10 parts. The pH was adjusted to 10.5 to 11.

Synthesis of colloidal silica sols by ion exchange

Starting with a sodium silicate solution, the Na+ ions are replaced by H+. As intermediate product thus silicic acid $Si(OH)_4$ is formed, which then condenses to yield SiO_2 seeds. Further addition of freshly ion exchanged silicate leads to a successive seed growth until the desired size is reached.

Finally the pH is adjusted in the alkaline region using NaOH. This high pH increases the zeta potential to high absolute numbers and helps to stabilize the silica dispersion (compare "Properties and processing of nanoparticles", page 35ff).

A low solid content of the silicate solution is crucial to prevent premature aggregation and a polymodal particle size distribution.

Many variations of this process exist. Instead of a continuous addition of silicate, a batch-wise temperature treatment can help to narrow the particle size distribution. A following controlled growth can then be realized by the addition of freshly exchanged silicate solution.

The concentration and temperature determine the number of seeds, their size and the final particle size distribution.

Besides spherical colloidal silica sols, also chain and barbell shaped SiO_2 particles are possible. By introducing small amounts of divalent ions like Ca^{2+} or Mg^{2+} particles, which were generated in a first step of the synthesis are "glued" together to pearl-chain like structures [38, 39].

Production of water glass

Alkali silicates are produced by melting a mixture of alkaline carbonates and sand. As a general rule, with decreasing alkaline and solid content, the degree of polymerization of the silicates is increasing. Standard types of water glass with SiO_2/Na_2O ratios of 2 to 3 consist only to a lesser extent of oligomers and to a large extent of big anionic SiO_2 clusters.

If a water glass solution is diluted with water, a shift of the equilibrium starts, which increases the size of these clusters more and more until precipitation occurs. The stabilizing high concentration of sodium hydroxide has been diluted to an extent, where the SiO_2 aggregates are no longer stable [40, 41]. Water glass can also be regarded as a colloidal dispersion of very fine SiO_2 particles smaller than 5 nm in size with a higher amount of alkali ions as stabilizers.

2.2.2.5 High temperature liquid phase synthesis of nanoparticle

A method, which has become quite popular in the last years, is the nanoparticle synthesis via the thermal decomposition of salts or metal-organic compounds. Solvents with a high temperature stability act as reaction media and also stabilize the resulting particles as surface modifiers. The educts are heated until thermally labile organic groups decompose or leave and the oxides result [42, 43]. If the product crystallizes readily and the temperature is high enough, by this reaction nanocrystals can be obtained without an autoclave step.

During the precipitation of mixed compositions, the resulting products often precipitate at different pH values leading to inhomogeneity. In the high temperature synthesis process, the determining factor is the decomposition of the anion, which is less dependent on the cation.

If a salt mixture with identical anions is used, in many cases even multicomponent compositions can be realized in a very homogeneous manner. Examples of thermolabile groups are acetates, nitrates, carbonates, carbonyls or peroxides [44, 42, 43]. By using fluoride or sulfur containing raw materials, the corresponding fluorides or sulfates can be synthesized (compare "Tailoring of the refractive index of sol-gel coatings", page 186)

➤ Example 6: Preparation of water-dispersible TiO_2 nanoparticles [45]

Under vigorous stirring, 600 ml anhydrous diethylenglycol were charged into a 1l-three neck flask. Then 20 ml titanium tetrachloride (0.182 mol) and 10 ml distilled water (0.556 mol) are added under nitrogen. The temperature is increased to 160 °C and the reaction mixture is heated 4 h under reflux. Thereafter the reaction product is separated by centrifugation. The solid obtained thereby is washed twice with acetone and dried under oil pump vacuum overnight. The resulting TiO_2 particles can be dispersed in amounts of more than 70 wt% in water without any additives. The primary particle size is about 5 nm and the particles essentially do not agglomerate in their aqueous dispersion. As crystalline phase, anatase is observed in XRD analysis.

Water dispersible titanium dioxide nanoparticles

When polyols are used as solvents in high temperature sol-gel processes, they not only moderate the reaction speed of hydrolysis but also act as surface modifiers and bind to the particle surface (compare "Sterical stabilization", page 46). Thereby they influence the nanoparticle growth and act as sterical stabilizers.

If diethyleneglycol reacts with both OH groups, the resulting nanoparticle surface will be rather hydrophobic. If a reaction only takes place with one group, thereby leaving a free OH^- group, then the particle will be more hydrophilic. During the synthesis, further surface modifiers can be added to shift the polarity of the resulting particles in a desired direction (e.g. silanes, phosphates, benzylalcohol).

In the Example 5, $TiCl_4$ is used as educt to yield a water soluble product. Considering the reaction conditions, residual chloride ions will remain in the product and on its surface and will provide an electrostatic stabilization of the particles in water. The example shows how important it is to choose the right educt. $TiCl_4$, $TiOCl_2$, $TiO(NO_3)_2$, $Ti\text{-}(OR)_4$ … all of these educts yield titanium dioxide after hydrolysis, but the particle morphology and stability will differ to a great extent.

When using thermally degradable anions in high boiling solvents, the synthesis strategy can be varied so that no water at all is used and the oxide formation takes place only by decomposition of the anions. This is a suitable way especially for multicomponent compositions with different iso-electric points (compare Example 19, page 73).

But even with small amounts of water, in the presence of polyols a homogeneous multicomponent mixture can be synthesized. This process is also suitable for the synthesis of transition metal nanoparticle compositions, where it often is difficult to find the corresponding alkoxides. Smart combinations of educts even yield phosphates or chalcogenides. The main selection criteria are the decomposition temperature and the decomposition products related to the temperature stability of the solvent.

2.2.2.6 *Electrolysis*

If the anion of a metal salt is decomposed by electrolysis in aqueous solutions, OH^- takes its place and the corresponding hydroxide is generated, which then condensates to the oxide. Zirconium dioxide and titanium dioxide dispersions were realized by electrolysis of $ZrOCl_2$ or $TiOCl_2$ solutions using iridium electrodes [46, 47]. Both processes have never gained a broad industrial acceptance.

If a metal anode is oxidized via electrolysis, also nanoparticles are generated. The synthesis of zinc oxide and ITO particles by using this principle has been described [48, 49].

In the esoteric media, some authors recommend home-made silver oxide nanoparticle dispersions as dietary supplements. Many cases of severe argyria, a permanent bluish colorization of the skin have been described as a severe side effect [50, 51].

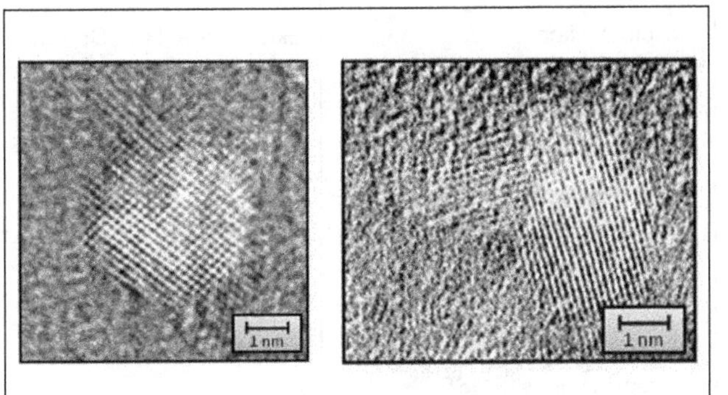

Figure 2.14: ZrO_2 and Al_2O_3 particles, produced by electrolysis of $ZrOCl_2$ and $AlCl_3$ solutions

Source: Merck KGaA

2.2.2.7 *Micro emulsion processes*

A micro emulsion consists of one phase which contains the three components water, oil and surfactants. In contrast to emulsions, micro emulsions are thermodynamically stable. The water-soluble educts, together with surfactants are dispersed in a hydrophobic phase like e.g. heptane. Usually the surfactant system is adjusted to the polarity of the water/hydrophobic phase mixture by small amounts of co-surfactants like partly water-soluble alcohols (butanol, pentanol, hexanol).

A stable dispersion results in which the droplet size limits the available reaction space for the educts, thereby limiting the possible particle size [52]. Precipitation then is induced by a change of pH or addition of further reactants.

The surfactants which are present in the micelles can act immediately as surface modifiers for the synthesized particles, thus preventing agglomeration. A significant disadvantage however is that the amount of surfactants relative to the product is very high and that a removal is often not possible with a reasonable effort.

2.3 *Commercial sources of nanoparticles*

Many compounds are already available in nanoparticulate form. Table 2.3 gives an overview about typical applications and commercial sources.

In order to realize the advantages of their nano-particulate nature, these products have to be transferred to an organic matrix like solvents, polymers, or coatings formulations.

The problems which have to be overcome when nanoparticles are processed will be covered in the following chapters. The central point which will be discussed is the tendency of nanoparticles to agglomerate, how it can be avoided and how a compatibility with the surrounding matrix can be achieved (compare "Sterical stabilization", page 46).

Table 2.3: Examples of commercially available nanoparticles

Nanoparticles	Source	Application
Pyrogenic silica	Evonik, Wacker	Rubber additive (tires), thixotropy modifier for paints
Carbon black	Evonik	Rubber additive (tires), paints
Layer silicates (bentonite, hektorite, laponite)	Rockwood	Paint additive, inorganic binder
Colloidal silica (SiO_2 in water)	Eka, Grace, H.C. Starck, Nyacol	Refractory binder, coating/ impregnation clarification of wine, beer
Organosilica sol	Nissan Chemical	Paint additive, inorganic binder
Boehmite, hydrotalcite	Sasol	Filler, additive for barrier coatings, fire protection
ZrO_2, Sb_2O_3 dispersions	Nissan Chemical, Nyacol	High refractive index coatings, fire protection catalysis
Polyhedralorgano-silsesquioxanes, POSS	Hybrid Plastics	Polymer modifier
Metals	Nanophase, SDC	Printable electronics, catalysts, explosives, high temperature resistant colors

3 Properties and processing of nanoparticles

In the first two chapters, we focused on the different possibilities to synthesize nanoparticles. Nanoparticles however, except in a purely academic environment, are rarely produced for their own sake. In an industrial environment, the objective in most cases is the production of a bulk material or a coating, which contains nanoparticles for a certain reason and function.

If the addition of nanoparticles is taken into account for a certain formulation, then usually a high expectation is built up with regard to realizing a special property profile. Sadly, in many cases these expectations are unreasonably high.

A participant of a practical seminar once asked: "Now that I have synthesized nanoparticles, how come that they are not transparent?" The transparency of the particles and the composites made thereof is a typical example of the expectations users have when dealing with nanoparticles. But which factors influence the transparency of a nanocomposite?

Light scattering evolves when light hits particles of a sufficiently large size, which show a sufficiently high difference in their optical density compared to their environment. The Raleigh equation shows the dependencies determining the degree of scattering Is. The decisive parameters are the particle size, the wavelength of the light and the difference in refractive index between matrix and particle. Further parameters are the viewing angle ϑ and theviewing distance x [303].

Equation 3.1: The Rayleigh equation describes the correlation between the intensity of scattered light and particle size, scattered wavelength and the difference of the refractive index between particle and medium

$$I_s \propto K \cdot \frac{(1+\cos^2 \vartheta)}{x^2} \cdot \frac{r^6}{\lambda^4} \cdot (\varepsilon - 1)^2$$

r^6 : Particle radius
λ^4 : Wavelength
$(\varepsilon-1)^2$: Difference in refractive index

The particle size is the dominating factor of equation. With increasing particle size, the scattered light increases to the sixth power. However not every wavelength of the light is scattered equally. Blue and ultraviolet light (shorter wavelengths) are scattered to a greater extent than red and infrared light.

This is the reason, why colloidal silica dispersions look bluish at incident angles and reddish when looking through the dispersion. An application of this effect can be found in paints and coatings with "frost effect" or a deliberate "color shift effect" caused by nanoscaled titanium dioxide particles of a size which scatters only the blue part of the spectrum (compare "Hiding power, transparency and particle size", page 72) [53]. An often neglected factor is the difference in refractive index between the particle and the surrounding matrix.

The rule "no difference-no scattering" can be demonstrated with halogenated solvents. If inorganic matting agents like silicates are dispersed in halogenated solvents which exhibit a comparable refractive index, then almost transparent dispersions can be realized. Trans-

Gerhard Jonschker: Sol-Gel-Technology in Praxis

ISBN: 978-3-86630-886-2

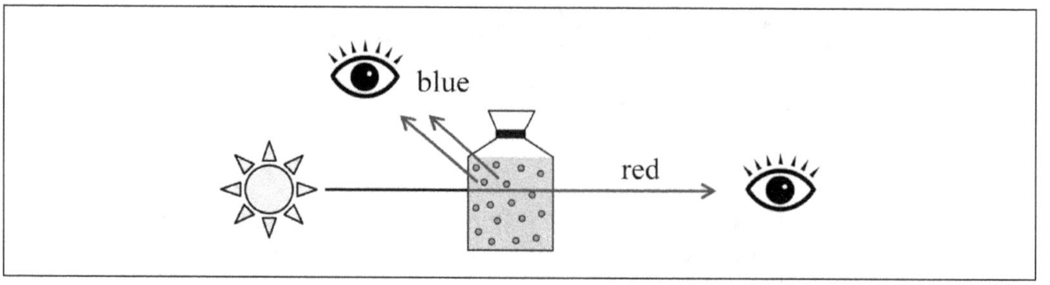

Figure 3.1: Practical application of the Rayleigh equation. Blue light (short wavelength) is scattered by colloidal dispersions to a greater extent than red light (longer wavelength). Thus colloidal dispersions look red when looking through and blue when looking at them.

ferred to paint applications, the consequence of this rule is that by choosing particles and resins with regard to their refractive index, the transparency of the coating can be optimized.

3.1 Agglomeration of nanoparticles

The production and processing of nanoparticles in practical applications is connected with the aspect of agglomeration and how to prevent it. What is the driving force for agglomeration?

Colloids are very small particles and exhibit a large surface to volume ratio. Whereas the properties of large particles which are bigger than 1 µm are determined mainly by their bulk properties, the influence of the surface increases significantly when the particles get smaller and smaller.

This can be illustrated by a thought experiment. A cube of 1 cm has a surface area of 6 cm^2. If this cube is separated into smaller cubes of 1 nm size, 10^{21} cubes result which exhibit a surface area of $6 \cdot 10^{21}$ nm^2, larger than a soccer field.

Atoms sitting on the surface are different in terms of their energy state, binding angle and surrounding than atoms in the bulk. With decreasing particle size they become more and more determining for the macroscopic material properties. Finally an interface determined system results.

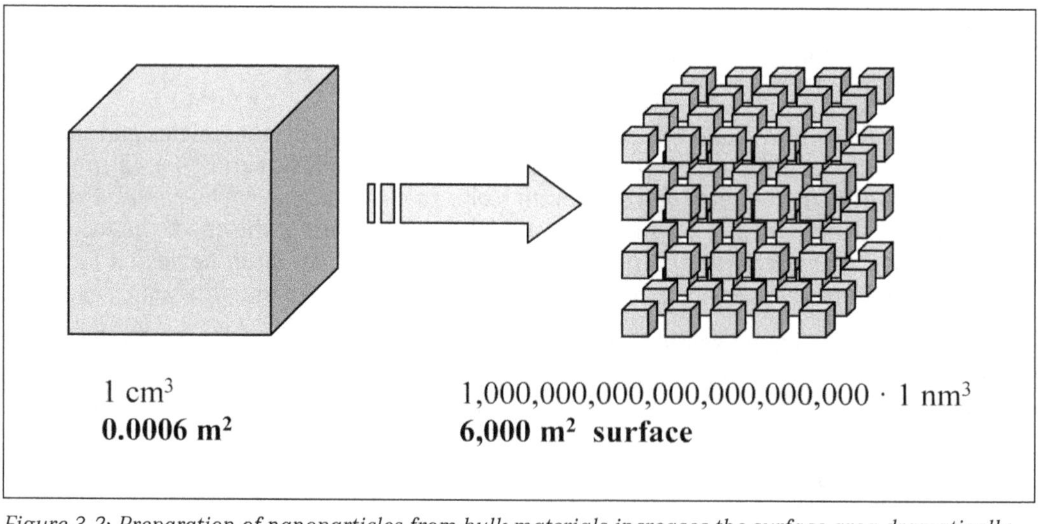

Figure 3.2: Preparation of nanoparticles from bulk materials increases the surface area dramatically

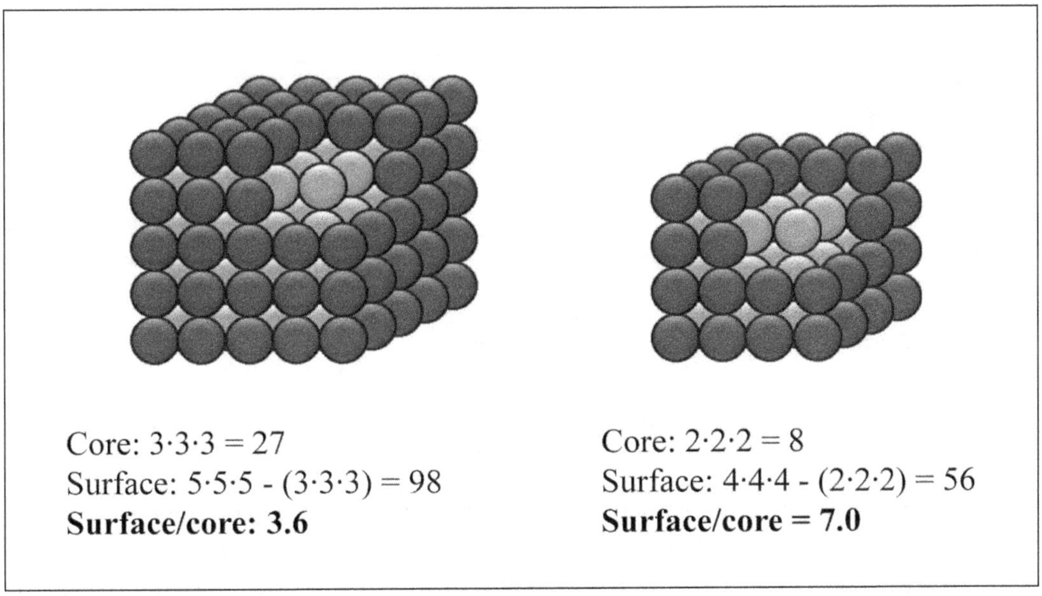

Figure 3.3: Illustration of the correlation between particle size and the surface-to-volume ratio

The energy content of colloids, due to the contribution of the surfaces and their surface free energy is remarkably higher than that of bulk materials. As a consequence, colloids are thermodynamically unstable and show a tendency to minimize their surface area by agglomeration.

This can only be prevented, if the energy gain by interaction with the matrix is higher than the energy gain by a reduction of the surface area, or formulated a little bit sloppy, if the nanoparticles "feel themselves more comfortable" in their surrounding matrix than agglomerated with each other.

This is the reason why surface modification and adaption of the surface modification to the properties of the matrix are crucial steps when processing nanoparticles.

The higher energy content of colloids in comparison to micrometer-sized particles also shows itself in a significantly reduced melting and sintering temperature, which is used in industrial processes to synthesize non-classic glasses and ceramics.

Green bodies or coatings made of nanoparticles can be sintered at significantly lower temperatures than the melting and sintering temperatures of the bulk materials may suggest. This is used besides other applications for high temperature and anti-corrosion coatings (compare "High temperature resistant coatings", page 139).

3.2 Shear thinning – useful agglomerates

Agglomeration has to be prevented in most cases when nanoparticles are processed. Sometimes however, it is especially their tendency to agglomerate which makes the use of nanoparticles attractive. When the viscosity of paints and coatings has to be adjusted, nanoparticles due to their large surface area can be used quite efficiently.

If they interact with each other e.g. via hydrogen bonding, they build three-dimensional networks, which can be destroyed again if shear forces are applied, a property called shear thinning. By shear thinning also the settlement of pigments in paint dispersions can be prevented. Pyrogenic silica is used for this kind of application in a large scale.

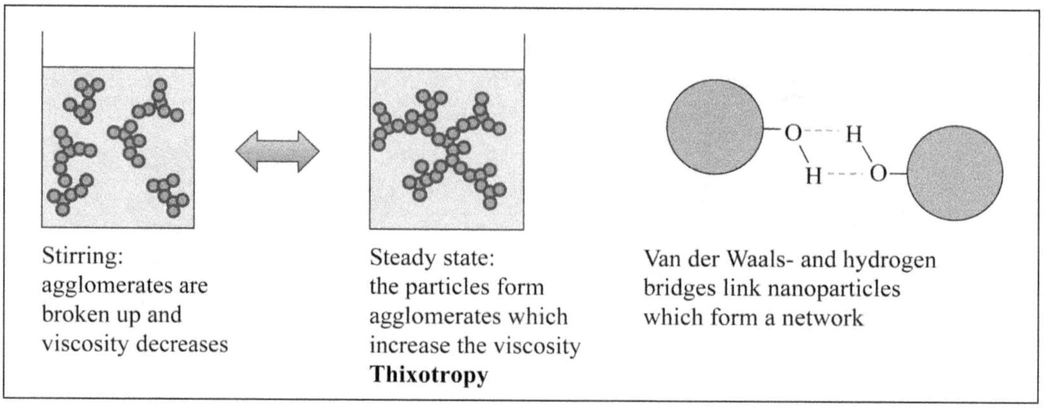

Figure 3.4: Illustration of thixotropic behavior. Reversible interparticulate interactions lead to the formation of fragile inorganic networks in the dispersing liquid

Whereas silica nanoparticles interact mainly via hydrogen bonding, nanoclays attract each other mainly via electrostatical interactions.

Clays, or better: "layer silicates" are a subdivision of silicate minerals and can be found in nature. For highest requirements concerning color and transparency, layer silicates can also be synthesized artificially. From a chemical standpoint, they consist of aluminum-magnesium silicates, which exhibit a negative charge in plane and a positive charge on their edges.

When dry, the 25 to 100 nm long and wide platelets of only 1 nm thickness are agglomerated to stacks. When wetted, water is intercalated between the platelets and leads to a widening of the platelet-to-platelet distance. Shear force, e.g. by stirring, then separates the platelets, which is called exfoliation.

As long as the shear forces are applied, the dispersion shows a low viscosity. As soon as stirring stops, the charges of planes and edges attract each other and a house-of-cards structure

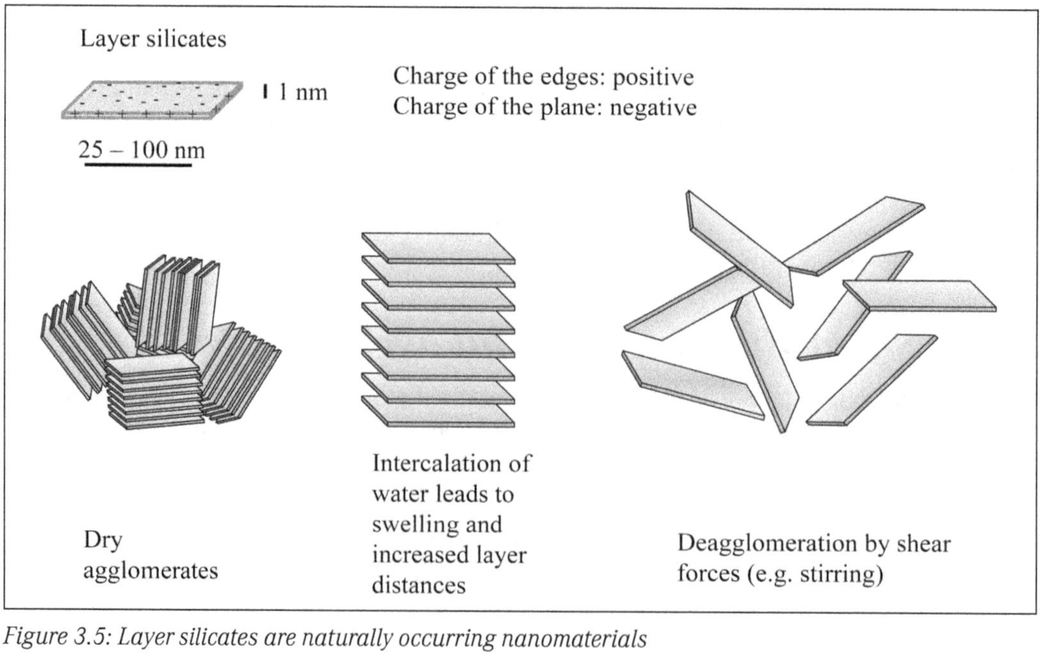

Figure 3.5: Layer silicates are naturally occurring nanomaterials

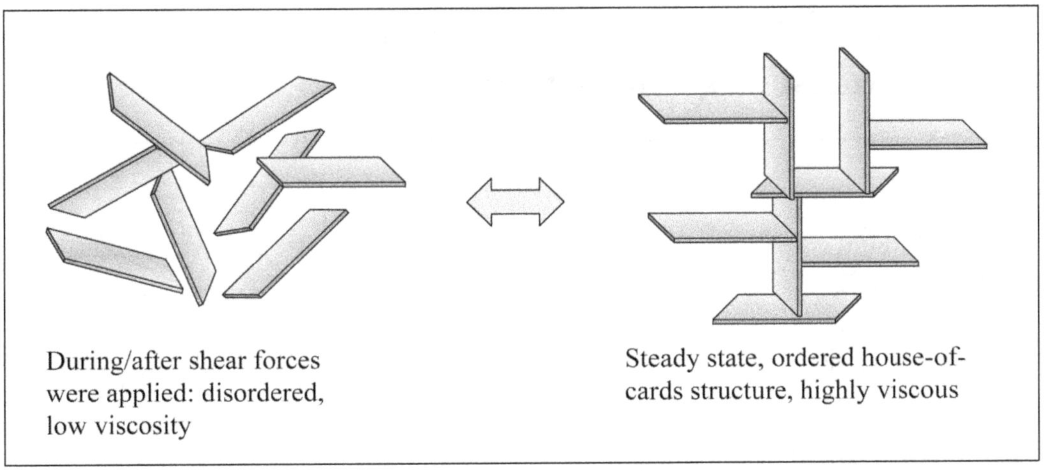

Figure 3.6: Attraction between the charges of the edges and the plane leads to the reversible formation of a house-of-cards structure, which increases the viscosity of the dispersion

is built, which increases the viscosity or even leads to gelation. If shear forces are applied, the structure is destroyed and the viscosity drops again.

Applications for layer silicates are ranging from cat litter to sophisticated water-based coating formulations. The synthetically produced Laponite or Hectorite is preferred for coating applications due to its transparency and color neutrality.

The opposite of shear thinning is shear thickening or dilatancy. Viscosity increases, as soon as shear forces are applied. The effect is known for concentrated dispersions of nanoparticles, when the dispersed phase cannot move fast enough when stirred. The same effect can be observed during walking on wet sand on the beach. When the sand contains just the right amount of water, it becomes dry and hard under the pressure of the foot.

For demonstration purposes, a dilatant dispersion can be formulated easily from starch or fumed silica. To water which contains 20 % glycerol, the powder is added under manual stirring until lumps show up which flow together again when stirring stops.

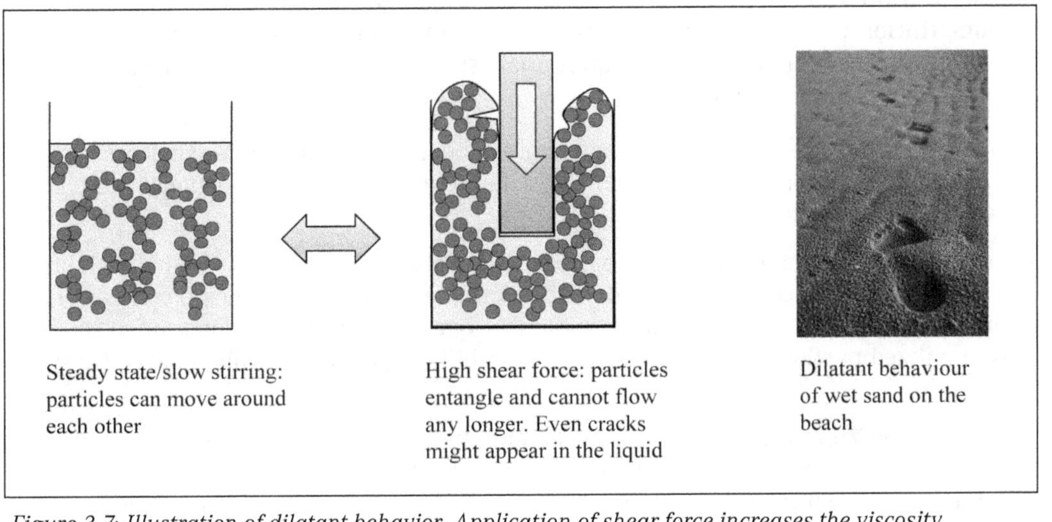

Figure 3.7: Illustration of dilatant behavior. Application of shear force increases the viscosity
Source picture: Franziska Elsner

3.3 *Stabilization of nanoparticles against agglomeration*

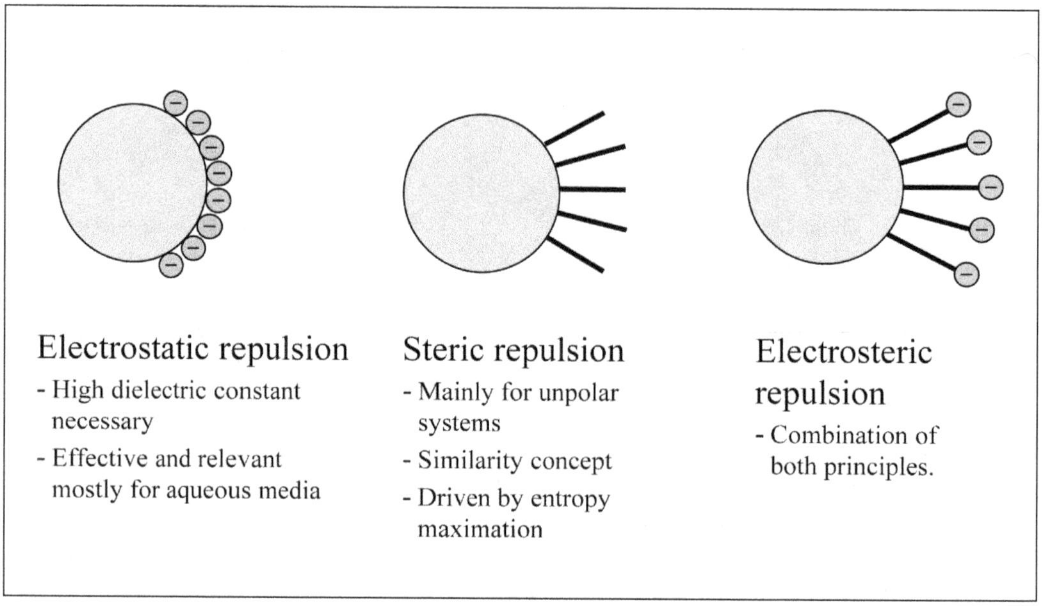

Figure 3.8: Schematic picture of the stabilization principles of colloids

In spite of their usefulness to adapt viscosity, the focus during formulation with nanoparticles in most cases is to avoid agglomerates. The agglomerates can cause an increased viscosity, precipitation and a decreased shelf/pot life.

A general rule is: "No formulation without stabilization". Nanoparticle dispersions are thermodynamically unstable and this becomes manifest in many ways [54].

- **Gelation,** by building of three-dimensional networks.
- **Coagulation** or precipitation of agglomerated or aggregated particles.
- **Flocculation** by reversible bonds between open and voluminous structures.
- **Coacervation,** if particles with an appropriate surface modification separate themselves from the rest of the solvent.

Dispersions which are not stabilized agglomerate, because the van der Waals forces attract the particles if they approach each other. These forces are inversely proportional to the sixth power of the particular distance.

Three principles of stabilization are possible, which also can be combined with each other. The repulsion of two equally charged surfaces is the force which keeps electrostatically stabilized particles at distance. Hard or flexible barriers prevent agglomeration of sterically stabilized particles and electrosterically stabilized particles exhibit charge carrying barriers.

All three stabilizing principles provide an energetic barrier, which prevents the particles from reducing their surface energy by agglomeration. This barrier however has to be higher than the kinetic energy of the particles to keep the dispersion in a stable condition.

In the following pages, the basics of the different stabilizing principles are discussed.

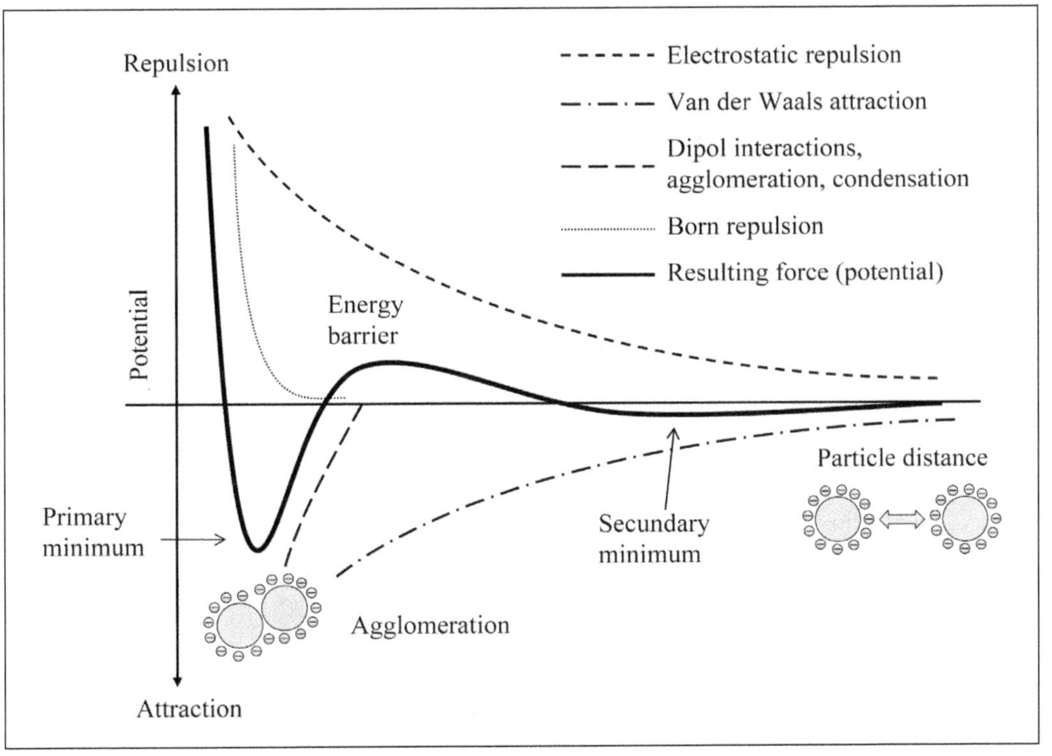

Figure 3.9: Repulsive and attractive forces develop when electrostatically stabilized particles approach each other

3.3.1 Electrostatical stabilization

The electrostatical stabilization is a common measure to prevent agglomeration of aqueous nanoparticle dispersions. The DLVO theory is the foundation of this stabilization method [55–57].

Inorganic particle surfaces act as acids or bases in contact with water. The existence of ionic groups on their surface thus is a logical result.

The surfaces of charged particles are partially neutralized by small amounts of counter ions, which build up a diffuse layer around the particles. This setup leads to an electric potential, which decreases with increasing distance to the surface. A negatively charged colloid thus is surrounded by a diffuse layer of positively charged ions.

When looking at the colloid from a distance, it seems to be neutral, because the particle charge is neutralized by the diffuse layer. When the particle moves, the diffuse ion layer cannot move at the same speed and a part of the layer is sheared off. This results in a measurable charge.

So a first layer of immobilized counter ions (Stern-layer) is distinguished from the mobile diffuse layer. The electric potential at the intersection from Stern to diffuse layer is called zeta potential, a very valuable indicator to determine the colloidal stability of dispersions [58, 71].

Table 3.1: Relative permittivity ε_r (dielectric constant) of different solvents

Particle	IEP	Polarity at pH 7
SiO_2	2.0 to 2.5	negative
Al_2O_3	8.5 to 9.5	positive
ZrO_2	6 to 7	neutral
MgO	11 to 12	strongly positive
TiO_2	5 to 6.5	weakly negative

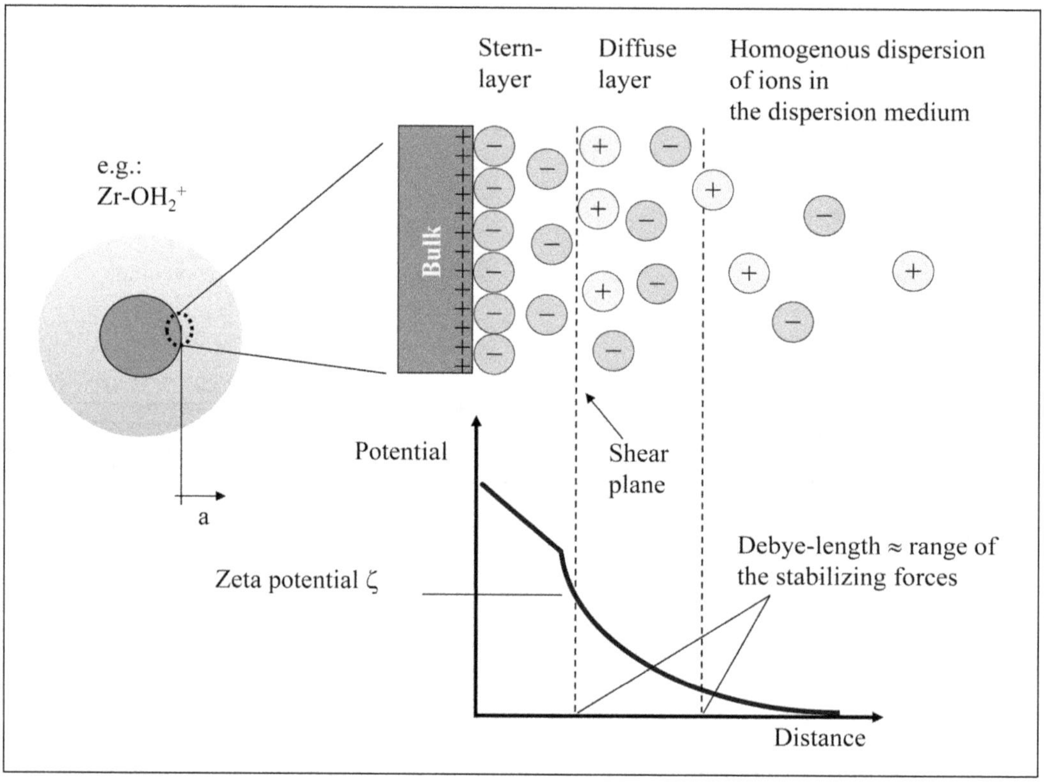

Figure 3.10: Illustration of the electrical double layer on the surface of charged colloids

The so-called Debye-length is dependent on the ion concentration and the temperature. It is an indicator for the range of the stabilizing force.

Figure 3.11 shows, that the range of the repulsive force, besides by the ion concentration is influenced also by the dielectric constant of the solvent and its temperature.

The higher the dielectric constant and the temperature, the larger the range of the repulsive force becomes. However raising the temperature also increases the kinetic energy of the particles, so in summary an increase in temperature decreases the colloidal stability.

The repulsive forces get stronger with decreasing ion concentration and increasing zeta potential. Therefore colloidal synthesis is usually carried out at the lowest possible ion strength and far away from the isoelectric point [5].

Proper control of the surface charge density of colloids is essential for the colloidal stability, a direct measurement however is hardly possible. What can be measured, is the zeta

$$\zeta \propto \exp(-\kappa)$$

The higher the temperature,
the higher the dielectric constant DC and
the lower the ion strength of a dispersion,

$$\kappa \propto \sqrt{\frac{\sum c \cdot z^2}{DC \cdot T}}$$

the higher is the electrically repulsing
electrical potential

$1/K$ = Debye-length = „thickness of the ion cloud“

Figure 3.11: The zeta potential depends on the ion strength and the dielectrical constant of the medium

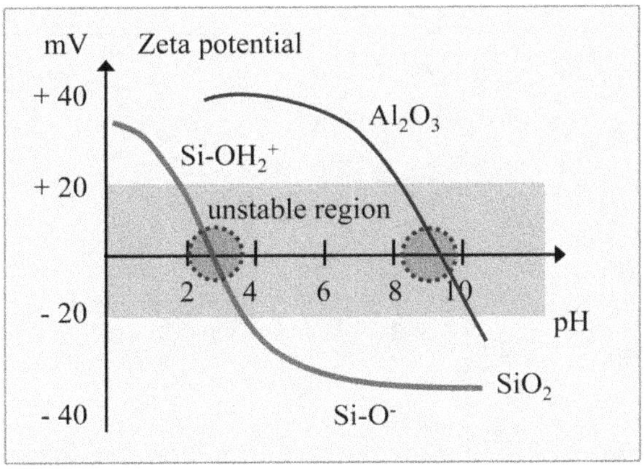

The zeta potential gives a first hint of the stability of an electrostatically stabilized colloidal dispersion at a given pH. Values < 20 mV in general are not sufficient to prevent agglomeration.

Figure 3.12: Zeta potential of SiO_2 and Al_2O_3 particles in water. The circles mark the isoelectric point

potential of moving particles. By the movement, a part of the loosely associated ion cloud is deformed and the particle no longer seems to be neutral. By observing the movement in an electrical field, thus the zeta potential can be measured.

Figure 3.12 shows the typical zeta potential curve of SiO_2 and Al_2O_3 particles in water at different pH values. The zeta potential can be positive or negative, depending on which species is dominating the colloidal surface. A first estimate of the colloidal stability can be made by the absolute value of the zeta potential. At values higher than 20 mV and lower than -20 mV, colloidal dispersions are considered as stable.

The zeta potential can be increased or decreased by a shift of pH and/or modification of the surface. For the purpose of surface modifications, ions or charge carrying organic molecules can be used (compare Figure 3.13).

Figure 3.12 shows the isoelectric point IEP as dotted circle. This is the pH value, at which a surface shows no electric charge; positive and negative charges are perfectly balanced.

Usually the IEP marks the pH value of highest instability of dispersions. SiO_2 particles however show an exceptional high stability at their IEP. One reason is that the Si-OH groups cover the particle surface with a hydrophilic layer which minimizes the interface energy. But this alone would not be sufficient to explain the extraordinary stability of SiO_2 in contrast to the other OH-group bearing oxides. One decisive difference can be found in the reactivity of the Si-OH groups compared to metal hydroxides. Without a catalytic activation, they are very inert and thus lead to a kinetic stabilization [5, 57, 137].

Another aspect of the SiO_2 anomaly is the exceptionally low Hamaker constant of SiO_2, which describes the van der Waals forces between two particles. Compared to e.g. titanium dioxide, the attracting forces between SiO_2 are about 35-fold smaller and therefore even adsorbed water can act as an efficient barrier to prevent agglomeration.

Table 3.2: Isoelectric point IEP of different oxides [59]

Medium	εr
Vacuum	1.0
Paraffin	2.2
Benzene	2.3
n-Propanol	18.3
Methanol	32.6
Glycerol	42.5
Water	80.1

Mixing of colloidal dispersions

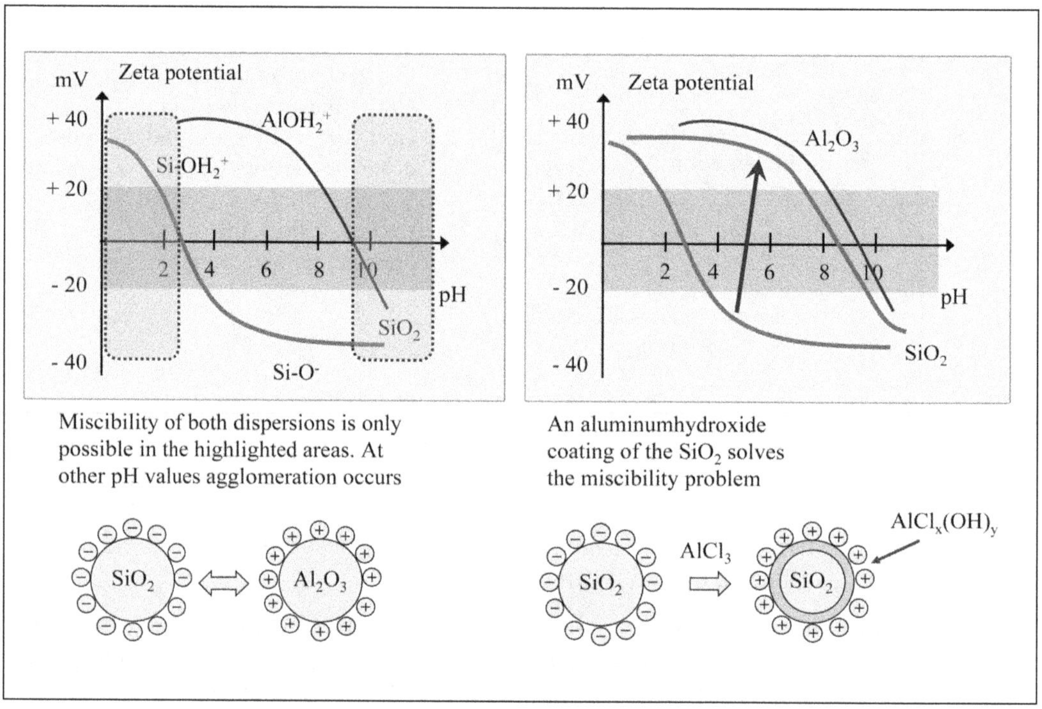

Figure 3.13: The zeta potential is an important parameter which has to be controlled when colloidal dispersions are to be mixed

Two colloidal dispersions can only be mixed if they show a comparable zeta potential, otherwise agglomeration occurs. Figure 3.13 shows a concept to improve the miscibility of SiO_2 und Al_2O_3 by a surface modification of SiO_2 with polymeric aluminum chloride.

For practical applications, two concepts can be used to realize an effective electrostatical stabilization of colloids.

- Increasing the charge density and thereby the zeta potential
- Increasing the range of the repulsive forces (the Debye length)

The charge density can be maximized by adjusting a suitable pH value. Via acid/base reactions with the particle surface, the pH determines the amount of charged species at the surface. The adsorption of counter ions on the particle surface is, besides other factors, limited by the available space. Multivalent ions like phosphates can lead to a higher charge density than it would be possible with monovalent ions.

A practical application can be found in the liquidification of clay mineral dispersions with phosphates [60] or in the stabilization of SiO_2 dispersions with aluminum chloride. It is important to avoid an excess of multivalent ions because according to the Schulz-Hardy rule, they can also cause flocculation [27].

A further possibility to increase the charge density and the zeta potential is the incorporation of dopants, which cause charge carrying lattice defects. Examples are colloidal silica sols, which were modified with sodium aluminate during the production process. This leads to a permanent negative charge and a significantly improved stability over a wide pH range (compare “Electrostatical stabilization”, page 41).

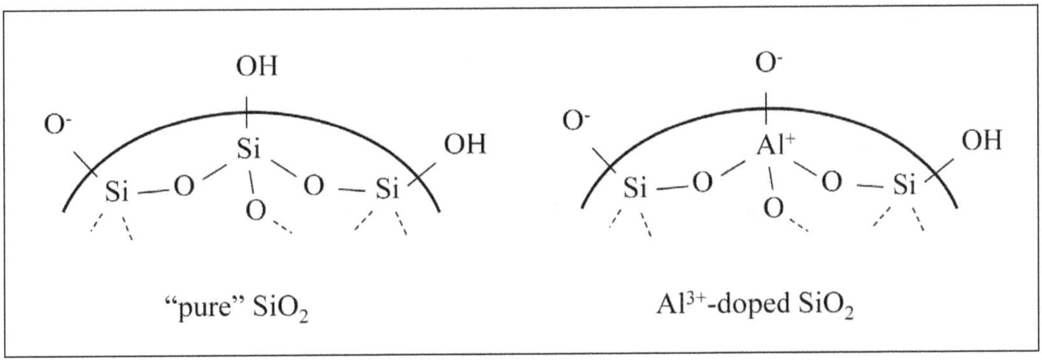

Figure 3.14: An aluminate-doped colloidal silica sol exhibits a higher surface charge density over a wider pH range, which leads to better electrostatical stabilization

The range of the repulsive forces (Coulomb repulsion) is determined by the dielectric constant and the concentration of electrolytes in the dispersion medium. The range can be extended by avoiding or removing ionic impurities and the selection of media with a high dielectric constant.

This point also explains the problems which have to be faced when water-based dispersions shall be transferred to organic media like solvents or resins. Usually these media exhibit a very low dielectric constant and so the range of the repulsive Coulomb forces shrink below the range of the attractive van der Waals forces. The consequence is agglomeration (compare "Sterical stabilization", page 46).

➤ ***Example 7: Preparation of cationically stabilized nanoparticles [61]:***

2.78 parts by weight of boehmite nanoparticles ("Disperal P3", registered trademark from Sasol Germany) were added with stirring to a mixture of 25 parts by weight of 1N acetic acid and 2.5 parts by weight of deionized water. The resultant mixture was treated in an ultrasound bath for three minutes until the boehmite nanoparticles had dissolved.

Boehmite nanoparticle dispersion

Boehmite is a needle shaped nanoparticle which can be bought in acetate-stabilized form. Acetic acid forms a chelate with the aluminum ions on the surface and builds up a sterical stabilization. Residual free Al-OH groups carry a positive charge at low pH values, which stabilizes the particle via Coulomb repulsion. Propionic acid or stearic acid with their long and hydrophobic alkyl groups would not be suitable for water, but would be the right choice for polar media (compare "Sterical stabilization", page 46).

A dispersion with the aid of an ultrasonic bath is only possible in cases, when the particles can be dispersed very easily. If agglomerates have formed already, the week shear forces in the ultrasonic bath are not sufficient to individualize the particles. Bead mills or a 3-roll mill are more suitable measures to disperse such particles.

In this example, a significant amount of acetic acid is added. It is only fair to ask the question, where an acetate-stabilized nanoparticle ends and an aluminum acetate solution begins.

Some commercially available nanoparticle dispersions contain up to 25 % by weight surface modifier (e.g. citric acid, nitrate, chloride, tetraalkyl ammonium, ...). Prior to a practical application of such dispersions, the user has to evaluate possible side effects of these amounts of salts or organic modifiers.

3.3.2 *Sterical stabilization*

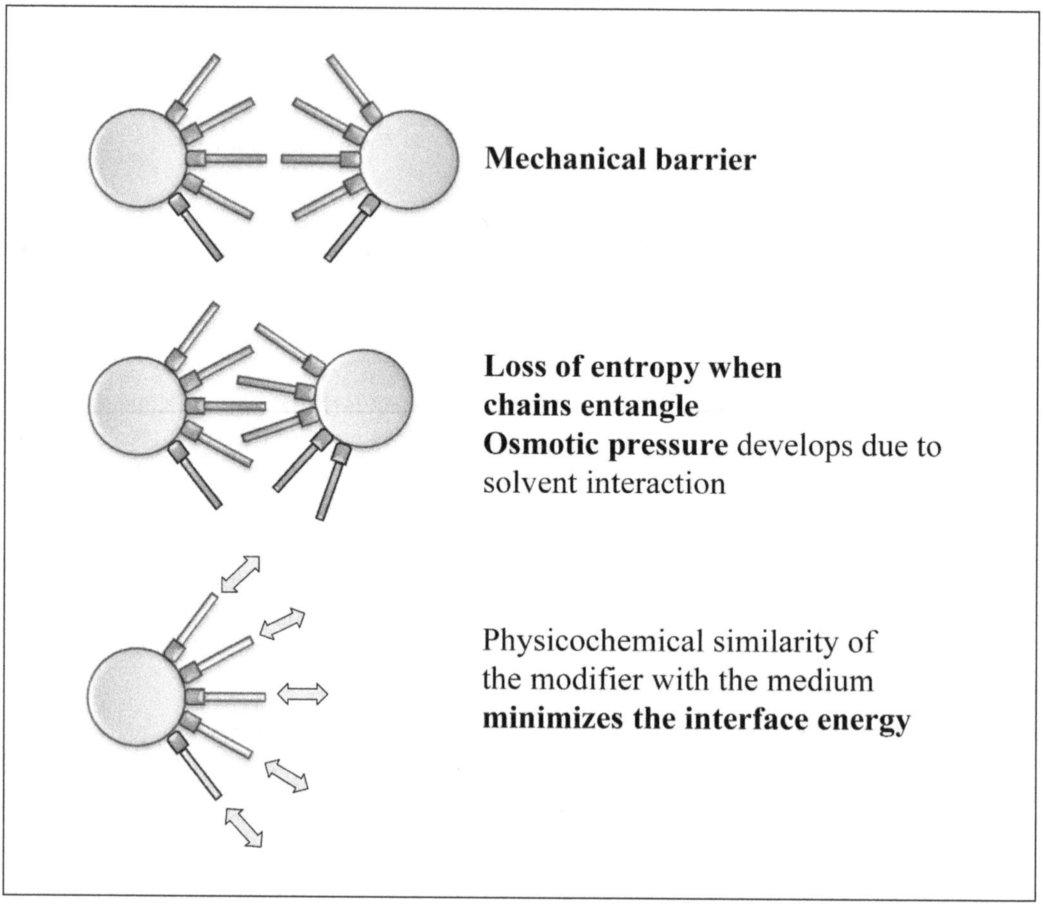

Figure 3.15: Illustration of the working principles of a sterical stabilization

Via electrostatic repulsion, it is possible to produce stable colloidal dispersions in water. To synthesize nanocomposites however, organic solvents are needed as compatibilizers for the transfer of nanoparticles to the organic resins.

Electrostatic repulsion as a stabilizing mechanism is almost useless in organic media, because the low dielectric constant limits the range of the repulsive forces. The high interfacial energy which develops between hydrophilic particle and hydrophobic resin is a massive stimulus towards agglomeration. Thus another stabilizing principle has to be chosen to prevent nanoparticles from agglomeration in organic media.

Since the early days of colloidal chemistry, organic protective colloids like gelatin or bone glue are used to stabilize aqueous particle dispersions. They adsorb onto the particle surface and act as hydrophilic barriers [7]. Instead of gelatin or bone glue, today mostly synthetic polymers like polyvinylpyrrolidone (PVP), polyethylenimine (PEI), polyvinylalcohol (PVA), hydroxypropyl cellulose or proteins are used [5, 7].

Besides their stabilizing properties, sterical stabilizers can also be used to tailor the particle size during the synthesis of dispersions via precipitation reactions. With increasing amount of stabilizers relative to the product, smaller particles are generated (compare Figure 3.18, page 49).

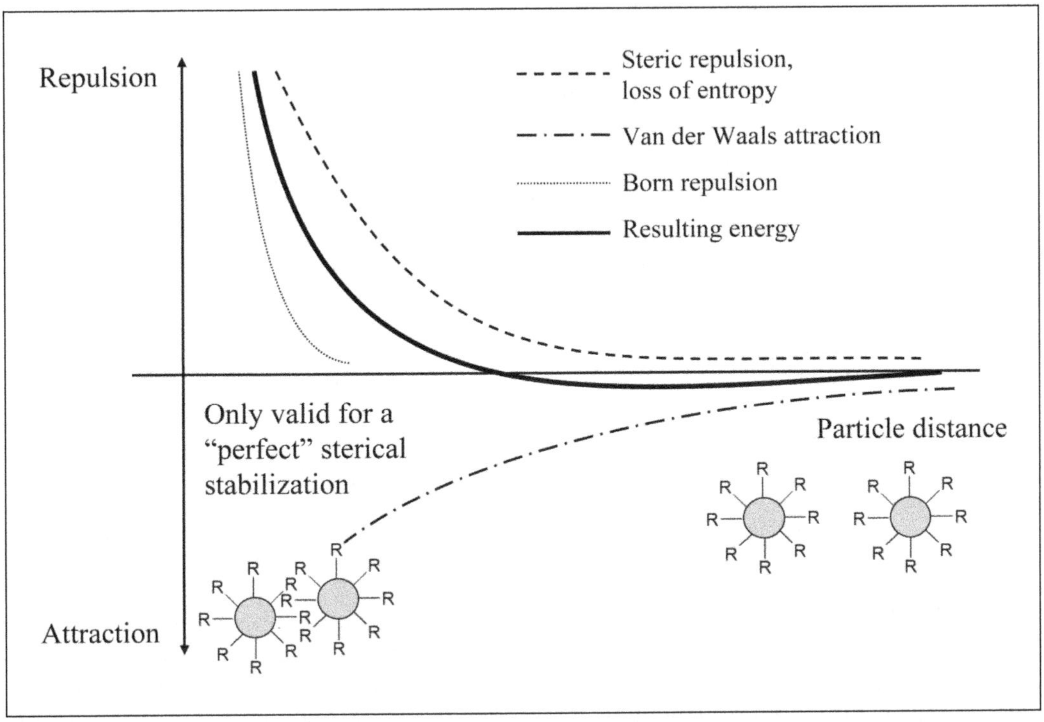

Figure 3.16: Repulsive and attractive forces determine the interaction of sterically stabilized particles

Especially for organic solvents, the surface modification of particles with sterical stabilizers is of great use. Their action is the sum of multiple properties, which are summarized in Figure 3.15. Organic molecules of a significant length can act as spacers and prevent a reaction of the surface groups of particles. Sterical stabilizers are usually chosen by considering both the reactivity of the particle surface and the polarity of the medium. According to the "similarity principle", sterical stabilizers work best if they show a similar chemical structure like the dispersion medium.

Between similar materials, the interfacial energy is low and no driving force towards agglomeration besides van der Waals forces develop. When two sterically stabilized particles approach each other, the stabilizing molecules interpenetrate and thereby reduce their mobility. As a consequence their entropy sinks. The tendency of the system to maximize entropy and the osmotic pressure of the solvent generate a counteractive force which drives the particles apart and prevents agglomeration.

In literature, the total energy curve of two approaching sterically stabilized particles in most cases is drawn as if these particles are shielded to 100 % by the modifier and thus are completely inert (compare Figure 3.16). This however is not true in the real world.

An idealized picture like Figure 3.16 means, that these particles could approach each other until they touch without leading to irreversible agglomeration. It can be observed however, that if sterically stabilized particle dispersions are dried, in most cases they cannot be redispersed to primary particle size again.

The reason for this behavior is the presence of residual reactive groups (Me-OH) on the surface which attract each other via hydrogen bridges and form covalent bonds. The acid test for the quality of a sterical stabilization therefore is the drying and redispersing of a colloidal dispersion. If the dispersion can be reduced to primary particle size again, the stabilization is excellent.

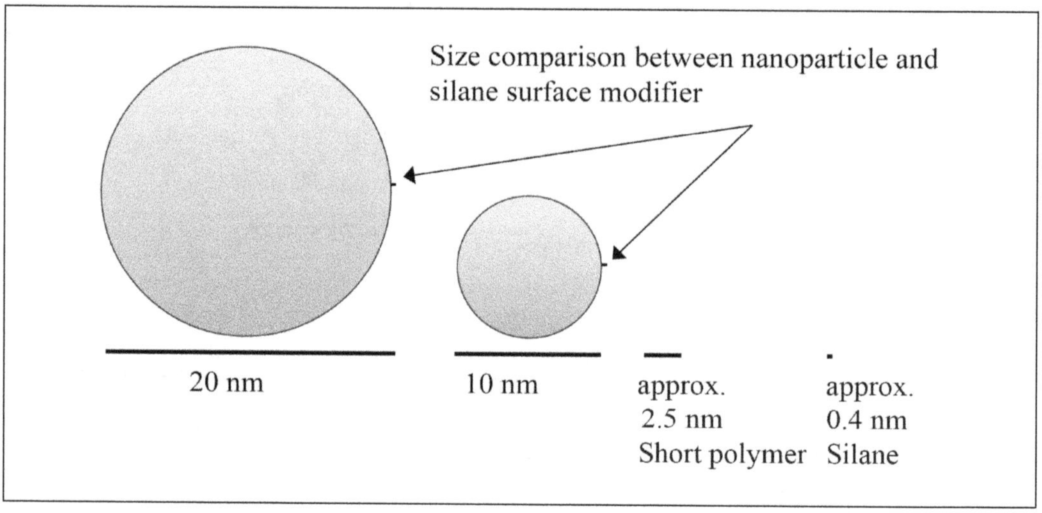

Figure 3.17: The sketch suggests that monolayers of typical silanes are not suitable to prevent nanoparticles from agglomeration. Reactive groups on the surface of the particles have to be covered completely

Figure 3.17 illustrates, why monolayers of conventional organosilanes are hardly suitable to provide a good sterical stabilization. Their small size and often incomplete reaction with the surface groups leave reactive sites which can lead to agglomerates. Only under harsh reaction conditions, a quantitative shielding of all reactive groups is possible (compare Example 13, page 58). To be effective, surface modifiers should exhibit a minimum length.

The requirements of a sterical stabilization can be summarized as follows:

- strong, if possible covalent bonds to the particle surface
- high coverage, leaving no free reaction sites
- sufficient length to maximize entropy and to be able to act as mechanical barrier
- intense interaction with the dispersion medium
- compatibility with other components of the dispersion

The compatibility of sterically stabilized particles can be influenced by a change of the composition of the dispersion medium (e.g. solvent exchange). If agglomeration occurs, the particles can be redispersed in a more suitable solvent if the sterical stabilization was complete. A broad variety of surface modifiers are known, which in general use five binding principles to particle surfaces.

Sterical stabilizers can be added to colloidal dispersions or even used during their synthesis. If chosen correctly, self-stabilizing colloidal dispersions are generated, in which the particle size can be controlled by the modifier-to-educt ratio [62, 63]. This can help to prevent

Table 3.3: Binding principles of sterical surface modifiers β

Principle	Example
Organic complex formers	Polyethyleneoxide, β-diketones, carbonic acids
Organic salts	Carboxylates, amines, tetraalkylammonium compounds, organophosphorous compounds
Silylation	Organofunctional silanes, inert and reactive
Esterification	Long chain alkyl or aryl alcohols like octanol, stearylalcohol, benzylalcohol
Polymer encapsulation	Adsorption, entanglement, often in combination with complex formation

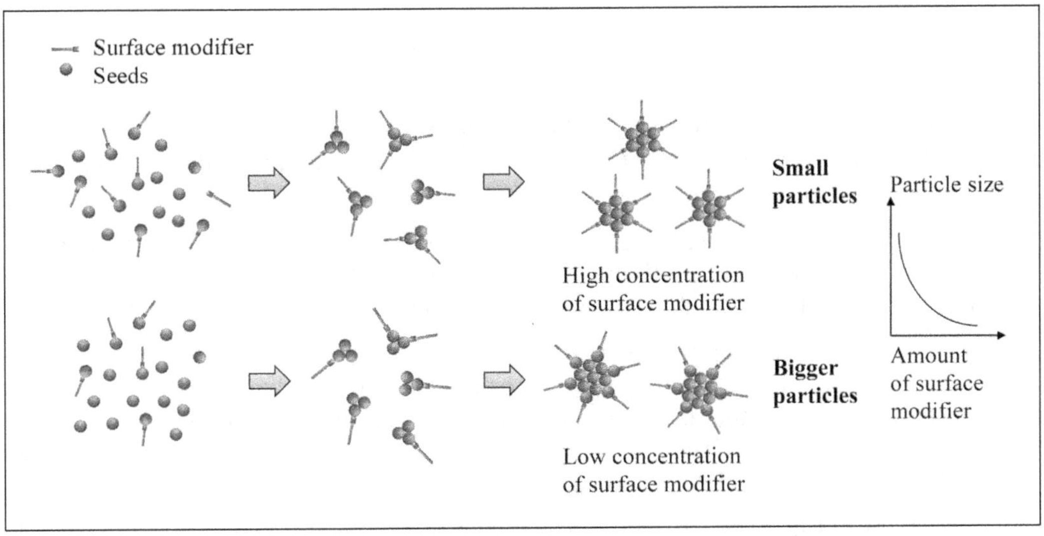

Figure 3.18: The particle size is influenced by the amount of surface modifier which is present during the reaction [62, 63]

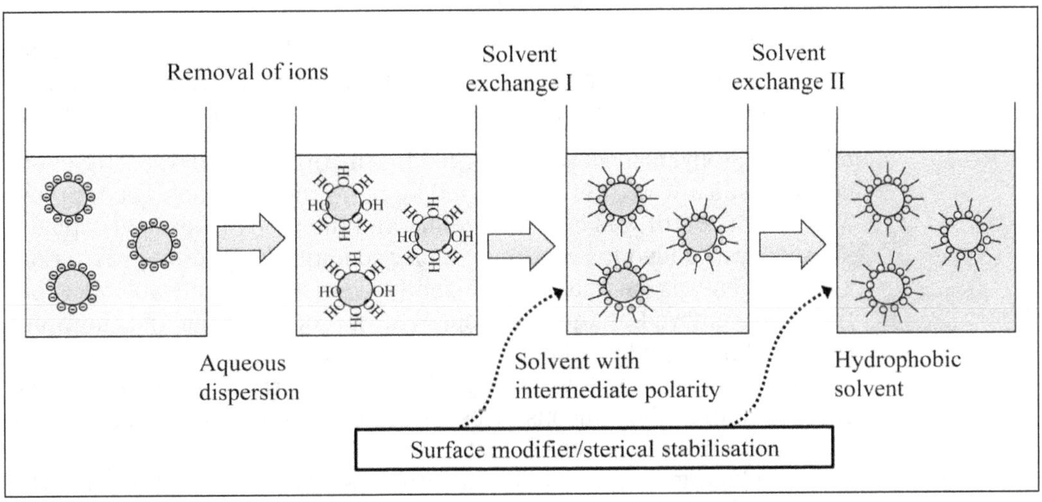

Figure 3.19: Best practice when transferring aqueous nanoparticle dispersions to organic media

agglomerates and to realize homogeneous, monomodal particle size distributions. Figure 3.18 illustrates the principle.

If compositions tend to crystallize, surface modifiers can help to influence the morphology (e.g. spheres, fibers, platelets, cylinders) of the resulting crystals. This principle is also used to prevent lime deposits by the addition of polycarboxylates to water [64-67].

The main use of sterical stabilizers is the transfer and stabilization of nanoparticles in organic media. Often starting from aqueous dispersions, the question arises how to replace the electrostatical stabilization by a sterical stabilization without causing agglomerates during the solvent exchange process. Because many organic dispersion media are not miscible with water, cosolvents which are both miscible with water and the organic medium can be of use. Figure 3.19 and Example 8 show the procedure for SiO_2.

In a first step, the electrostatically stabilized SiO_2 dispersion is deionized to shift the pH to the isoelectric point (approximately pH 2 to 3). The electrostatical stabilization by Si-O$^-$ Na$^+$

groups thereby is replaced by a Si-OH stabilization. For other colloids, this approach cannot be recommended, because only SiO_2 shows the exceptionally high stability at the isoelectric point which is called "SiO_2 anomaly".

A water-miscible solvent can be used in the following step to remove water by distillation. A prerequisite for this process is that a suitable azeotrope exists, which allows an efficient distillation process.

During this step, a first surface modification of the colloids makes sense. After the water has been removed, the target medium can be added without causing agglomeration if the surface modification has been chosen accordingly to the already discussed principles.

When choosing the intermediate compatibilizing solvent, it is important to use solvents which have a lower boiling point compared to the target medium so that they can be removed easily in a final process step. Isopropanol, n-propanol, tetrahydrofurane and acetone are suitable and also allow the fast removal of water by distillation. Decisive however is the colloidal stability during and after mixing them with the water-based colloidal dispersion.

Many variations of this basic procedure are known and some of them are used in the following examples.

➤ *Example 8: Preparation of a SiO_2 alcosol in n-propanol* [68, 69]

The alkali ions of an aqueous dispersion of colloidal silica (17 nm in diameter, 30 % SiO_2 by weight, stabilized with alkali) were removed with an acidic ion-exchanger until the pH of this sodium-free sol was about 3.5.

6 kg of n-propyl alcohol were charged into a still, separately, 3.2 kg of the collodial silica were diluted with 8.0 kg of n-propyl alcohol. The n-propyl alcohol in the still was heated to reflux and the propanol-diluted colloidal silica sol was added to the alcohol under conditions of thorough agitation. The addition was carried out sufficiently slow, so that the water was continuously removed as the azeotrope with n-propyl alcohol and the water content of the liquid was kept low (less than 15 %), by distilling water from the still as rapidly as it was being added in the form of the aqueous alcosol. When all of the aqueous alcosol had been added, the distillation was continued until the water content in the sol was reduced to below 0.5 %, thus producing an essentially anyhdrous silica alcosol in the still. The anhydrous alcosol prepared in this manner contained 19.9 % solids, which was practically all colloidal silica, and the water content was 0.53 %.

Preparation of SiO_2 dispersions

Aqueous silica sol is a cheap raw material for the preparation of organosilica sols. The cation exchange removes traces of sodium and thereby the electrostatic stabilization. The ion exchanged silica sol now exhibits a pH value near the isoelectric point IEP at which it shows an unusual stability ("SiO_2 anomaly").

In the following solvent exchange process, n-propanol reacts with the Si-OH group and forms Si-O-pr groups which act as sterical stabilizers for the organosilica sol (compare Example 14, page 58). The procedure can also be adapted for longer chain alcohols, but the reaction conditions to achieve complete esterification can be quite harsh.

Typically, the temperature of the dispersion is increased gradually up to the boiling temperature of the surface modifier. Water and lower boiling solvents leave the dispersion. It takes up to some hours at boiling temperature until the esterification is complete. Maintaining an inert gas atmosphere during this process is mandatory to avoid partial oxidation of the surface modifier. Even redispersable nanopowders are accessible via this method (compare Example 13, page 58).

➤ *Example 9:* *Preparation of a SiO_2 alcosol in butyl acetate* [70]

920 g of a commercially available silica sol dispersed in methanol (trade name: "MA-ST", manufactured by Nissan Chemical Industries, Ltd., SiO_2 concentration: 40 % by weight, water content: 2.3 % by weight, average particle diameter: 22 nm) was placed in a reactor and 103 g of n-butyl trimethoxysilane (Tama Chemical Co., Ltd.) was added dropwise over 10 minutes with stirring.

Then, the temperature was increased to 50 °C and the aging was carried out for 1 hour at this temperature. Thereafter, the solution was transferred to a 2 l flask and distillation was carried out at a pressure of 150 mm Hg in a rotary evaporator while n-butyl acetate was charged. Further, the pressure was gradually reduced to 70 mm Hg, and distillation was continued with addition of butyl acetate. When the total addition amount of n-butyl acetate reached 1,820 g, solvent replacement was completed. The resulting silica sol dispersed in n-butyl acetate had SiO_2 concentration of 30 % by weight, viscosity of 3.3 mPa·s, and n-butyl alcohol concentration of 1.2 % by weight.

SiO_2 alcosol in butyl acetate

The transfer of an aqueous silica sol into hydrophobic organic solvents needs an intermediate solvent which is compatible with both water and the hydrophobic solvent. Organosilica sols in alcohols, like methanol or isopropanol can be bought or prepared quite easily and are compatible with a variety of more hydrophobic media due to their Si-O-alcohol groups.

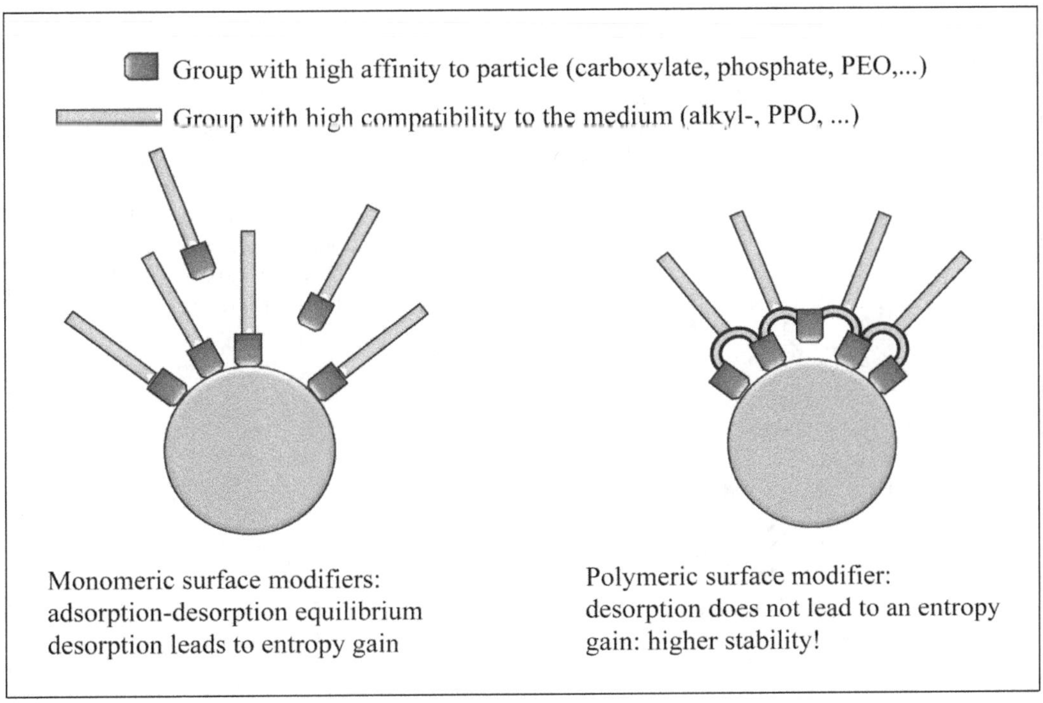

Figure 3.20: Monomeric and polymeric surface modifiers for particles

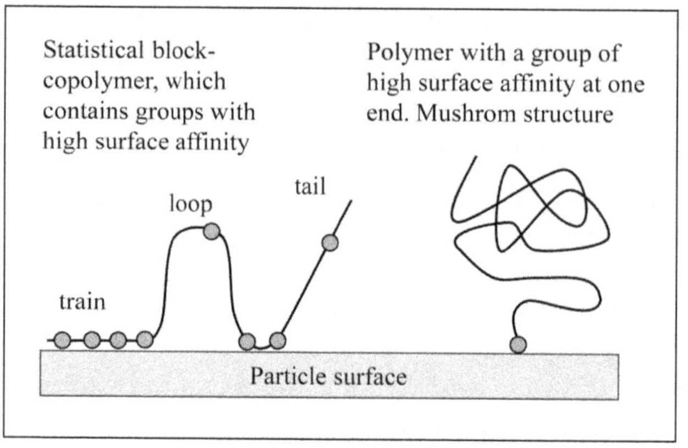

Figure 3.21: Arrangement of polymer chains on surfaces depending on their content of groups with high surface affinity

If necessary, additional surface modifiers can be used which should be chosen accordingly to the similarity principles for sterical stabilizers.

For the target solvent butyl acetate, the n-butyl group is a reasonable choice. For a reactive sterical stabilization, e.g. methacryloxypropyl trimethoxysilane would have been a good modifier. The residual free Si-OH groups and traces of water are sufficient to ensure a reaction of the silane with the organosilica sol. Even if some alkoxy groups remain unreacted, they will participate in the stabilization of the dispersion (compare "Hydrolysis and condensation", page 94).

Polymeric sterical stabilizers can show advantages compared to monomers. Figure 3.20 illustrates the entropy gain of monomers when they desorb. The entropy gain of polymers, when just one binding site desorbs is much smaller, therefore they can stabilize particles more effectively than monomers.

There are different ways possible how polymers adsorb to surfaces. A high density of binding sites leads to "trains", whereas parts of the polymer with fewer binding sites orient themselves more to the dispersion medium and form "loops", "tails" and "mushrooms".

Depending on their length and binding site density, polymers can lead to agglomeration via bridge-flocculation or mosaic-adhesion like shown in Figure 3.22.

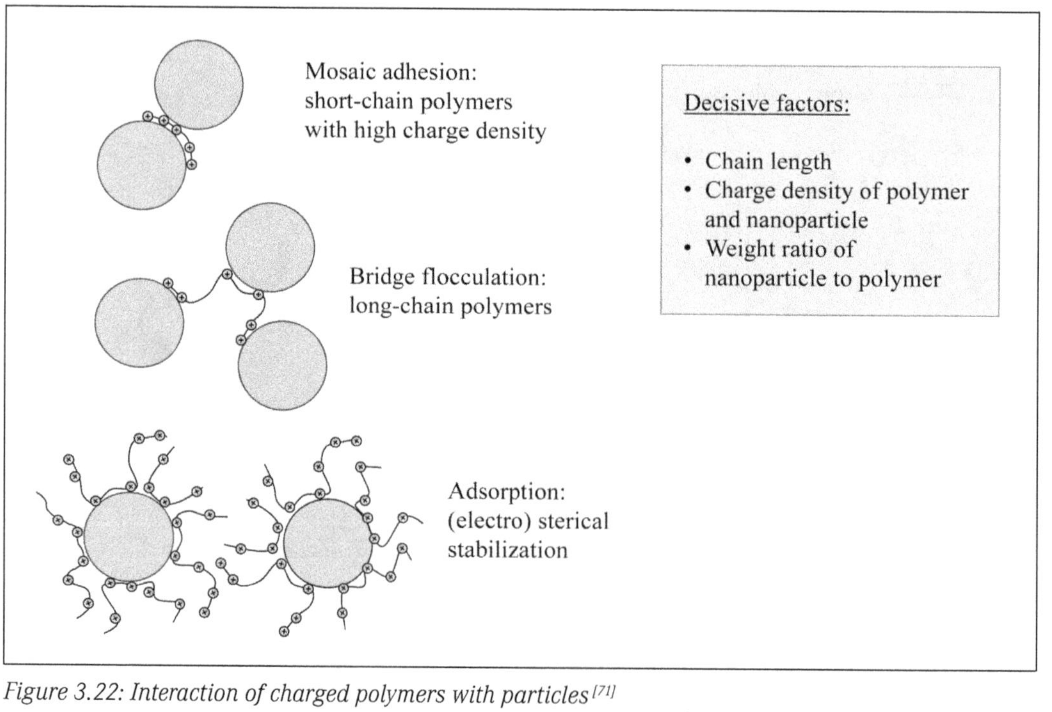

Figure 3.22: Interaction of charged polymers with particles [71]

If the polymers encapsulate the nanoparticle completely, a core/shell structure is formed. This polymer shell can be generated by three different grafting methods (grafting-to, grafting-through and grafting-from) and is considered to be an effective measure of stabilization. In a grafting-to process, reactive polymers are brought into contact with the nanoparticles and react with their surface.

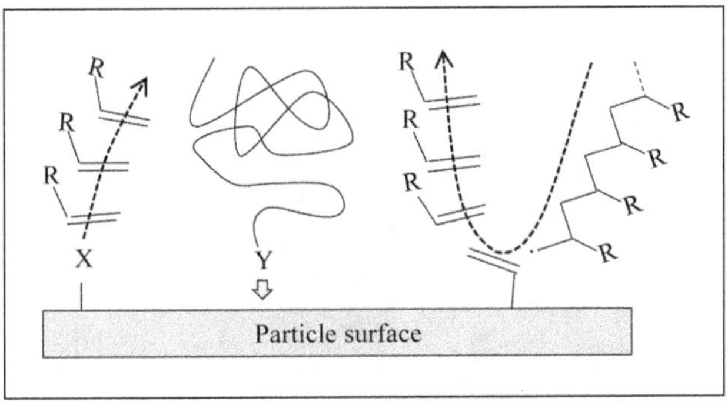

Figure 3.23: Surface modification by grafting-from, -to and -through processes. X: radical starter, Y: group with high affinity to particle

During a grafting-from process, the monomers are polymerized in the presence of nanoparticles which carry on their surface initiator groups. Because monomers need less space, as a general rule a more complete coverage of the surface reactive groups of the particle can be achieved compared with the polymer grafting-to process.

In a way which is comparable to the grafting-from process, in the grafting-through process, the polymer is synthesized in the presence of surface modified particles. In this case however, polymerizable groups are used as surface modifiers, which take part in the polymerization process [5, 72].

➤ *Example 10: Preparation of polymer-encapsulated SiO_2 particles* [73]

A mixture of 8.55 g 2-(ethylhexyl) methacrylate, 3.88 g (polyethylene oxide)-methacrylate, 0.132 g mercaptoethanol and 12 g toluene is heated to 70 °C. 0.066 g AIBN, dissolved in 12 g toluene is added dropwise and thereafter the reaction mixture is stirred for 24 h at 70 °C. After the reaction mixture has cooled down to room temperature, the produced copolymer is precipitated by adding methanol and dried under vacuum.

300 mg of the copolymer is dissolved in 6 ml n-heptane and 66 mg of an aqueous SiO_2 dispersion (30 % by weight SiO_2, particle size: 15 nm) is added to 2 ml ethanol. Both solutions are mixed under agitation. After 3 minutes, 0.2 ml H_2O are added and the mixture is cooled down to 0 °C to induce a phase separation. Polymer-encapsulated agglomerate-free SiO_2 particles are found in the unpolar phase. Their diameter was determined to be 37 nm.

Polymer-modified SiO_2 particles

The example describes a two-step grafting-to process, in which first a polymeric surface modifier is synthesized and in a subsequent step then is precipitated onto SiO_2 nanoparticles. The adsorption of polymers to nanoparticle surfaces can be used to hydrophobise hydrophilic nanoparticles and thereby compatibilize them with the organic phase. The nanoparticles then move to the organic phase to minimize their surface free energy and are sterically stabilized against agglomeration.

The example describes a classical radical polymerization in toluene, starting with a mixture of the unpolar ethylhexyl methacrylate and the polar PEG-methacrylate. Mercaptoethanol serves as radical chain transfer agent and helps to adjust the chain length of the resulting polymer to foster optimal adhesion and prevent bridge flocculation.

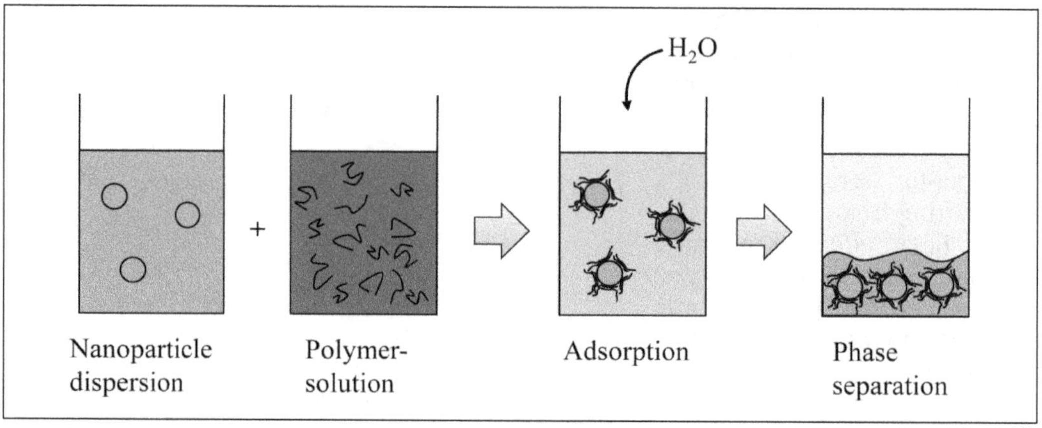

Figure 3.24: Grafting-to process, described in Example 10. By shifting the polarity of the solvent, a phase separation takes place. The surface modified particles move to the unpolar phase.

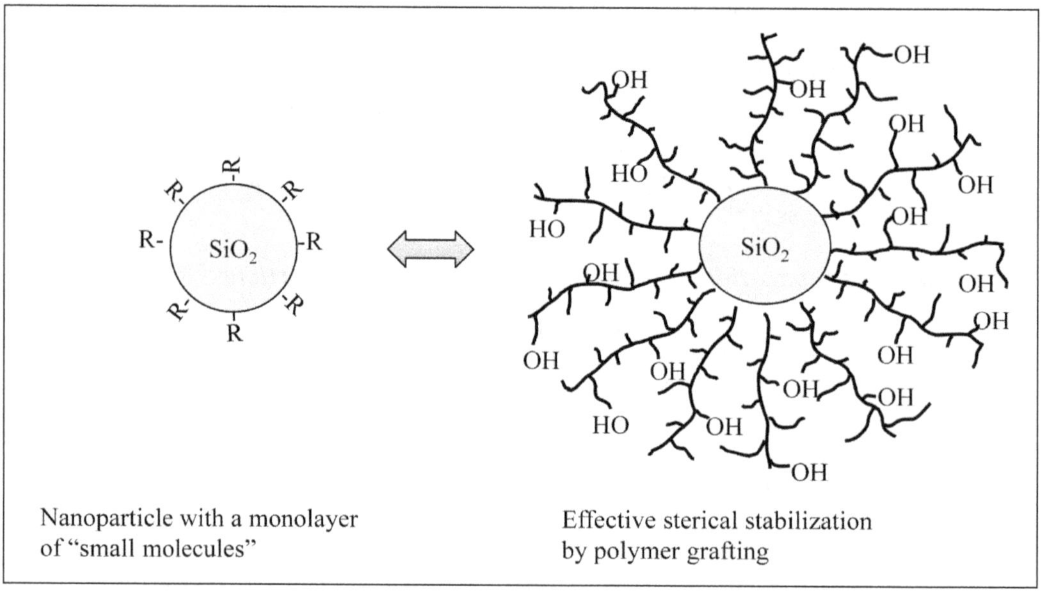

Figure 3.25: Core/shell particle in comparison to the conventional stabilization method with small molecules (e.g. silanes) [74]

In the statistical copolymer, the two components have different functions. The PEG moiety adsorbs via its oxygen brigdes to the particle surface and the ethylhexyl-moiety orientates itself towards the solvent phase and is responsible for the good compatibility with unpolar solvents. The branched ethylhexyl group is known for superior stabilizing properties compared to linear alkyl groups. A precipitation in methanol is finally used to purify the polymer from educts and other byproducts.

Aqueous silica sol can only be diluted with ethanol if it has been ion exchanged before or if it is otherwise stabilized appropriately (compare "Nanoparticles in coating resin", page 59). If brought into contact with a solution of the surface active polymer, the PEG moiety adsorbes to its surface within seconds and the unpolar rest of the polymer orientates itself towards the unpolar solvent.

A sudden increase in the polarity of the solvent during the addition of water initiates a phase separation into a water/alcohol and a heptane phase. The now hydrophobic particles try to

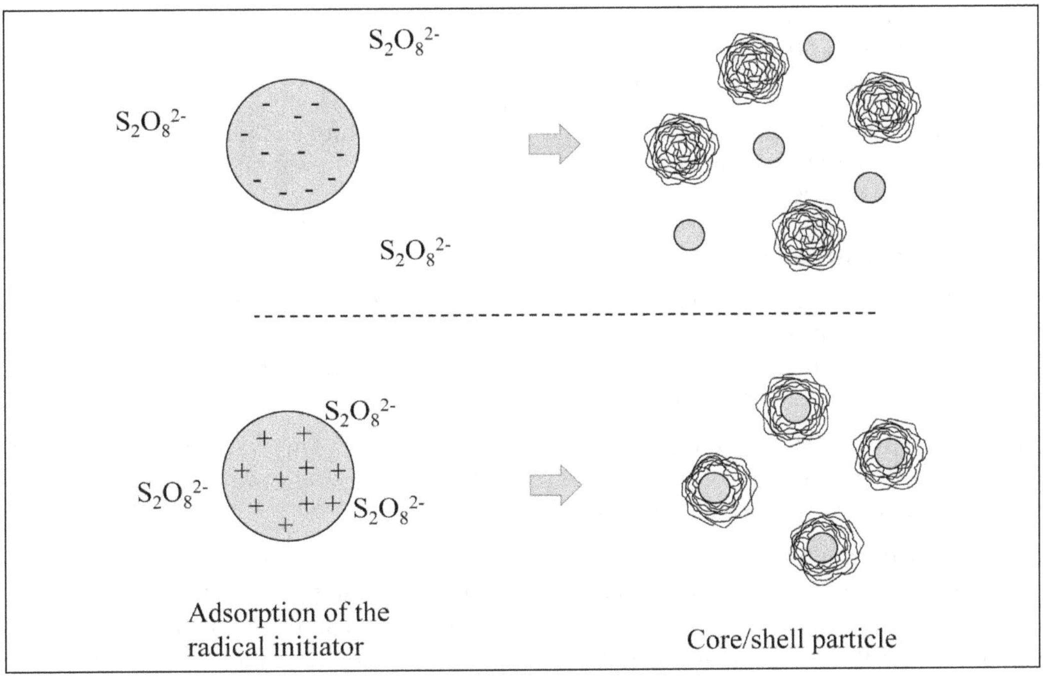

Figure 3.26: Emulsion polymerization in the presence of nanoparticles yields different results, depending on the (surface) charge of the particle and the radical initiator

minimize their interfacial energy and stay in the heptane phase. The result of this operation is a dispersion of sterically stabilized particles in a hydrophobic solvent which can be used as additives for paints or adhesives.

Polymeric steric stabilizers resemble conventional coating resins and thus are ideally suitable to stabilize inorganic nanoparticles in conventional coating resins (compare "Sterical stabilization", page 46). Such core/shell particles can exhibit excellent stability against agglomeration and good compatibility with the resin. Figure 3.25 illustrates the difference between conventional surface modifiers which are based on small molecules and a polymer- based core/shell concept. Such systems have been commercialized as additives to improve scratch resistance of coatings (compare "Improving the scratch resistance of coatings", page 82) [74].

In principle every polymerization method can be used to synthesize the polymer shell, however radical polymerization methods, either statistically or controlled, are state of the art [5, 75].

A method which is used commercially for latex binders (e.g. façade paints) is the emulsion polymerization in the presence of surface modified nanoparticles [88]. During the polymerization process, the nanoparticles are more and more encapsulated by the polymer. Depending on the weight ratio of polymer to nanoparticles, the system can be described either as polymer-modified nanoparticles or nanoparticle-modified polymers.

All these processes have in common that the polymerization takes place in proximity of the particle surface in order to ensure the incorporation of the nanoparticle into the polymer droplets. To foster this incorporation, polymerizable groups, radical initiators or chain transfer reagents can be attached to the particle surface. One important additional possibility is to adapt the surface charge of the particle to the surrounding polymer and the charge of the radical initiator.

Particles with a positive surface charge show an affinity towards negative ions like e.g. peroxodisulfate S_2O_{82-} which are adsorbed and can start the polymerization in the proximity of the particle surface. The polymer grows and encapsulates the particle. A negatively charged particle hardly adsorbs $S_2O_8^{2-}$. Polymerization then happens primarily in solution and rather separate polymer droplets are generated.

The initiators transfer their charge to the polymer. So depending on their charge, positively or negatively charged polymers are generated, which adsorb preferably to particles with the opposite charge [5].

3.3.2.1 Redispersable nanopowders

Covalent bonds between particles make redispersing of unmodified nanopowders a hard task. By common measures like stirring or ultrasonic treatment, the bonds between the particles cannot be broken up again. The smaller the particles are, the smaller are also the applicable mechanical forces to the bond.

An electrostatical stabilization is useless without a solvent, so only a sterical stabilization can be used. In common textbooks, sterically stabilized particles often are described as hard inert spheres with a surface modification.

If these spheres approach each other, repulsive forces grow and finally drive both spheres apart again (compare Figure 3.16, page 47). This picture is hardly true in the real world. If sterically stabilized particles are dried to a powder, they usually cannot be redispersed again. The reasons are insufficiently covered reactive groups of the particle surface which build strong covalent bonds or hydrogen bridges if the particle surfaces approach each other.

So drying and redispersing nanoparticle dispersions turn out to be an excellent test of the quality of the sterical stabilization. Literature proposes turns out various methods to prepare redispersable nanopowders, which basically refer to the principles, introduced in Table 3.3 like core/shell particles, organosilylation or alkyl ammonium ions [76-80].

➤ *Example 11: Production of a methacrylate filled with nanoscaled particles [81]*

1,050 g of silica sol (Levasil 200/40 %, BET=200 m^2/g, 40 % SiO_2, Na^+ removed with ion exchanger) were stirred with 62.58 g of γ-methacryloxy propyltrimethoxysilane for 1 h. The material was then diluted with 1,250 g of THF and, while stirring, 63 g of chlorotrimethylsilane were added.

After an hour, two phases have formed. The top phase contained no solid and was discarded. The bottom phase was diluted with 150 g of THF, 63 g of chlorotrimethylsilane were added with stirring and, after an hour, a further phase separation was carried out. The top phase was again discarded. The bottom phase was diluted with 400 g of toluene, and THF/water were distilled off with addition of further toluene.

The resulting toluene sol, which was still acidic, was heated under reflux and the distillate flowing back was passed via a column filled with sodium carbonate. After six hours of reflux, the sol no longer gave an acidic reaction. The toluene sol obtained was freed from volatile fractions under reduced pressure at 60 °C which yielded a white powder. By stirring with a magnetic stirrer, 50 % dispersions of this powder in methyl methacrylate (MMA) can be produced which have a viscosity of 18 mPa·s and are optically clear.

Nanoscaled particles for methacrylates

Ion exchanged colloidal silica sols show a pH of 2.5 to 3.5 which roughly corresponds to the IEP of SiO_2. In this state, they are excellent catalysts to promote hydrolysis in the sol-gel process and allow working without an external catalyst, which might disturb the final product.

The example describes the hydrolysis of methacryloxypropyl trimethoxysilane in the presence of colloidal silica, which leads to a surface modification of the silica nanoparticles with a reactive polysilsesquioxane shell. The use of chlorotrimethylsilane then leads to a further hydrophobisation of the particles which induces the phase separation into a solvent-rich coacervate (compare Figure 3.4, page 38).

Chlorotrimethylsilane is an interesting surface modifier, because if it binds to the particle surface, it has no further reactive group like Si-OR left and thereby shields effectively. Even more, the only byproduct if an excess of chloromethylsilane reacts with traces of water is hexamethyldisiloxane, a volatile silicone fluid, which is miscible with alcohols and becomes part of the solvent mixture.

During the solvent transfer to toluene, residual HCl from the hydrolysis of the chlorosilane is removed by heating. The THF which is used in the first step is an excellent water miscible solvent for nanoparticles, however it should be controlled for hydroperoxides on a regular basis.

The almost complete conversion of the reactive groups with trimethylchlorosilane yields an inert particle, which can be dispersed without applying high shear forces in unpolar media. An impressive example of an effective sterical stabilization!

➤ *Example 12: Preparation of ZnO nanopowders which can be redispersed in organic media* [82]

40.88 g zink chloride are dissolved in 150 ml methanol. Separately, 24 g of NaOH are dissolved in 200 ml methanol. After the NaOH solution in methanol has been cooled down to room temperature, both solutions are united under vigorous stirring. After 60 minutes of stirring, the resulting precipitate is separated by centrifugation and washed with deionized water. The wet ZnO cake is dispersed in 650 ml of water and heated up to 80 °C. Thereafter carbonic acid derivatives are added as surface modifiers (e.g. 13.7 g "Akypo RO 90", R-O- (CH_2CH_2O-)nCH_2COOH n = 12-16). A stable dispersion of ZnO is yielded.

Redispersable zinc oxide

In analogy to Example 4 (page 30), in the first step a precipitate of agglomerated nanoparticles is formed. With the removal of the generated salt in the next step, the crucial prerequisite for a redispersing of the particles is realized. The polarity of the carbonic acid derivative determines the properties of the final product. Alkyl carbonic acids like stearic acid are suitable rather for hydrophobic media, whereas alkyl polyethylene oxide carbonic acids like "Akypo RO 90" are more suitable for polar media.

The precipitation in methanol is the main reason for the formation of small particles. The same amounts of zinc chloride and sodium hydroxide, dissolved in water instead yield coarse precipitates which cannot be redispersed again down to primary particle size. Special attention has to be paid to the washing step before the peptisation with the carbonic acid is possible. Without removal of the salt which results from the reaction, the zinc oxide cannot be peptized because the high ion strength limits the debye length of the repulsive forces.

An alternative to a peptisation after precipitation is using a precipitating agent, which already contains a surface modifier. Thereby during nanoparticle formation the newly generated surfaces can be covered directly with the modifier (compare Figure 3.18, page 49).

Instead of sodium hydroxide, e.g. a solution of K-O-Si$(CH_3)_3$ can be used. The resulting zinc oxide particles are then covered with a trimethylsilyl-group [83].

➤ *Example 13: Preparation of a SiO_2 nanopowder which can be redispersed in organic media* [84]

400 ml milliliters of the anhydrous alcosol of Example 8 were mixed with 300 ml of a branched chain octadecyl alcohol (e.g. "Fine Oxocol", Fa. Nissan Chemicals). n-Propanol was then distilled from the mixture at atmospheric pressure, leaving the colloidal silica as a relatively clear, slightly viscous solution in the octadecyl alcohol. This colloidal solution was then heated to elevated temperature to bring about an exchange of octadecyl groups for propyl groups. Thus, distillation of the mixture was continued at atmospheric pressure until the temperature in the distillation flask was 140 °C. Nitrogen was then blown through the flask and the temperature was further raised to 200 °C for 3 h.

The pressure in the distilling equipment was then reduced to between 12 to 15 millimeters, and the free octadecyl alcohol was removed, the boiling point at this pressure being about 170 °C. The powdery residue in the flask was then further dried in a vacuum oven at 10 millimeters pressure and at a temperature of 180 °C for a period of 2 days. The dry powdery product thus obtained was readily soluble in kerosene, to give a clear colloidal solution.

The colloidal solution obtained by dissolving the product in kerosene could be dried to a powder and this could be redissolved in kerosene. A colloidal solution of this product in kerosene containing 20 % by weight of SiO_2 appears to be permanently stable toward gelling. Electron micrographs show that the silica particles in the kerosene sol are of the same size as in the original aquasol. However, they are no longer hydrophilic, and when the kerosene is evaporated, the surface-esterified silica powder is highly hydrophobic.

Surface modification of redispersable SiO_2 nanopowders

Like in Example 8 (page 50), SiO_2 particles are sterically stabilized by esterification of their surface Si-OH groups with alcohols.

The esterification with bulky alcohols is a suitable measure to stabilize colloids. Without a catalyst like titanium alkoxides, this esterification needs high temperatures and long reaction times to be complete. To achieve redispersable powders, an almost complete coverage of the surface Si-OH groups is necessary. Branched alcohols excel by a very good stabilization and compatibility with organic media (e.g. ethylhexyl, isopalmityl, …).

➤ *Example 14: Production of ZrO_2 (zirconium dioxide) nanoparticles which can be redispersed in organic media* [85]

7.6 g (70 % in n-PrOH) of Zr(OPr)4 were combined with 136 g of n-hexanol and, after stirring at room temperature for 5 min, 0.90 g of 37 % HCl in 6 g of n-hexanol was added. The entire mixture was then treated in an autoclave at 250 °C and 300 bar for 7 h.

The amphiphilic ZrO_2 nanoparticles formed, having hexoxy groups on the surface, were centrifuged off and taken up in 5 ml of i-PrOH. Subsequently, they were dried at 50 °C and 10 mbar. High-resolution transmission electron micrographs show that the particles are crystalline and the particle sizes are from 3 to 5 nm.

Redispersable zirconium dioxide particles.

During the reaction, a trans-esterification combined with a hydrolysis and condensation reaction takes place. Zirconium propoxide, like most metal oxides, is very sensitive to humidity. Even during the weighing or filling or the flasks, significant amounts of alkoxides can hydrolyze if they come into contact with traces of humidity in the air or which is adsorbed on the walls of the reaction vessel. In general, handling under dry inert gas atmosphere is highly recommended.

By adding a sub-stoichiometric amount of water (compare Figure 5.6, page 95) an uncontrollable precipitation is avoided and the formation of small particles is favored. These particles which consist of zirconium oxide/hydroxide are sterically stabilized by unhydrolyzed zirconium propoxide Zr-O-Pr and transesterified Zr-O-Hex groups.

HCl is not necessary for promoting the hydrolysis and condensation, because the metal alkoxides already show a very high reactivity towards water. HCl here serves as a source for chloride ions, which stabilize zirconium oxide via salt formation. To promote crystallization, this example uses a solvothermal treatment in an autoclave. If it is not necessary to achieve the highest possible density and crystallinity, this step can be replaced by refluxing the reaction mixture for some hours at ambient pressure.

3.3.3 Electrosterical stabilization

Electrosterical stabilization combines the principles of electrostatical and sterical stabilization. Because the electrostatic part of the stabilization in an organic media only shows a short range, the electrosterically stabilization is often used in water/organic solvent mixtures or simply to add some extra stability in water-based dispersions due to the sterical barrier (compare "Stabilization of nanoparticles against agglomeration", page 40).

Amino-, carboxy-, sulfonate, phosphate-containing organic polymers or silanes can be used for electrosterical stabilization of nanoparticles. An example for an electrosterical stabilization can be found on page 183 (Example 52).

3.4 Nanoparticles in coating resins

Inorganic fillers are used since a long time in organic resins as extenders or as an active ingredient. Nanoparticles due to their large surface area and potential to realize transparent nanocomposites now open up the possibility to transform conventional polymers into new hybrid materials with extraordinary properties. The modification of conventional organic resins with nanoparticles for sure is an exciting topic with no end in sight [86].

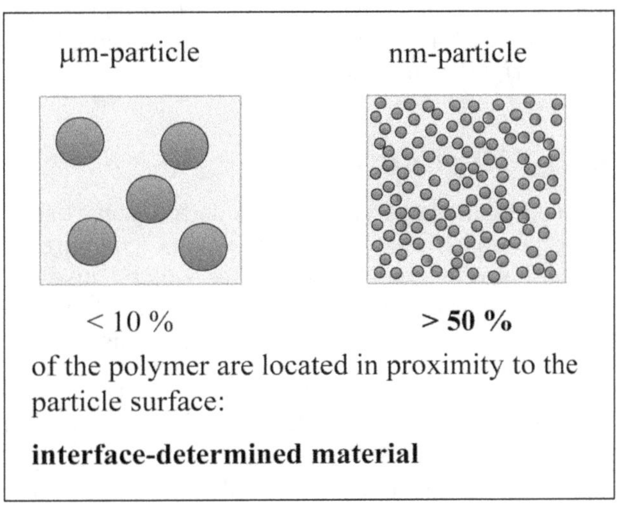

Figure 3.27: Illustration of the importance of the interface in relation to the size of the particles

Different to micrometer sized fillers, the action of nanoparticles is not limited to their bulk proper-

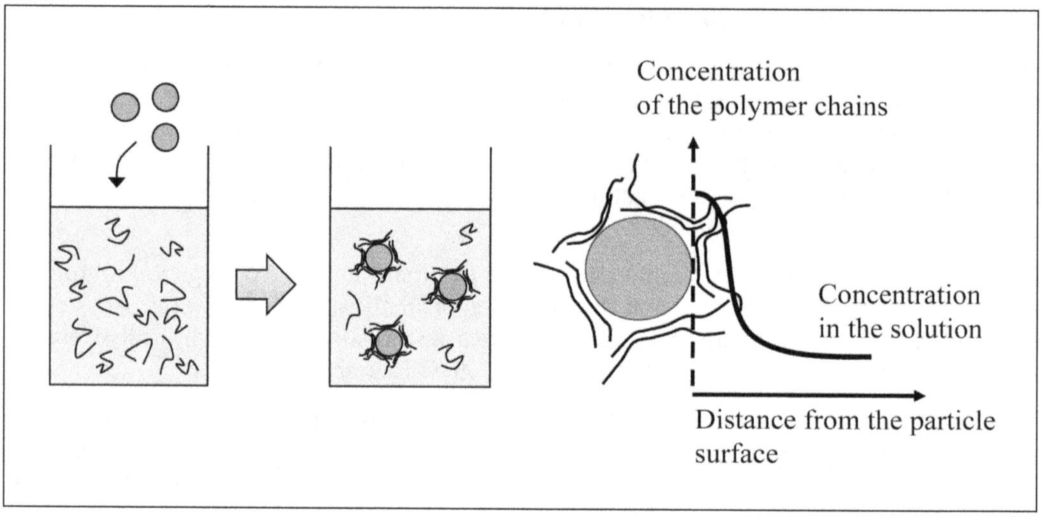

Figure 3.28: Polymers adsorb to nanoparticle surfaces. Adsorption leads to a denser packing of the polymer chains on the particle surface [71]

ties but dominated by their interface properties. They are able to change the conformation and density of the surrounding polymers in their proximity, an action which has a range of several nanometers into the polymer. Figure 3.27 illustrates how a filler-modified polymer in this way becomes an interface-determined nanocomposite.

The volume fraction of the polymer which is exposed to the nanoparticle surface accounts for such a significant part of the total volume, that it becomes the third component besides particle and polymer and can be decisive for the properties of the composite.

The orientation of the polymer chains is changed in the proximity to the nanoparticle surface. Thereby nanoparticles can change e.g. T_g, E-modulus, temperature resistance and density of the polymer disproportionally high compared to their addition level [87].

An application of this principle can be found in polymer lattices for façade paints. By modification with nanoparticles, hard coatings can be realized with polymers having a low T_g. The high hardness combined with the hydrophilicity of the SiO_2 particles leads to a reduced soiling (compare "Hydrophilic coatings" [88], page 179). Figure 3.28 illustrates how polymers can adsorb to nanoparticles surfaces.

3.4.1 *Organic resins*

The modification of organic resins with nanoparticles follows a similar logic like the transfer of nanoparticles to organic solvents (compare "Sterical stabilization", page 46). The crucial point to avoid agglomerates which cause haziness and increased viscosity is to choose an appropriate surface modification. Figure 3.29 shows three different possibilities to incorporate nanoparticles in organic resins.

In order to disperse nanoparticle powders, high shear forces are necessary. Only if the sterical surface modification of the particles is nearly perfect, the particles can be dispersed to primary particle size again. For applications with an emphasis on transparency, this approach is hardly suitable. In analogy to milling, the dry particle agglomerates have to be wetted, dispersed and stabilized with suitable modifiers (compare "Top-down processes", page 20).

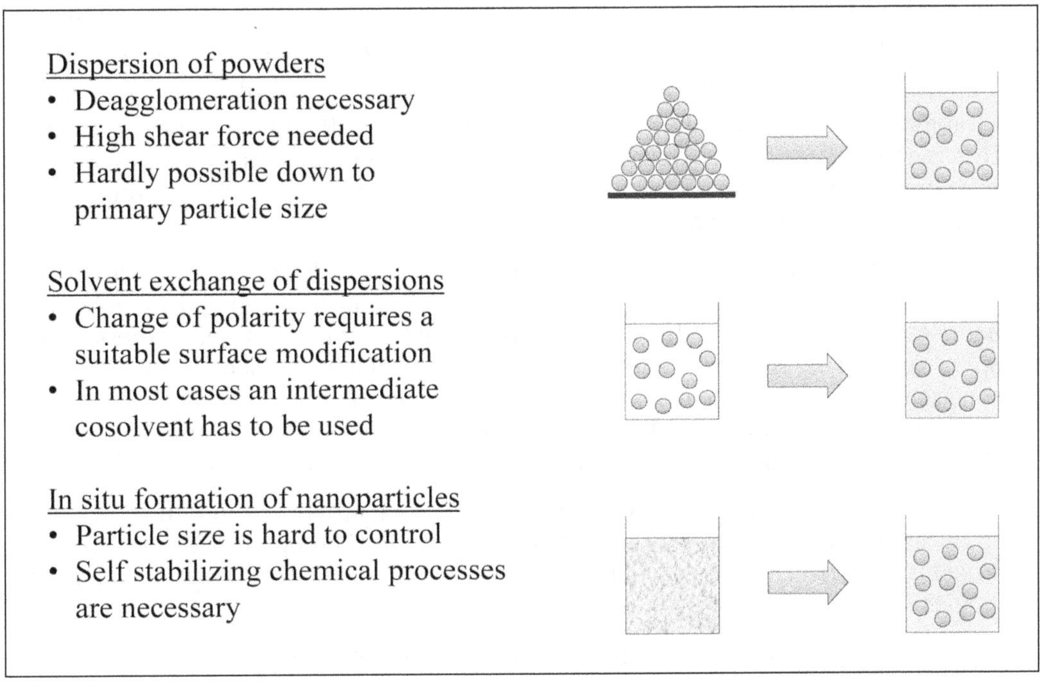

Figure 3.29: Possibilities how to introduce nanoparticles in resins

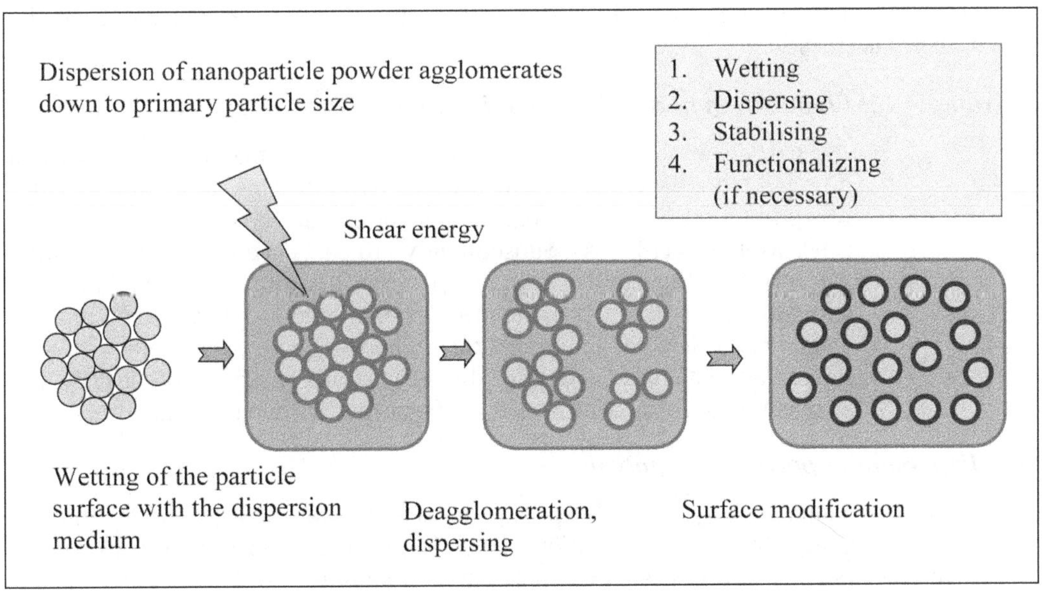

Figure 3.30: Necessary working steps to disperse nanoparticle powder agglomerates

If it is planned to react the nanoparticles with an organic resin in a further process step, then a tailored reactive surface modification has to be applied. A lesser known method is the in situ generation of nanoparticles in resins. Liquid reactive precursors are dissolved in the resin and reacted with either water or a separate reagent in a following step, either in the liquid phase or after curing the resin.

For example literature can be found describing the generation of Ag0 or titanium dioxide nanoparticles. in resins [89]. Reduction of silver salts by organic components of the coating or induced by UV light can be used to prepare silver [90]. The polymer network limits the mobility of the seeds and thereby prevents the generation of bigger agglomerates.

Starting with silver halogenides, the reaction can even be reversible. Silver and the halogen, both trapped in the network can react back to the colorless silver halogenides, a reaction which is the basis for phototropic coatings and glasses. The organic components in the resin however limit the number of cycles due to a reaction with the aggressive halogen. Almost inorganic sol-gel coatings are a good basis for phototropic coatings.

The method which is most often applied to transfer nanoparticles to organic resins starts from nanoparticle dispersions. First, the nanoparticles are transferred to a suitable intermediate solvent, which is capable of both solving the organic resin and dispersing the nanoparticles. During solvent removal in most cases a visual inspection indicates whether the surface modification of the particles had been suitable and complete. The occurrence of haziness, precipitation or gelation is a hint that the particles agglomerate and are not "comfortable" with their environment (compare "Sterical stabilization", page 46).

The used surface modifiers can be differentiated into compatibilizers and functionalizers. Whereas compatibilizers focus on the shielding of surface-OH groups, functionalizers allow a covalent bond between particle and resin. This difference can be crucial for the mechanical properties of the composite.

Functionalized nanoparticles can be regarded as cross-linkers whereas inert nanoparticles generate a cavity filled with an inorganic particle (compare Figure 4.16, page 85) [91, 92]. This especially is the case for surface modifications containing silicones. So, identical nanoparticles can yield composites with different mechanical properties, depending on the type of their modification.

➤ *Example 15: Preparation of a SiO_2 dispersion in a polyacrylatepolyol* [93]

100 parts of an aqueous colloidal silica sol were admixed with 3.9 parts of n-propyl-trimethoxysilane, which was stirred in. Thereafter this mixture was stirred into 620 parts of isopropanol and at 40 °C and 85 mbar was concentrated to 113 parts. Subsequently 110 parts of a hydroxyl-containing polyacrylate ("Desmophen A 870 BA", Bayer AG) were added. The volatile constituents were subsequently removed by distillation at 40 °C and 58 mbar in a manner sufficiently gentle that the higher-boiling butyl acetate present in the polyacrylate remained in the dispersion. This gave a water-clear dispersion having a SANS-determined diameter distribution of 8 ± 2.5 nm.

SiO_2 dispersion in polyacrylatepolyol

High dilution and ion exchange are prerequisites for the transfer of SiO_2 particles into the polyacrylate (compare "Sterical stabilization", page 46). By the surface modification with propyltrimethoxysilane, in a first step an inert, sterical stabilizer which acts as compatibilizer with the resin is reacted with the nanoparticle. Then the water is removed via its azeotrope with isopropanol, so that finally the resin dissolved in butylacetate can be added.

As a general rule of thumb for the removal of water via an azeotrope, a 5 to 7 fold excess of solvent (e.g. isopropanol or n-propanol) has to be used. The boiling point of the intermediate solvent should be lower than that of the target solvent or medium to ensure that it can be removed easily. This is especially important, if during the following production steps side reactions of the intermediate solvents can be expected. This is the case in this example, when isocyanates are used to cure the polyacrylatepolyol binder. A side reaction would reduce the isocyanate content and threaten the curing process.

➤ ***Example 16: Preparation of a SiO_2 dispersion in a UV-curable reactive thinner*** [94]

The following were mixed under stirring at room temperature:

480 g of an acidic silica sol, freshly prepared, containing by weight 50 % of silica with a mean diameter of 50 nm, and having a pH of 2, 1,860 g of isopropyl alcohol, 206 g of 1,6-hexanedioldiacrylate, 111.5 g of vinyltrimethoxysilane, i.e. 0.46 g of vinyltrimethoxysilane per gram of initial dry silica.

The reaction mixture thus obtained is next subjected to distillation at reduced pressure of approximately 150 mbar whilst keeping the temperature of the external bath at approximately 35 °C for approximately 12 hours total distillation time. Then the reaction medium is cooled to room temperature.

In this way a solution is obtained which is clear, limpid, transparent and stable over time, containing by weight 50.2 % silica and 0.65 % water and having a Brookfield viscosity of 84 mPa·s.

SiO_2 dispersion in reactive thinners

The colloidal silica sol used in this example can be synthesized, starting from sodium silicate or alkaline silica sol by ion exchange. Aqueous silica sol shows a good compatibility with organic solvents at pH 2 to 3, which corresponds to its isoelectric point (IEP). At this point, the stabilization has changed from electrostatic repulsion to stabilization with Si-OH groups (compare "Stabilization of nanoparticles against agglomeration", page 40).

In this example vinyltrimethoxysilane is used as a reactive sterical stabilization. As an alternative, also methacryloxypropyl trimethoxysilane could be used, which generally shows a higher reactivity during UV curing.

Special attention should be paid to the high amount of modifier relative to the colloidal silica sol. For a complete coverage of the surface, nanoparticles need up to 20 weight% surface modifiers due to their huge surface area which has to be covered. The synthesis process in this example is a remarkable "one pot" approach. Silica sol, reactive thinner, intermediate solvent and surface modifier are mixed together at the same time. This approach is only possible because the organic resin HDDA is highly soluble in the isopropanol/water mixture.

A more conventional approach would be to first modify the silica sol and remove the water and then to add the organic resin. This approach would also minimize the losses due to the high volatility of HDDA during vacuum evaporation.

3.4.2 Waterborne organic binders

The incorporation of nanoparticles into waterborne resins is of unequal complexity compared to solventborne resins. Whereas solventborne resins consist of only one liquid phase, which is the solution of the resin, waterborne resins consist of an emulsion of nanoscaled polymer droplets surrounded by surfactants in water [5, 95].

The nanoparticles introduce a third phase into the system and can arrange themselves with the resin droplets in a variety of ways. Nanoparticles can be present in the polymer droplets, on their surface or only in the water phase. Figure 3.31 illustrates some possibilities.

If the surface modification of the nanoparticles is somehow between hydrophobic and hydrophilic, they can add colloidal stability to the polymer droplet by arranging themselves as a shell around the droplet, sinking partly into the resin. This effect is known under the term "Pickering stabilization" and can also be observed for exfoliated clays [96, 97].

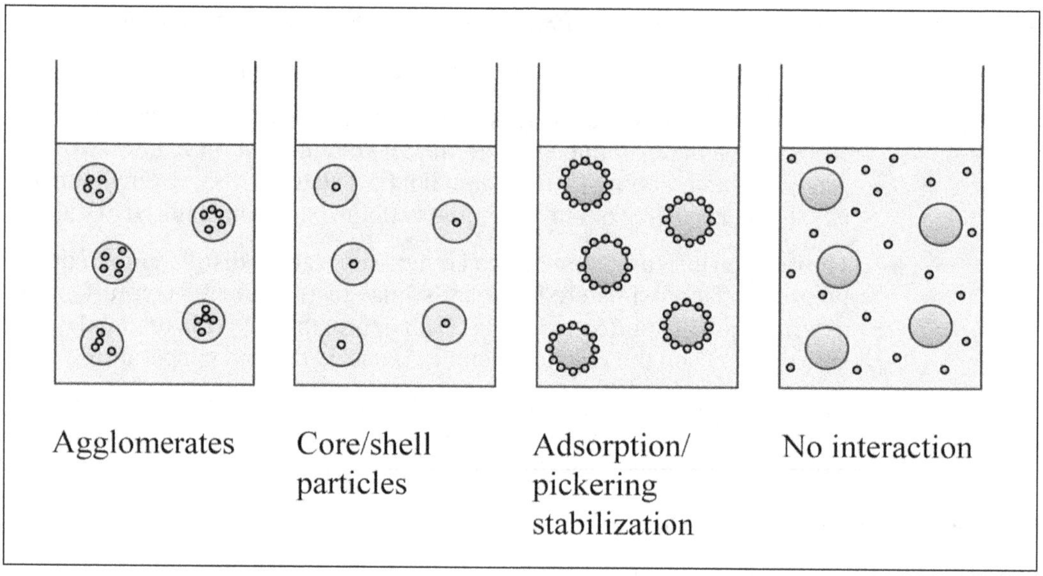

Figure 3.31: Possibilities how nanoparticles can be dispersed in polymer latices

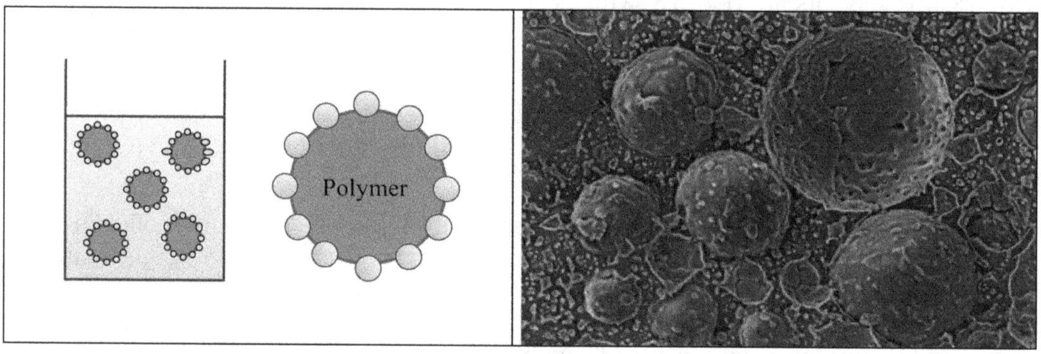

Figure 3.32: Pickering stabilization of polymer emulsions *Source: BASF SE*

If nanoparticles are dispersed in a resin emulsion, they arrange themselves depending on their surface modification either in the water, in the polymer droplets or at the interface between water and polymer droplet.

Another way is to introduce nanoparticles at an early stage of the resin preparation. Waterborne polymers are synthesized either by radical emulsion polymerization out of monomers or by dispersion/chain-elongation of a pre-modified oligomer in water.

During emulsion polymerization, the monomer is solubilized by surfactants in the form of micelles and only a small portion is water soluble. The radical initiator starts the polymerization in the water phase and first polymer seeds are generated which are stabilized by the surfactant. Monomer diffusion drives the growth of the polymer until all the monomer has been polymerized.

Nanoparticles can be introduced at various stages of this process [98-102].

- dispersion of the nanoparticles in the monomers prior to the polymerization
- presence of nanoparticles with a reactive surface modification during the emulsion polymerization
- addition of the nanoparticle dispersion after the emulsion polymerization is finished

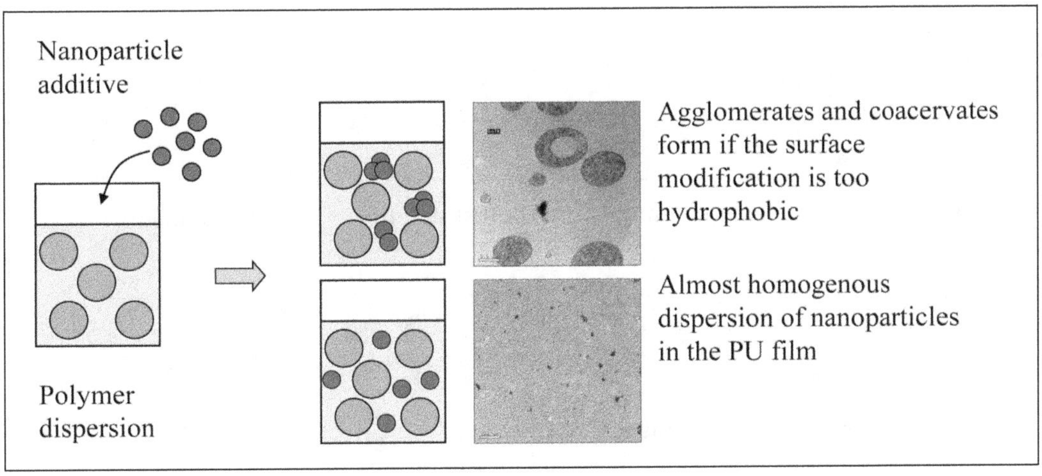

Figure 3.33: Matching the polarity of the surface modification and the polymer resin is essential when adding nanoparticles to waterborne polymer dispersions Source: Merck KGaA

- dispersion of nanoparticles in a polymer oligomer prior to the dispersion step (for PU-emulsions)

The point of time and way of addition of the nanoparticles significantly influences the distribution and location in the aqueous binder and the cured resin coating.

The homogeneous dispersion of nanoparticles in polymer lattices after their production is not trivial. The nanoparticles have to be dispersed in water or a water/solvent mixture to be compatible with the polymer emulsion. For these products, solvents are only tolerated to a very low amount to keep the benefits of water as a dispersing medium.

The surface modification of the nanoparticles therefore has to be sufficiently hydrophilic to use water as dispersion medium, but also sufficiently hydrophobic to ensure compatibility with the polymer during coalescence and drying. The polymer usually is very hydrophobic to maximize chemical resistance and water resistance of the coating after drying. So only a very narrow bandwidth of polarity exists, in which the nanoparticles are stable in both phases.

During coalescence and film formation of the latex, the nanoparticles should remain distributed homogeneously throughout the polymer phase to optimize transparency and mechanical properties of the film. If the surface modification is too hydrophobic, a dispersion in water is not possible, if it is too hydrophilic, the dispersion in the polymer film is disturbed.

During film formation of a nanoparticle-modified latex, different structures result, depending on the distribution of the particles. If the nanoparticles are homogeneously dispersed in the resin droplets, the water can evaporate and the resin droplets can form a coating film by coalescence without agglomeration of the nanoparticles.

If the nanoparticles are distributed in the water phase, they concentrate in the voids between the resin droplets during drying and form agglomerates or even an interpenetrating network which affects transparency of the resulting hybrid. Figure 3.34 illustrates this behavior.

Besides radical emulsion polymerization, also other polymerization methods can be used to produce waterborne resin binders. Polyurethane-based dispersions are synthesized by dispersing NCO-functionalized oligomers in water. Acetone or N-ethylpyrrolidone are often used as water miscible aprotic solvents for this process [103].

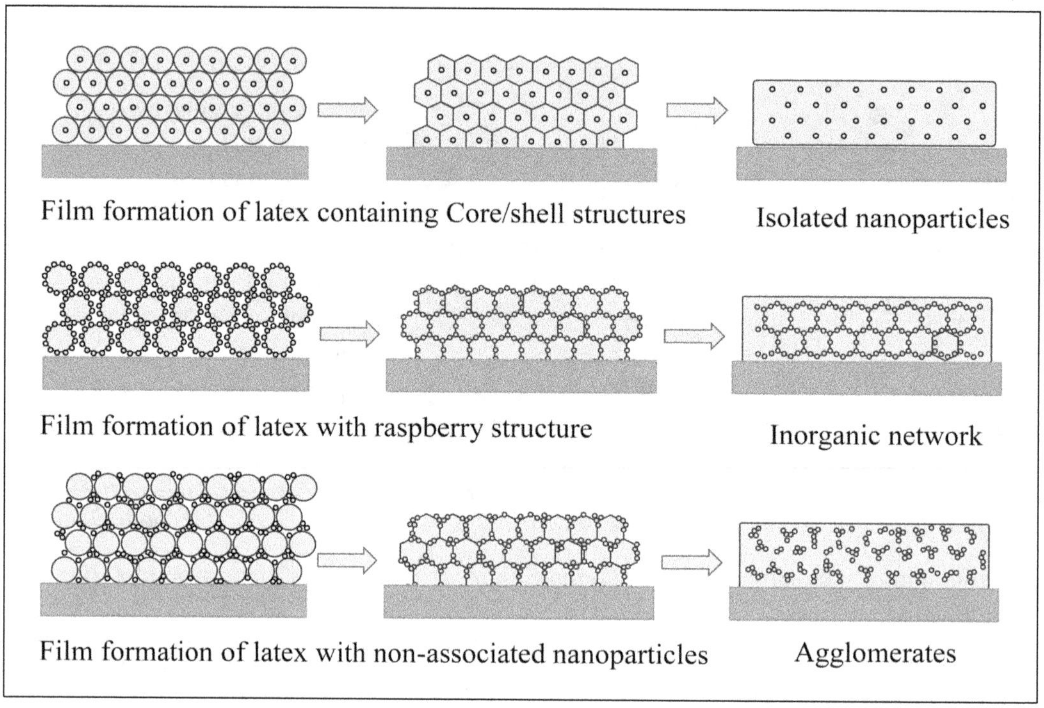

Figure 3.34: Different structures can result during the film formation of latices containing nanoparticles (idealized illustration)

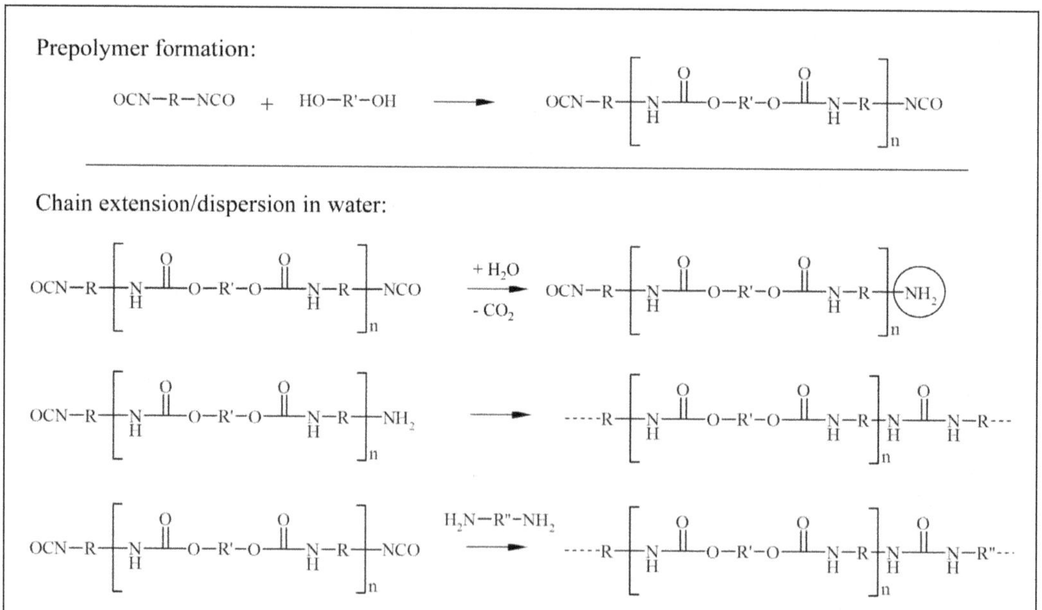

Figure 3.35: Production of PU-dispersions. A prepolymer, dissolved in a water-miscible solvent is dispersed in water while adding amines

By the reaction of the isocyanate group with water, CO_2 and an amino group is produced which readily reacts with another isocyanate to build a carbamate. Thereby the chain length of the prepolymer is rapidly extended.

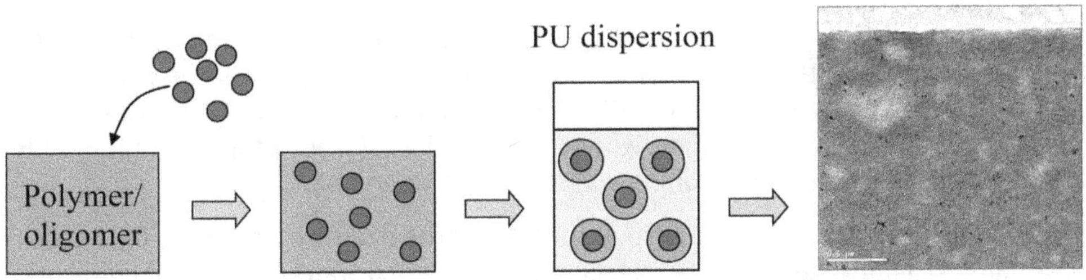

1. Dispersing the surface-modified nanoparticles in a prepolymer/oligomer
2. Dispersing the composite in water

Figure 3.36: 2-Step production of a nanoparticle dispersion which contains homogenously dispersed nanoparticles

Additionally, other amines can be added to the water to modify the resin or to speed up the polymerization reaction. Ionic and/or non-ionic groups induce the compatibility and stability of the dispersion in water. After the organic solvents have been removed by distillation, a stable emulsion of a polyurethane/polyurea in water is received.

The terms "dispersion" and "emulsion" often are used imprecisely in the coatings business. Dispersions describe solid particles which are distributed more or less stable in a solvent, whereas emulsions describe liquids which are distributed in another liquid. Waterborne resin binders are emulsions, because the resin in the droplets is liquid, but usually these resin emulsions are not used without inorganic particles, e.g. pigments or fillers and so, one can speak of a dispersion as well.

Basically nanoparticles can also be added to the final PU dispersion. The special production process, starting from a solventborne prepolymer however opens up the possibility to introduce the nanoparticles into the prepolymer before or during the chain extension step in water. Thereby the separation of polymer and nanoparticles is prevented and homogeneous dispersions of PU droplets containing nanoparticles can be synthesized.

Example 17 shows one possibility to introduce nanoparticles into PU dispersions.

➤ ***Example 17: Preparation of a UV-curable PU-dispersion containing nanoparticles*** [104]

187.4 g polyesteracrylate with a hydroxyl number of 80 mg KOH/g ("Laromer LR 8800", BASF) and 31.5 g N-ethylpyrrolidone were placed in a glass beaker. To this initial mixture was added 0.06 g 3-tert-butyl-4-hydroxy-anisole. Subsequently, via a drop funnel, 83.9 g 4,4›-dicyclohexylmethanediisocyanate ("Desmodur W", Bayer AG) were added dropwise.

The reaction mixture was agitated at 70 °C with a pass-over of compressed air and converted up to an NCO content of <=5.0 %. Thereafter, 9.5 g dimethylol propionic acid and 7.2 g triethylamine were added. The mixture was converted further at 75 °C to an NCO content of 2.8 %.

The resulting NCO-terminated prepolymer was dispersed with vigorous agitation in a mixture comprising 469 g demineralized water, 7.8 g monoethanol amine and 206 g "Bindzil 305 FG" (aqueous silica sol containing 30 % SiO_2, EKA Chemicals). A dispersion with the following characteristics was obtained: Viscosity: 25 mPa·s, pH value: 8.2, solid content: 35.4 %

Nanoparticle-containing UV-curable PU dispersion

The production of the PU dispersion in this example is done via the emulsion of a NCO-terminated polyester acrylate in water. By reaction of the NCO group with water in a first step the amine is generated which then rapidly reacts with another NCO group to a carbamate. Thereby a chain extension takes place, which turns the low viscous prepolymer into a highly viscous polyurea/polyurethane (compare Figure 3.35, page 66).

The NCO-terminated low viscous prepolymer is synthesized by addition of an excess of diisocyanates to a polyester acrylate. The compatibility with water is ensured by incorporation of the ammonium salt of dimethylol proprionic acid as hydrophilic component.

If during the chain extension emulsion step nanoparticles are present, the synthesized polymers can be adsorbed and thereby the particles are completely enclosed by the polymer. In this example, unmodified SiO_2 particles are used, which show a high affinity to amine groups and form salt-like structures (SiO^{-} $^{+}NH_2$-R). The formation of covalent bonds between particles and polymer however can be further promoted, if suitable surface modifiers are added (e.g. aminosilanes).

Following another synthesis route, it would also be possible to add the nanoparticles prior to the emulsion step to the solventbased solution of the oligomer. Then, nanoparticle dispersions which are free of water and protic solvents have to be used (compare Figure 3.36).

The water-miscible solvent N-ethylpyrrolidone stays in the final product and can hardly be removed due to its high boiling point. If a solvent-free dispersion is desired, N-ethylpyrrolidone can be replaced by a protic water-miscible solvent with a lower boiling point like acetone, which can be readily removed by distillation.

The resin contains unsaturated components. To protect these components from oxidation, butylhydroxyanisol, a conventional antioxidant is used.

➤ *Example 18: Preparation of a waterborne polymer resin which contains nanoparticles [105]*

220 g of "Dowanol PnB" (propyleneglycol n-butylether) were initially introduced into a 5 l reaction vessel with a stirring, cooling and heating device and were heated up to 138 °C. A mixture 1) of 4 g of di-tert-butyl peroxide in 4 g of "Dowanol PnB" was added dropwise at this temperature in the course of 30 minutes. Immediately thereafter, a mixture 2) of 298.3 g of i-bornyl methacrylate, 292.0 g of hydroxyethyl acrylate, 169.8 g of butyl methacrylate, 126.5 g of styrene, 90.4 g of 2-ethylhexyl acrylate and 12.5 g of "CoatOSil 1706" (vinyl-tris- i-propoxysilane) was metered in during the course of 3.5 hours, and immediately thereafter a mixture 3) of 63.8 g of styrene, 90 g of hydroxyethyl acrylate, 50 g of butyl acrylate and 28.7 g of methacrylic acid was metered in during the course of 1.5 hours.

In parallel with mixture 2) and 3), a mixture 4) of 14.5 g of di-tert-butyl peroxide in 14.5 g of "Dowanol PnB" was metered in over a period of 5 hours. A mixture 5) of 4 g of di-tert-butyl peroxide in 4 g of "Dowanol PnB" was then metered in over a period of 1 hour. The mixture was then cooled to 70 °C. 436 g of a 36 wt.% strength colloidal dispersion of silicon dioxide in methyl ethyl ketone (organosilicasol "MEK-ST", Nissan Chem. Am. Corp.) were introduced into this melt. 31.2 g of N,N-dimethylethanolamine were then added, the mixture was homogenized for 30 minutes and dispersing was carried out with 1,790 g of water at approx. 60 °C over a period of

2 h. The methyl ethyl ketone was then distilled off in vacuo. A copolymer dispersion having the following data was obtained: OH content 4.0 %, acid number 13.5 mg of KOH/g, solids content 40.3 %, viscosity 370 mPa·s, pH (10 % strength in water) 8.4, average particle size 125 nm.

Nanoparticle-containing waterborne resins

In a first step in this example, a solution of a silylated polyacrylate in a glycolether is synthesized, which is rendered water-compatible by carbonic acid groups. The silyl-groups of the polymer react with the nanoparticle surface when the SiO_2 organosol is added so that the polymer encapsulates the silica particles.

The product “MEK-ST” is stabilized by methoxy-groups which quickly hydrolyze upon contact with water and generate Si-OH groups which then can react with the silyl-groups of the binder. Water is added slowly to the nanoparticle-modified binder, so that enough reaction time is given to incorporate SiO_2 homogeneously into the resin droplets. Such nanocomposites often show a better scratch and abrasion resistance compared to unmodified resins.

The incorporation of nanoparticles in polymers can also show undesired side effects. Besides increased viscosity, brittleness or haze, a degrading of the polymer by photocatalysis or chemical interaction is possible (compare “Hydrophilic coatings”, page 179). Si-OH groups show an IEP of 2.5 and thus react as acids (compare Figure 3.12, page 43). They can degrade polymers which leads e.g. to yellowing. This problem can be addressed by a complete reaction of the surface Si-OH groups with organosilanes and/or by neutralization with alkali ions [106].

Lewis acids, like aluminum hydroxide or zirconium hydroxide can catalyze the polymerization of epoxides. Even traces like can be found on cationic silica sols can lead to problems in the processing of modified polymers [24].

4 Application of nanoparticles in paints and coatings

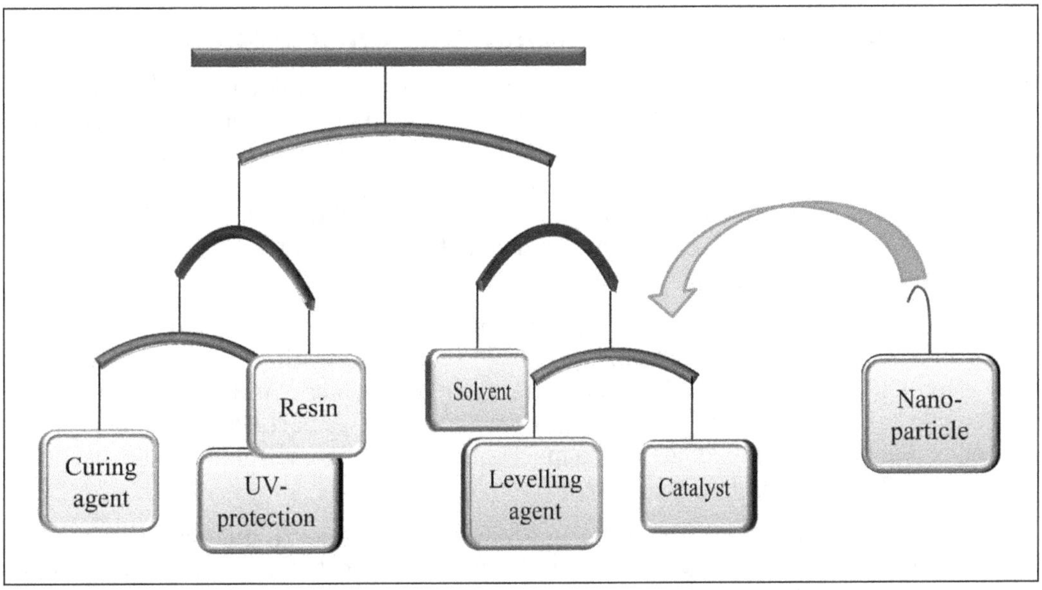

Figure 4.1: Nanoparticle additives must not disturb the delicate balance of coating ingredients

When trying to incorporate nanoparticles into paints and coatings, in most cases the basic questions are not different compared to other more conventional coating additives. Paints and coatings are complex mixtures of resins, curing agents, solvents and other additives which were composed in a tedious time-consuming process, often based on trial and error.

Adding a further additive to this sensitive equilibrium often causes unpredictable side effects.

The cause of those effects in most cases is the large surface area of the nanoparticles, which adsorbs some of the components of the coating. The change of properties caused by this adsorption process can hardly be foreseen and has to be investigated on an empirical base.

For example it has been reported, that the efficiency of SiO_2 nanoparticles which were added to improve the scratch resistance of coatings can be improved dramatically by polyethersiloxane leveling agents [107, 108]. On the other hand, stratification or agglomeration can occur, if the HLB value (hydrophilic/lipophilic balance) of the leveling agent is too high or too low.

HALS (Hindered Amine Light Stabilizers), which are used in coatings to support the action of organic UV absorbers are secondary amines, which show a high affinity to SiO_2 surfaces. Their adsorption can a) decrease the stabilizing action of the HALS and b) lead to a destabilization of the SiO_2 dispersion in the coating.

Again, the prediction of the effects of a nanoparticle addition to a coating formulation, due to the complexity of interactions is not possible in everyday lab work. The possible effects

Gerhard Jonschker: Sol-Gel-Technology in Praxis

ISBN: 978-3-86630-886-2

and side-effects of nanoparticles are manifold and maybe even a little bit scary to the unexperienced user, but exactly the same is the true for many classes of additives and should not be the reason for not working with nanoparticles.

4.1 Color and light

4.1.1 Hiding power, transparency and particle size

Transparency, respectively the scattering power of fillers is determined by the particle size as well as the difference in the refractive indices of resin and filler (compare "Agglomeration of nanoparticles", page 36). This is the reason, why pigments should have a high refractive index to increase their hiding power and nanoparticles with an emphasis on transparency should exhibit a low refractive index near that of the resin. Titanium dioxide, the classical white pigment, has a refractive index of 2.5 to 2.7 (rutile phase) and SiO_2 as a preferred transparent nanofiller 1.45 to 1.46.

Figure 4.2 shows, that for each resin-pigment combination an optimum particle size exists at which the hiding power shows its maximum [5, 109]. At the same time, a minimum particle size exists, at which the particle does not scatter the light anymore. As a consequence of the relationship between refractive index and scattering intensity, a particle with higher refractive index like titanium dioxide has to be much smaller than low refractive index particles like SiO_2 to avoid scattering.

If the particle size increases, according to the Raleigh equation (Equation 3.1, page 35) first the blue part of the spectrum is scattered, whereas the rest of the light still passes the particle unscattered. This effect in combination with reflecting metal flake pigments is used to generate an angle-dependent color impression called "Frost effect" [53, 110].

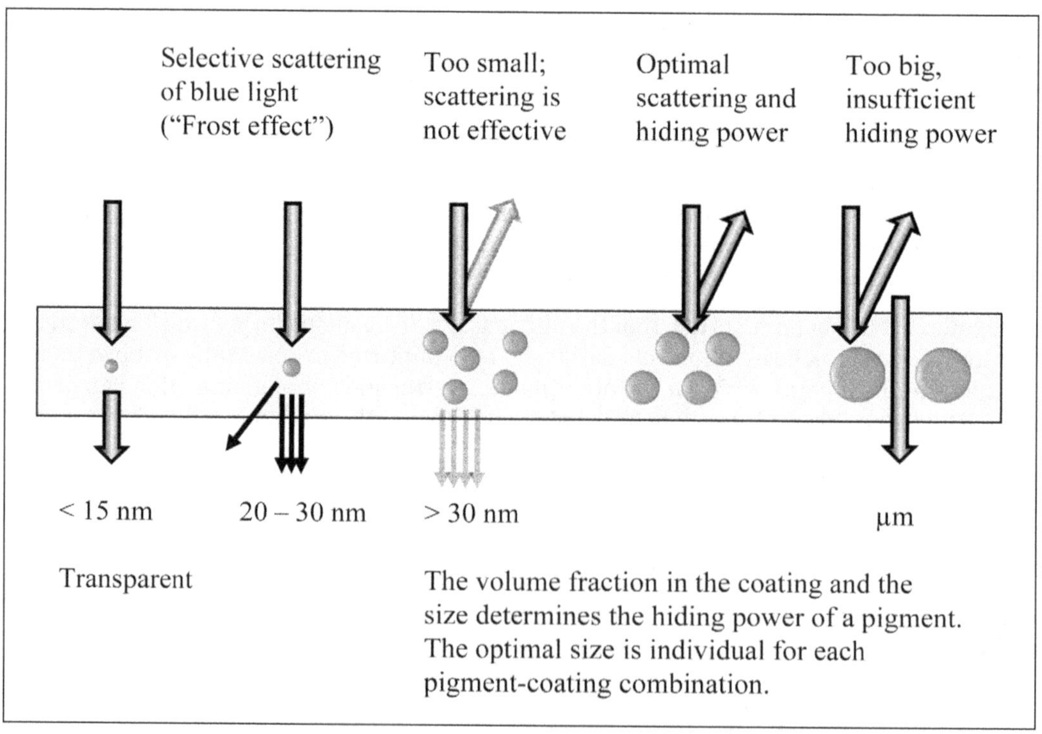

Figure 4.2: Illustration of the interaction of particles with different sizes with visible light

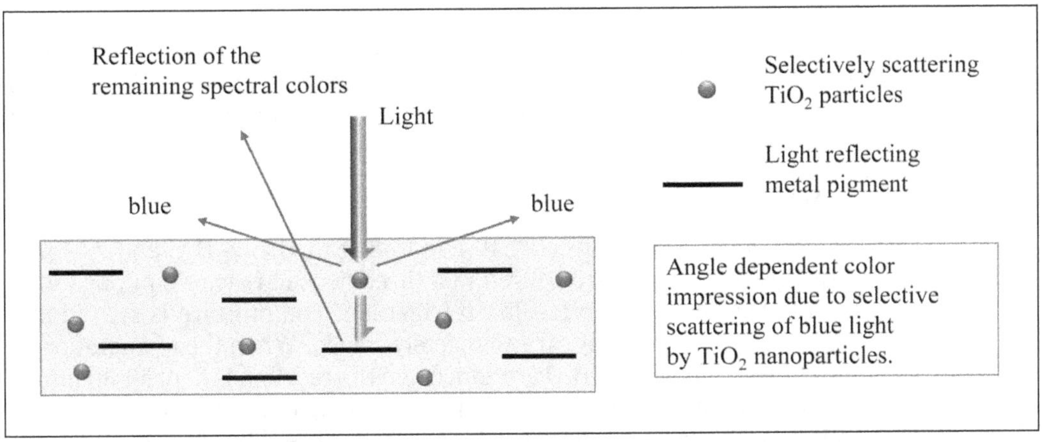

Figure 4.3: Realization of a "frost effect" by the combination of nanoscaled TiO_2 and reflecting metal pigments

When viewed at perpendicularly, the coating appears yellowish and when viewed from the side bluish, because only the blue part of the light is scattered. The same effect can result in an interesting color travel coating when the nanoscaled titanium dioxide is combined with traditional pigments.

4.1.2 Nanoscaled pigments

Inorganic pigments are available in a myriad of different colors and shades. As chromophors, typically e.g. $Fe^{2+/3+}$, Cu^{2+}, Co^{2+} or Cr^{3+}-ions are used. An advantage compared to organic pigments is their high thermal and chemical stability, but transparent colors usually are not realizable.

For UV resistant transparent colors, nanoscaled inorganic pigments might be a good starting point. Their small size predestines nanoscaled inorganic pigments also as coloring agents in thin sol-gel coatings (compare "Colored and pigmented sol-gel coatings", page 150).

If a smooth, glossy finish is desired, the size of the pigment has to be significantly smaller than the coating thickness. Therefore, most conventional pigments are not suitable for the only a few micrometer thick sol-gel coatings. For high temperature resistant coatings, inorganic oxide nanoparticles, besides noble metal colloids are the only possibility to realize intense transparent colors (compare "High temperature resistant coatings", page 72).

Nanoscaled oxide pigments usually are mixed oxides. The preparation of homogeneous multicomponent nanoparticles however is challenging. Only minor differences in the reactivity of the educts and their isoelectric points (IEP) can cause differences in their behavior during hydrolysis and condensation. If the reactions do not run in parallel, the consequences are inhomogeneity, seed formation and growth are disturbed, which finally leads to agglomeration and phase separation [111-113].

Example 19 reveals a possible solution for this problem by using the polyol-moderated high temperature solvent synthesis.

➤ ***Example 19: Preparation of nanoscaled color pigments*** *[114]*

Vanadium doped $ZrSiO_4$:

100 cm^3 diethylene glycol, 3.22 g of $ZrOCl_2 \cdot 8\ H_2O$, 2.08 g of tetraethoxysilane and 0.17 g of NH_4VO_3 are poured in a three necked reactor equipped

with stirrer, reflux refrigerator, thermometer. The mixture is stirred and heated up to complete dissolution of the reagents. The mixture is thereafter heated up to 180 °C and the solution becomes opalescent. The solution is left under vigorous stirring for 1 h at 180 °C and then cooled down; a transparent solution of $ZrSiO_4$:V is obtained.

$CoCr_2O_4$:

50 cm³ of diethylene glycol, 2,49 g of $Co(CH_3COO)_2 \cdot 4\ H_2O$ and 7,43 g of $Cr(CH_3COO)_3 \cdot 4\ H_2O$ are poured in a three necked reactor equipped with stirrer, thermometer and reflux refrigerator. The mixture is stirred and heated until the reagents are completely solved. When the temperature of 140 °C is reached, 2 ml H_2O are added and the reaction is kept for 30 min at the same temperature. Thereafter it is heated up to 180 °C and the solution is kept at the same temperature for about 2 h. The mixture is cooled down under stirring obtaining a green solution of $CoCr_2O_4$.

Synthesis of nanoscaled pigments

The production of multicomponent oxide nanoparticles via controlled precipitation processes is difficult, because when the pH changes gradually, the products precipitate according to their IEPs in the given order but not simultaneously. This can hardly be avoided if the precipitation is not carried out using high-speed methods.

An elegant alternative to high speed mixing devices is to use a precipitation principle, which is independent from the pH, like the synthesis in high boiling solvents using thermolabile anions or organic residues. The solvent is chosen with emphasis on its boiling point (> 180 °C) and solvation power for the educts. By heating, the anions are decomposed, expelled or eliminated via a chemical reaction (compare Table 2.2, page 25). Nanoparticles grow, which are stabilized by the solvent (compare “Sterical stabilization”, page 46). Ethyleneglycols with low molecular weight have proven to be especially suitable for this type of reaction.

Typical variations of this process are:

- Decomposition of thermolabile metal alkoxides
- Decomposition of nitrates, reaction of nitrates with the solvent
- Hydrolysis of acetates
- Elimination of NH_4Cl or $(NH_4)_2CO_3$ by combination of ammonium salts with carbonates or chlorides

Since nanoparticles usually are not processed as dispersions in glycols, the high boiling solvent has to be removed. Typically an incompatible solvent is added to the dispersion. The precipitated nanoparticles are filtered, washed and then redispersed in a more suitable solvent.

The adsorbed solvent molecules protect the nanoparticles against agglomeration and determine the compatibility of the nanoparticles with other solvents. Synthesis in glycols leads to water-dispersible particles. If a rather hydrophobic solvent like trioctylphosphate is used for the synthesis, the particles can only be dispersed in unpolar solvents.

Nanoscaled metal particles are ideal pigments due to their very high absorption coefficients. Even when added in small amounts, they lead to intense, durable colors (compare “Colored and pigmented sol-gel coatings”, page 150). Their synthesis follows the same principles like the synthesis and stabilization of oxide nanoparticles [115].

➤ *Example 20: Preparation of a gold nanoparticle dispersion* [116, 117]

8.1 mmol% (3.1896 g) tetrachloroauric acid trihydrate is dissolved in a 2-liter stirrer vessel in 270 ml water. Parallel thereto, 18 mmol (9.8426 g) tetra-n-octylammonium bromide (TOABr) is dissolved in 720 ml toluene. The dissolved TOABr is then added under a N_2 flow during stirring quickly to the tetrachloroauric acid solution. The toluene phase immediately turns red and the water phase becomes clear and colorless.

The reaction mixture then is stirred for five minutes under a blanket of N_2. During this, 90 mmol $NaBH_4$ is dissolved in 225 ml water. The $NaBH_4$ solution is added quickly to the mixture under vigorous stirring. The toluene phase now becomes dark red and the water phase remains colorless. The reaction mixture then is continuously stirred for 30 minutes. After that, the phase separation takes place in a separating funnel and the aqueous clear phase is discarded. The dark red toluene phase now is washed once with 90 ml 0.1 M H_2SO_4, thereafter once with 90 ml 0.1 M NaOH and thrice with 90 ml H_2O. The toluene phase is divided in half and filled into centrifugal vessels. To each vessel 20 mmole (2.45 g) 4-dimethyl-amino pyridine (DMAP) is added while stirring to precipitate the nanoparticles. Thereafter, the gold nanoparticles are separated in 15 minutes at 4350 r/min and the almost clear toluene is discarded. The gold nanoparticles, which are stabilized with 4-dimethyl amino pyridine are dried and thereafter can be dispersed in water.

Dispersion of gold nanoparticles

The synthesis of noble metal nanoparticles is realized via the reduction of the corresponding metal salt while protective colloids or other stabilizers are present. A broad variety of procedures, partly even from the early 19th century can be found in literature.

All possible reaction pathways have in common, that measures have to be taken to prevent aggregation of the metal clusters to larger particles. Therefore, a significant amount of stabilizers is added which adsorbs immediately on the newly formed surfaces to prevent uncontrolled growth.

By the sterical demanding stabilizer and phase transfer catalyst tetra-n-octylammonium bromide in a first step, the metal salt is transferred to the toluene phase. The following reduction with $NaBH_4$ leads to seed formation and growth in analogy to the sol-gel process for oxides.

The quaternary ammonium salt stabilizes the metal particles and prevents excessive growth and agglomeration. By washing, the high salt content of the reaction mixture has to be removed quickly to stabilize the colloids.

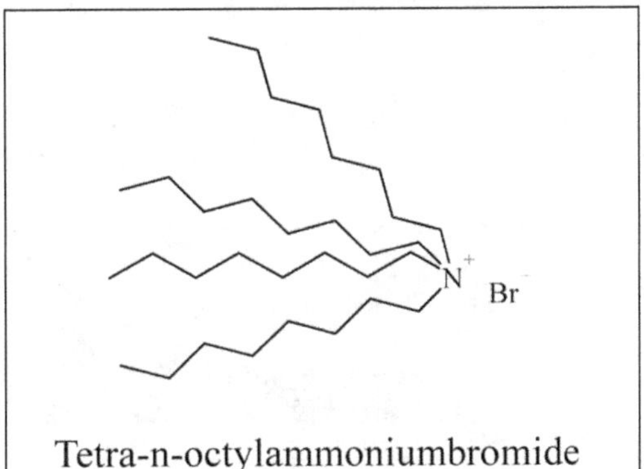

Figure 4.4: Tetra-n-octylammoniumbromide is an effective surface modifier

Tetra-n-octylammoniumbromide bonds comparably week to the metal surface. Thiols and amines are more suitable to ensure a strong bond to metal substrates. 4-dimethylamino pyridine is a hydrophilic modifier, which shows a strong affinity to metal surfaces. After addition to the toluene phase, it displaces the weakly bonded ammonium salt, hydrophilizes the nanoparticles and thereby leads to precipitation and compatibility with water as a dispersing agent.

As an alternative to precipitating and redispersing the nanoparticles, an aqueous solution of 4-dimethylaminopyridine can be added. The nanoparticles then move spontaneously from the toluene to the water phase, what can be observed visually by the color change.

If the nanoparticles are to be used for formulations based on organic solvents, a more hydrophobic surface modifier than 4-dimethylaminopyridine should be used. Typically alkane thiols like hexadecanethiol or mercaptopropyl trimethoxysilane can be used. The silane, besides the stabilization additionally offers the advantage of building a protective silicate shell around the metal particles.

4.1.3 Interference pigments

Even micrometer scaled pigments can consist of nanometer sized building blocks. Interference pigments are built from layered aggregates of titanium dioxide, silicon dioxide or iron oxide nanoparticles which were deposited on natural mica or synthetic inorganic platelets. Depending on the thickness of the different layers, the color impression develops due to interference effects of the reflected light [118, 119].

The production process involves the precipitation of nanometer-scaled titanium dioxide particles under conditions of minimal colloidal stability and in the presence of mica platelets. Because formation of a stable dispersion is not an option, the nanoparticles are deposited on the mica substrate. The thickness and the refractive index of the deposited layer determines the color impression for the eye.

➤ ***Example 21: Production of a blue interference pigment*** [120]

100 g of SiO_2 flakes (125 nm thick, particle size 5 to 50 μm) are heated to 75 °C in 2 l of deionised water. At a stirrer speed of 1,300 rpm, 17 g of a 50 % $SnCl_4$ solution are added. During this addition, the pH is kept constant at pH 1.4 using potassium hydroxide solution (30 % by weight).

The pH is subsequently lowered to pH 1.3 using hydrochloric acid (10 % by weight). At this pH, 1,200 g of $TiCl_4$ solution (400 g/l) are added. The pH of 1.3 is kept constant using potassium hydroxide solution (30 %). After

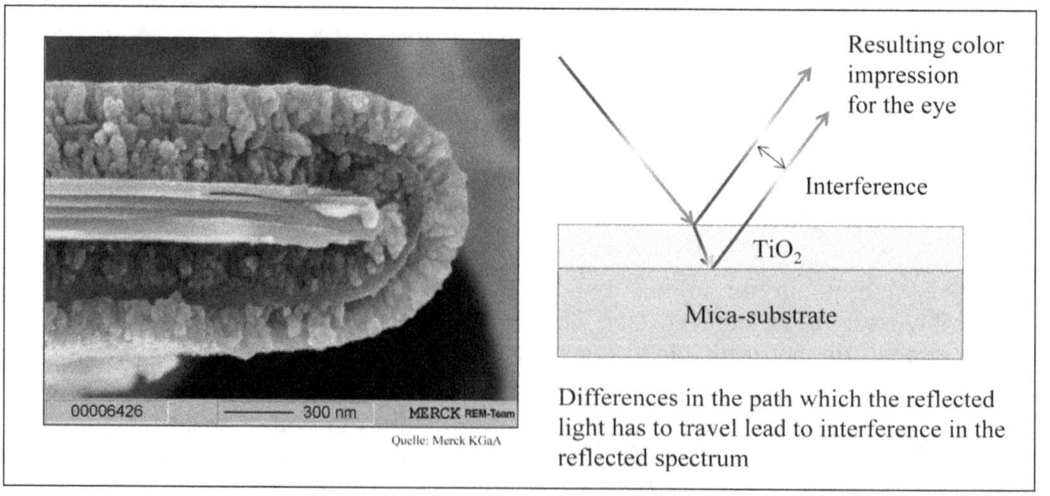

Figure 4.5: Structure and working principle of interference pigments *Source: Merck KGaA*

addition of the $TiCl_4$ solution, the pH is raised to pH 6 using potassium hydroxide solution (30 %), and the mixture is stirred for 15 min. The product is filtered off and rinsed with deionised water. After drying at 110 °C, it is calcined at 800 °C. An interference pigment having high luster and a blue interference color is obtained.

Production of a blue interference pigment

The production of interference pigments basically is a typical precipitation reaction. Starting from titanium chloride and potassium hydroxide, titanium hydroxide and potassium chloride are generated. To avoid formation of a stable dispersion of titanium dioxide and to foster the precipitation on the offered mica surface, special reaction conditions are necessary.

The colloidal stability mainly is dependent on the zeta potential of the particles and the ion strength of the solvent. The precipitation takes place at pH 1.3, a pH value at which titanium dioxide exhibits a positive particle charge and should have a sufficiently high zeta potential to form stable colloidal dispersions. The byproduct of the synthesis however is potassium chloride and this raises the ion strength of the reaction medium rapidly, thereby reducing the range of the repulsive forces continuously so that finally no stable colloids can be built anymore.

The concentration of titanium hydroxide species is kept very low during the continuous precipitation, so that the formed nanoclusters cannot aggregate to colloids, but rather are deposited onto the offered surface of the SiO_2 flake. The addition of tin chloride in the beginning of the process creates a thin tin oxide layer on top of the SiO_2 flakes which promotes adhesion and influences the crystallization process of the titanium oxide layer.

This principle of depositing nanoparticles on substrates by precipitation is also used to protect organic pigments. In literature the encapsulation of ITO nanoparticles and sensitive organic pigments by SiO_2 coatings in analogy to the principles of the interference pigment synthesis is described [121].

Other sources mention the encapsulation of organic pigments by heating them in an $AlCl_3$ solution [5]. Via hydrolysis, AlOy(OH)x is generated which forms an inorganic shell around the organic pigment (compare "Barrier coatings", page 198).

4.1.4 *IR absorption*

The desire for getting as much natural light as possible inside of vehicles and our homes goes parallel with the desire to keep excessive heat outside. Worldwide, a gigantic amount of energy is used to cool down interior space which has been heated up before by solar irradiation.

Glass, covered with conductive or semiconductive layers of sputtered ATO, ITO or FTO (antimony doped tin oxide, indium tin oxide, fluoride doped tin oxide) is capable of reflecting the IR part of the sunlight while maintaining a high degree of transparency. It is difficult however to coat complex geometries or small batch series via sputtering, therefore coating processes using conventional or sol-gel resins which contain dispersed semiconducting particles have been investigated [122, 123].

A prerequisite for the reflection of IR radiation is a (semi-)conductive area of at least the dimensions of the wavelength to be reflected. Therefore IR radiation (wavelength larger than 800 nm) can only be absorbed by dispersed nanoparticles but not reflected.

The nanoparticle heats up and dissipates the absorbed energy through the coating material. Even without reflection, by this mechanism of IR absorption, the coating can prevent

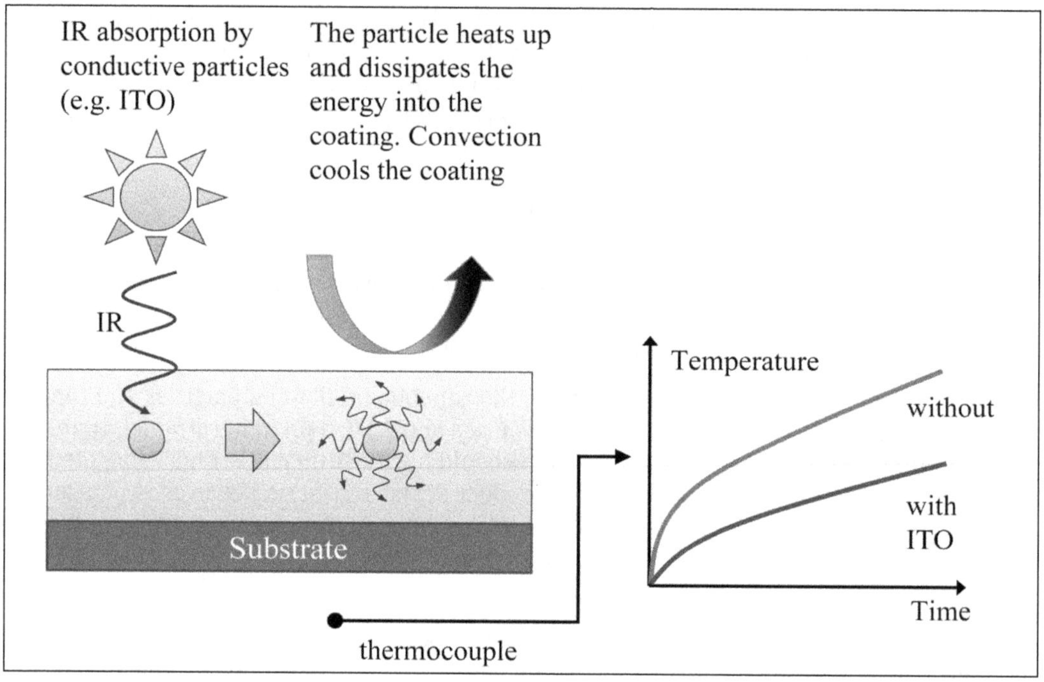

Figure 4.6: Function of an IR absorbing coating. Semi-conductive particles like ITO absorb IR-radiation and transform them to heat. Convection cools the coating

the heating-up of the room behind the coating (compare Figure 4.6). An example for the synthesis of ITO nanoparticles and the preparation of a stable dispersion can be found on page 21 (Example 2).

4.1.5 UV absorption

Organic materials quickly age under the influence of light (UV), air (O_2) and water, so that they have to be protected. UV absorbers based on organic small molecules are the state-of-the-art additives for polymers and coatings to prevent their quick outdoor deterioration. Since they are organic molecules, with time they are themselves destroyed by the UV radiation, and as softeners may lead to an undesired weakening of the mechanical strength of coatings and polymer parts.

Therefore it is desirable to develop inorganic UV absorbers which are UV stable and will rather improve the mechanical properties of coatings than to worsen them [124]. Table 4.1 shows a survey of the properties of several inorganic nanoparticles in comparison to organic UV absorbers.

Table 4.1: Assessment of inorganic and organic UV absorbers, + good, - mediocre

	TiO_2	ZnO	CeO_2	Organic absorber
Refractive index	2.7	2.0	2.2	approx. 1.5
UV stability	+++	+++	+++	0
Migration stability	+	+	+	–
No photo-catalysis	–	o	+	++
Absorption edge	340–370 nm	360–370 nm	360–420 nm	360–380 nm
Color	+ colorless	+ colorless	– yellow	+ colorless

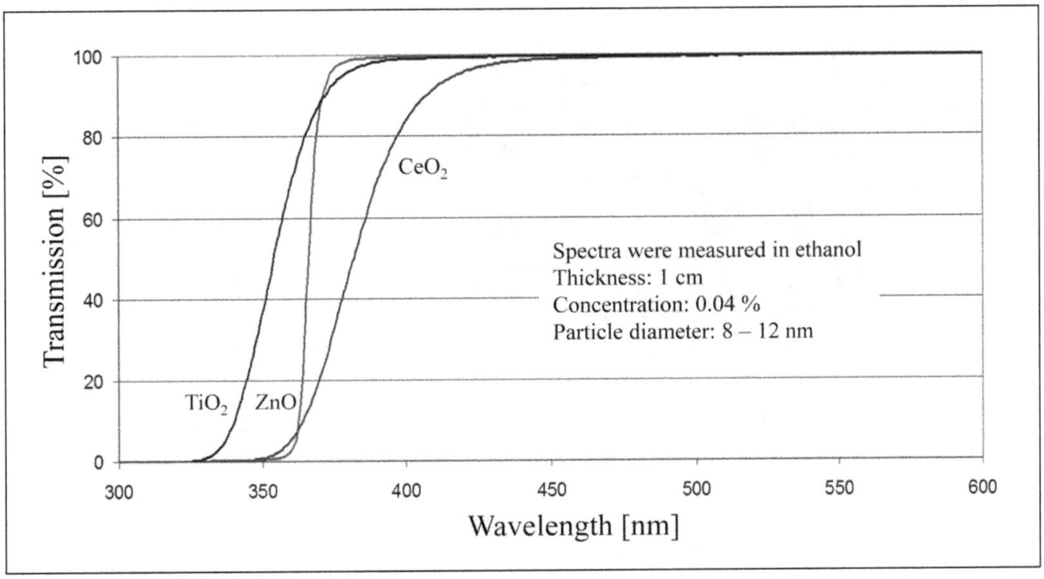

Figure 4.7: Transmission spectra of inorganic UV absorbers in ethanol Source: Merck KGaA

Titanium dioxide, zinc oxide and cerium oxide differ mainly in their absorption edge, which is the wavelength which separates the blocked part of the spectrum from the part which can pass. In order to maximize transparency and UV protection, it is desirable to have an absorption edge which is close to the visible part of the spectrum but without causing any coloration by absorbing parts of the visible spectrum. Cerium oxide as an example absorbs a significant part of the blue range of the spectrum, which results in a yellow color [125].

Titanium dioxide effectively blocks the UV-B range, but part of the UV-A range can pass. Among the inorganic UV absorbers, zinc oxide shows the most favorable absorption edge and blocks most of the UV-A and -B radiation without causing coloration.

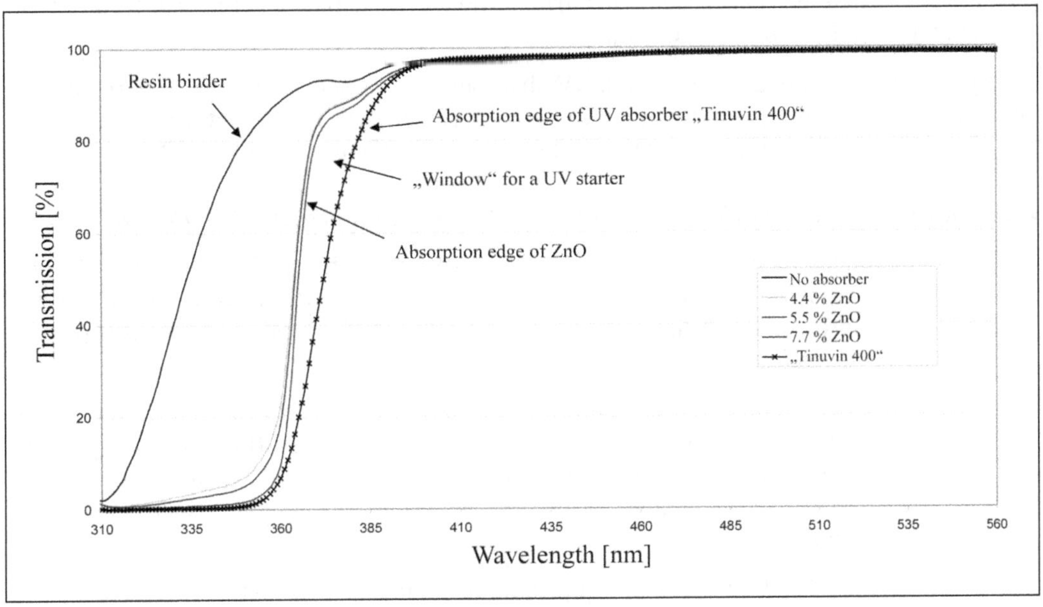

Figure 4.8: A UV-curable, UV resistant wood coating is possible by using ZnO as UV absorber
Source: Merck KGaA

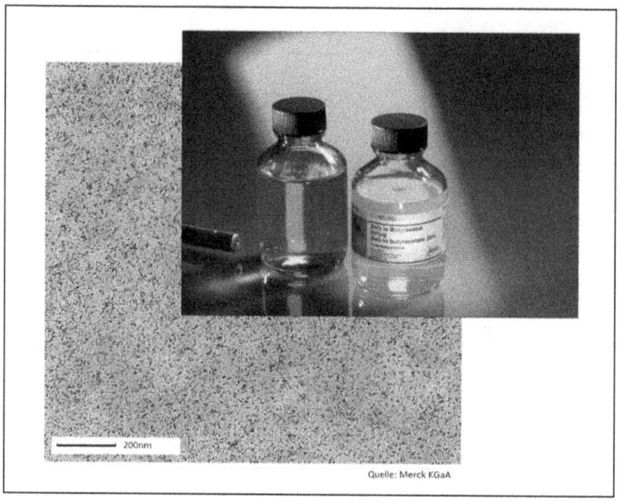

Figure 4.9: Electron-microscopical picture of ZnO particles dispersed in a PU-coating. As a comparison the picture also displays a bottle of water and the ZnO dispersion in butylacetate, showing a nice Tyndall effect *Source: Merck KGaA*

Even with an UV absorber like zinc oxide, UV curing coatings are possible. Figure 4.8 shows the transmission spectrum of a wood coating containing either an organic UV absorber or zinc oxide. The small area in the UV-A range which can pass the zinc oxide is enough to be used with a UV starter which is designed to operate at this wavelength.

If the particle size of the inorganic absorber increases, e.g. via agglomeration, then scattering causes haziness (compare Equation 3.1, page 35).

All inorganic UV absorbers have a rather high refractive index. To realize fully transparent coatings, therefore only very small particles in agglomerate-free dispersions can be used. 10 nm particle diameter, especially for titanium dioxide is the limit.

Nanoparticles of a size which does not scatter visible light might still scatter UV light since scattering is a function of both wavelength and particle size. So even nanoparticles which are not absorbing UV light like SiO_2, will still contribute to a UV protection of the coating by UV scattering.

Larger particles cause haziness. For the scattering, the size of the particles and agglomerates is decisive and not the size of the primary particles.

When the UV radiation is absorbed, the particles heat up and dissipate the heat through the coating (compare Figure 4.6, page 78).

Inorganic UV absorbers unlike organic UV absorbers do not need a HALS (hindered amine light stabilizer) to work properly. However it has been shown, that an addition of HALS is beneficial even for formulations which contain inorganic UV absorbers [126, 127].

➤ *Example 22: Combination of nanoscaled ZnO with an organic UV absorber* [127]

Firstly, two solutions “1” and “2” were prepared at 75 °C. Solution 1 comprised 13.6 g (0.1 mol) of zinc chloride (Riedel de Haën) in 500 ml of isopropanol. Solution 2 comprised 8.0 g (0.2 mol) of sodium hydroxide (Riedel de Haën) in 500 ml of isopropanol.

Solution 2 was metered into solution 1 with stirring. The reaction mixture was then stirred for 60 minutes at 75 °C. The resulting white suspension was cooled to room temperature and transferred to a rotary evaporator flask. Isopropanol was then distilled off at 70 °C and a pressure of approx. 150 mbar.

29.8 g of the residue, comprising 0.025 mol of ZnO was dispersed in 200 ml of toluene at 40 °C. A mixture of oleic acid (0.001875 mol) and 4-methoxycinnamic acid (0.00375 mol) was then added. The resulting mixture was heated to 70 °C and held at this temperature for 4 hours. The ZnO-comprising, slightly cloudy suspension was separated from solid NaCl in a separating funnel. Measurement of the UV spectrum revealed a lambda max (toluene): 294 nm.

Combination of zinc oxide with an organic UV absorber

The preparation of zinc oxide from isopropanol instead of from water results in much smaller particles, which can be more easily dispersed. If the reaction is carried out at low concentrations, uncontrolled agglomeration can be avoided to a large extent. Monomodal dispersions of zinc oxide can be prepared.

A further advantage of a precipitation starting from isopropanol is, that the formed NaCl byproduct is not soluble, crystallizes and can be filtered off easily at a later step of the reaction.

Freshly prepared zinc oxide can be deagglomerated easily by chelating surface modifiers like carbonic acids (compare Example 12, page 57). The hydrophobicity of the carbonic acid derivatives determines the compatibility of the surface-modified particle with the solvent. Long chain alkyl-, or even better branched alkyl carbonic acids have proven to be excellent compatibilizers with organic media.

This example makes use of the fact that a widespread UV absorber based on cinnamon acid exhibits a carbonic acid group. Used as a surface modifier, a ZnO particle is designed which carries an organic UV absorber as stabilizer. Both absorbers can act in synergy and the typical migration of the organic UV absorber is suppressed.

Inorganic UV absorbers show a photocatalytic activity. In the presence of UV radiation, oxygen and water they generate radicals, which aggressively destroy every organic material in their proximity (compare "Hydrophilic coatings", page 179).

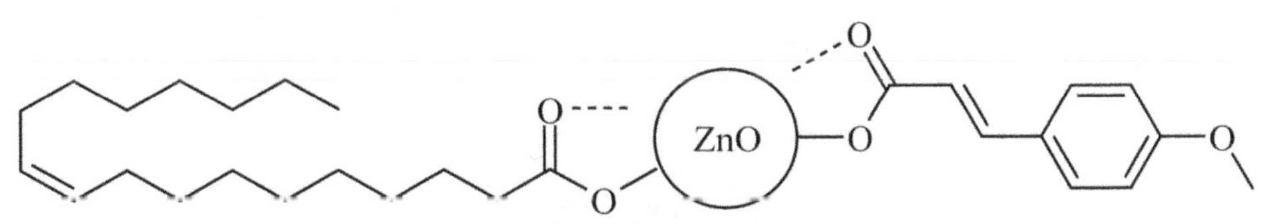

Figure 4.10: A combination of an organic UV absorber and an unsaturated carbonic acid can be used to modify the surface of ZnO-nanoparticles

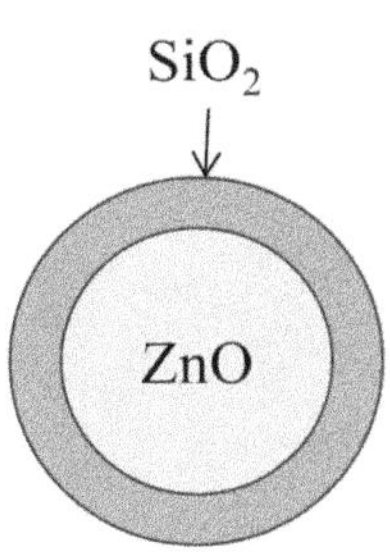

Photooxidation of isopropanol to acetone

- 40 min irradiation
- 15 mW/cm^2
- Identical amounts of isopropanol related to the BET surface of the particles were used

Material	Acetone [ppm]
ZnO 5 µm	47
ZnO 10 nm	77
ZnO/SiO_2 10 nm	2.9

Source: Merck KGaA

Figure 4.11: An inorganic layer of SiO_2 is suitable to suppress the photochemical activity of ZnO

This activity has to be suppressed, if the coating resin shall be protected. Besides a lattice doping with aluminum ions, inorganic coatings like silicon dioxide or aluminum oxide are used to separate the coating resin from the active surface of the UV absorber. The effectivity of these measures can be evaluated by investigating the photooxidative reaction of isopropanol to acetone in the presence of inorganic UV absorbers.

Figure 4.11 shows that the nanoscaled zinc oxide due to its larger surface area exhibits almost twice as much photoactivity compared to the micrometer scale zinc oxide. This photocatalytic activity can almost quantitatively be suppressed by a SiO_2 coating.

The SiO_2 shell can also enhance the chemical stability of zinc oxide. Being amphoteric in nature, zinc oxide is sensitive against contact with acids and bases, a behavior which can cause problems especially in outdoor applications.

4.2 Improving the scratch resistance of coatings

The scratch resistance of coatings can be significantly increased by the addition of inorganic particles. This has been recognized at the latest since automotive headlights made from polycarbonate could be protected with SiO_2 nanoparticle-modified acrylate resins. But even the assumed simple question "what is scratch resistance?" cannot be answered in a simple way. Scratch-, mar-, abrasion resistance, vandalism, wet and dry test methods mingle to a mind-boggling mess.

Looking at Figure 4.12 it becomes clear that there is no such thing as a universal scratch resistant coating! Different requirements and expectations demand different test methods

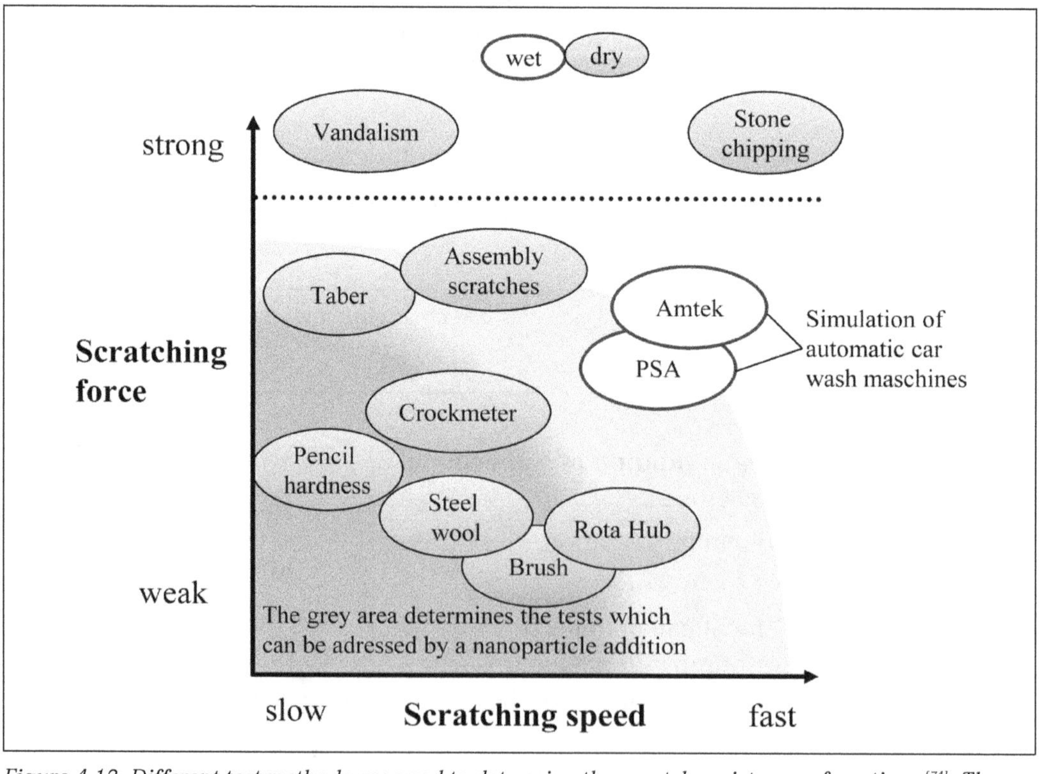

Figure 4.12: Different test methods are used to determine the scratch resistance of coatings[74]. The gray shading gives an impression which test results can be improved by the addition of nanoparticles to the coating

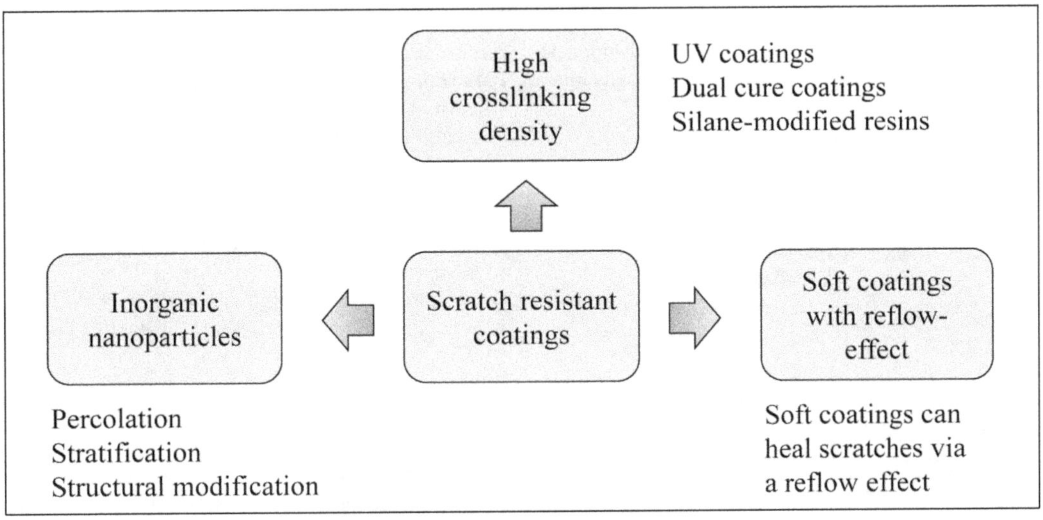

Figure 4.13: Different concepts to realize scratch resistant coatings are followed by the coating industry

and different material properties. Hence, the intuitive test with the car key is not a test for scratch resistance, but pure vandalism. Likewise stone chipping resistance cannot be compared to a car washing simulation.

Scratch resistance by no means is hardness. This statement especially proves right in current concepts which are followed to realize more scratch resistant car clear coats. Besides the addition of nanoparticles, self-healing, soft and elastic coating concepts based on UV curing, dual cure or silanized resins are investigated (compare "Scratch resistant coatings", page 155).

If a cured coating resin is described in a simplified way as a network of polymer-segments and -knots, the strengths and weaknesses of the different concepts can be illustrated.

A low cross-linking density, which means fewer knots and longer polymer segments describes a coating which can be scratched easily, but shows an enhanced ability to heal the scratches by a reflow mechanism.

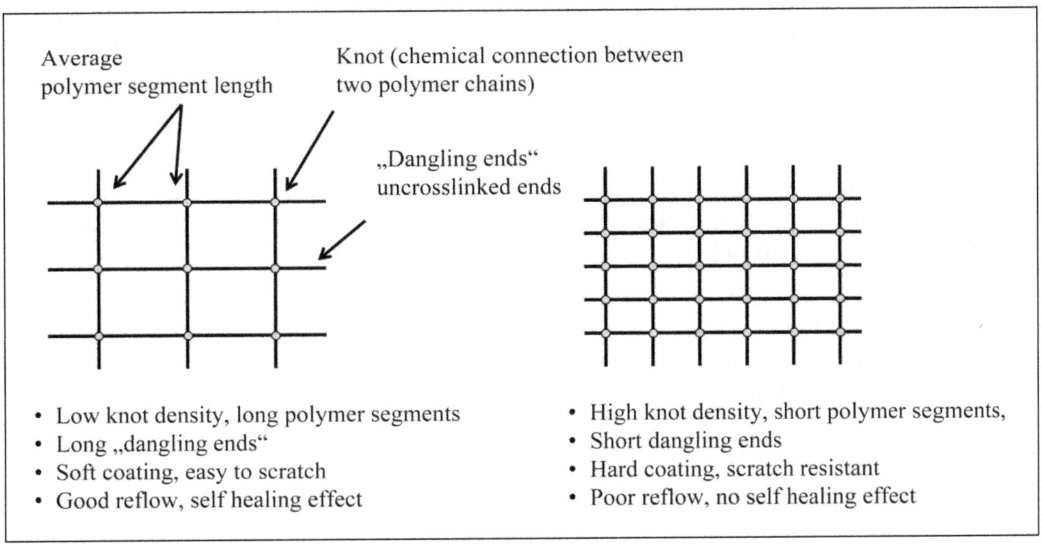

Figure 4.14: An idealized concept: Coatings can be described as a combination of polymer segment length and knot densitiy to understand their response to scratches

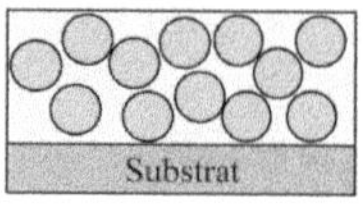	**Percolation**	Advantages: Persistent, good resistance against heavy duty use Disadvantages: Large amount of nanoparticles necessary. Viscosity and brittleness increases. Compatibility with the resin can be an issue
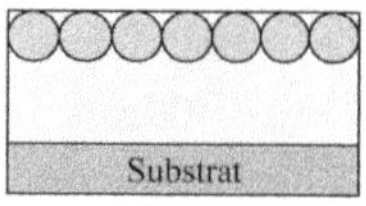	**Stratification/ leafing**	Advantages: Small amount of nanoparticles necessary, no increase in viscosity Disadvantages: Only temporary effect until the upper layer is gone. Possible problems with recoatability
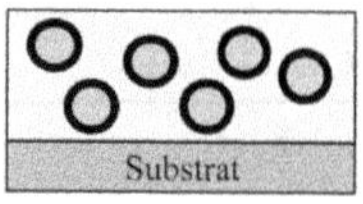	**Structural modification**	Advantages: Persistent effect, only small amount of nanoparticles is neccesary. No change in viscosity Disadvantages: Limited effect against heavy duty use

Figure 4.15: Strategies to increase the scratch resistance of coatings with nanoparticles

Many knots and short polymer segments describe a coating which is on the one hand very hard, but also very brittle. If the coating is damaged, a crack propagates easily through the material and cannot be stopped by plastic deformation. The degree of self-healing of such coatings is very poor.

None of both concepts seems to be the optimum solution. Nanoparticles open up new possibilities to increase scratch resistance of coatings by introducing a new component to the resin. Their homogenous distribution in the coating is essential for their effect (Figure 4.15).

Percolation concept

If the amount of nanoparticles in a coating is increased up to the point when on average every nanoparticle is in contact with another, then an inorganic network has been formed within the organic resin. This amount of nanoparticles is called the percolation threshold. The inorganic network is able to withstand mechanical impact and so the scratch resistance of the coating mainly is a function of the hardness of the particles and the intensity of their interparticulate interaction.

The surface modification of the nanoparticles usually is inert and compatibilizing. Since the percolation threshold requires addition levels of between 15 to 25 weight% of nanoparticles relative to the binder resin, the character of the coating material is changed significantly. The high hardness of the inorganic network induces brittleness and thus a tendency to crack formation and adhesion problems.

However, a coating not only has to be scratch resistant, but furthermore has to fulfill other criteria which are disturbed by a high loading with nanoparticles. Therefore it has been investigated, how to make use of the scratch resistance improving properties of inorganic nanoparticles while using significantly smaller amounts.

Leafing concept

Starting from the thought that only the surface of coatings has to be scratch resistant, the leafing concept has been developed. A very thin layer of nanoparticles, only a few nanometers thick, increases the scratch resistance of the coating, whereas the rest of the coating remains almost unchanged.

The realization of this concept uses the well-known idea of leafing pigments, which e.g. by a modification with stearic acid orient themselves to the surface. By the surface modification a slight incompatibility with the resin matrix is induced. The system minimizes the interfacial energy by orientation of the particles to the surface and to each other. Typically less than 5 % of nanoparticles are necessary to realize a complete coverage of the surface [128]. Since the nanoparticles are only present in the upper layer of the coating, the higher scratch resistance is only given until in this layer is destroyed by weathering.

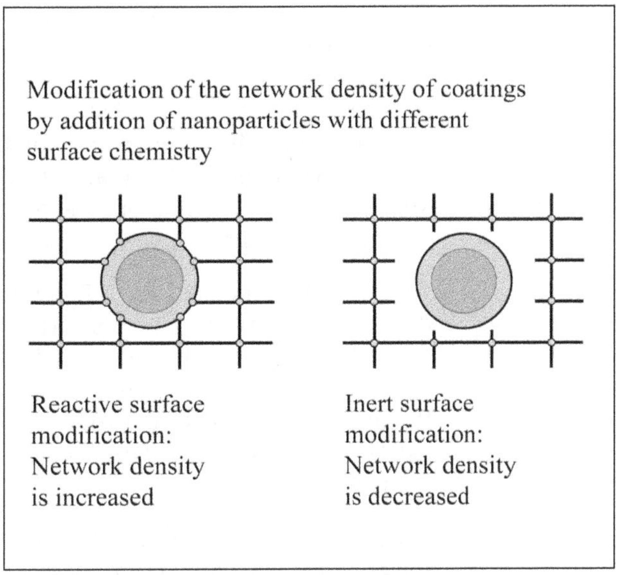

Figure 4.16: Structural modification of coating resins by addition of nanoparticles with different surface modifications

When such coatings systems have to be repaired, the nanoparticle containing upper layer has to be removed completely to ensure good leveling and adhesion. The leafing effect is also dependent on the processing and drying parameters which make body shop repairs quite challenging.

A recently proposed concept uses small amounts of nanoparticles without a leafing effect. The nanoparticles are homogeneously dispersed throughout the coating and lead to a structural modification of the coating resin. Figure 4.16 illustrates the two different concepts using this approach [74, 107].

If nanoparticles are coated with a low surface energy coating, they become inert towards the resin and disturb the formation of a continuous polymer network. A particle-filled cavity is generated, which improves the reflow capability of the coating by allowing plastic deformation.

If nanoparticles are modified with reactive groups on their surface, they act as cross-linking points and become part of the polymer network (compare Figure 3.25, page 54). By tailoring the composition and T_g, the polymer shell can also be designed to show viscoelastic properties. Via a reversible deformation, an impacting energy can be absorbed, which contributes to the resistance of coatings against scratches.

Both concepts can reduce the number of especially fine scratches significantly. Even addition levels of as low as 0.1 % can show positive effects. It should be clear however, that such small amounts of nanoparticles are not able to protect the coating against a rough treatment. The hardness of the coating is hardly changed.

➤ *Example 23: Preparation of a silicone-modified alkoxysilane and surface functionalization of nanoparticles [129]*

A 250 ml four-necked flask is charged with 100 g of Si-H-functional polysiloxane. This silicone is easy to prepare by means of an equilibration reaction as described by Noll (Chemie und Technologie der Silicone, Wiley/

VCH, Weinheim, 1984). The silicone is heated to 70 °C under N_2, and 16.3 g of vinyltrimethoxysilane (e.g.: "Geniosil XL10"; Wacker Chemie GmbH) are added. Then 10 ppm of hexachloroplatinic acid are added. After the exothermic reaction has subsided, the reaction mixture is stirred at 80 °C for 1 h. Then reduced pressure is applied and approximately 2 g of unreacted vinyltrimethoxsilane and volatile constituents of the polysiloxane are separated by distillation. The product is of low viscosity and possesses an amberlike colour.

A 250 ml four-necked flask is charged with 75 g of aqueous silica sol ("Köstrosol 2040 AS", Chemiewerk Bad Köstritz) and this charge is mixed with 75 g of 1-methoxy-2-propanol and heated to 60 °C. 1.6 g of propyltrimethoxysilane ("Dynasylan PTMO", Degussa AG) are added dropwise to the mixture, which is then stirred at 60 °C for 2 h. 80 g of methoxypropylacetate are added and 110 g of solvent mixture are removed under reduced pressure at 70 °C.

Subsequently, 1.25 g of the silicone-modified alkoxysilane are added and the mixture is stirred at 75 °C for 2 h. Following the addition of 3.75 g of wetting and dispersing additive ("Disperbyk 168", BYK-Chemie GmbH) the dispersion is homogenized and then a solids content of 35 % is adjusted by removal of solvent under reduced pressure at 75 °C. The resulting dispersion is of low viscosity and shows no tendency towards gelling or sedimentation after 28 days storage.

Surface modification of nanoparticles with silanes

The synthesis and use of application-specific silanes is an important tool of sol-gel- and nanotechnology. Unfortunately, commercially available silanes are not always the right answer to the actual material problem which has to be solved.

In this example, the platinum catalyzed addition of a Si-H group to an olefinic double bond is used to synthesize an asymmetric silicone with one alkoxysilane functionality. The following reaction with silicon dioxide nanoparticles yields a SiO_2-silicone core/shell particle. Obviously this form of a sterical stabilization is not sufficient, because in this example a significant amount of conventional dispersing aids is added to further stabilize the dispersion. The use of such core/shell dispersions as nanoadditive to coatings to improve their scratch resistance is discussed.

This example describes a similar reaction like shown in Figure 5.31 (page 117) and underlines that silicone derivatives can be a useful tool in the sol-gel process (compare "Network modifiers", page 109).

It has been shown in practical applications, that the mechanical properties of nanoparticle- containing coatings can be influenced to a great extent by other components of the formulation. Especially leveling-, slip-, degassing- or wetting additives, depending on their structure can act synergistic or deteriorating on the scratch resistance of the coating [130].

Nanoparticles introduce a large interfacial area into the formulation which is capable to interact with surface active additives. Per gram nanoparticles, a surface area of up to 350 m^2 per gram is generated, so the size of the inner interfaces exceeds by far the size of the surface of the coating which is exposed to the environment. So small changes in the chemical reactivity or polarity of the particles can have distinct consequences for the mechanical properties of the coating.

Crockmeter test for determining dry scratch resistance

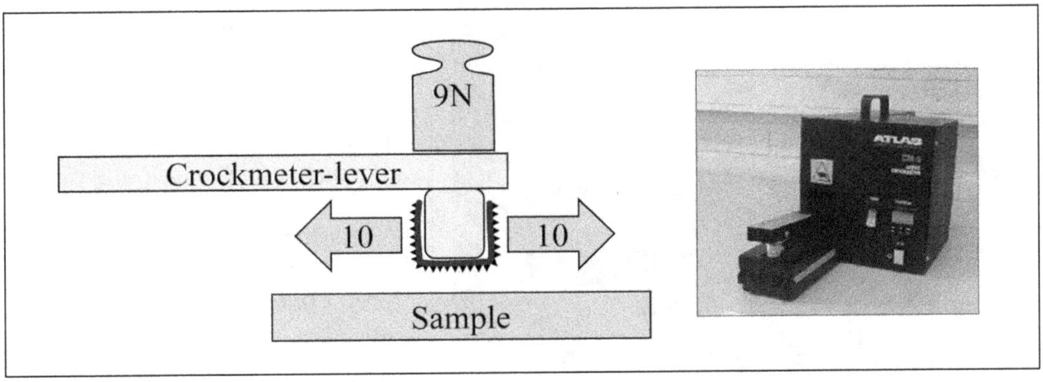

Figure 4.17: Crockmeter-Test to determine the scratch resistance according to Daimler test regime PBODC390 [132]

The Crockmeter test has been established as a tool to measure the scratch resistance of coatings [131]. In contrast to car-wash simulations like the Amtek Kistler test, this test simulates, how dry dirt particles are scratching the surface of a coating. The test results are compa-

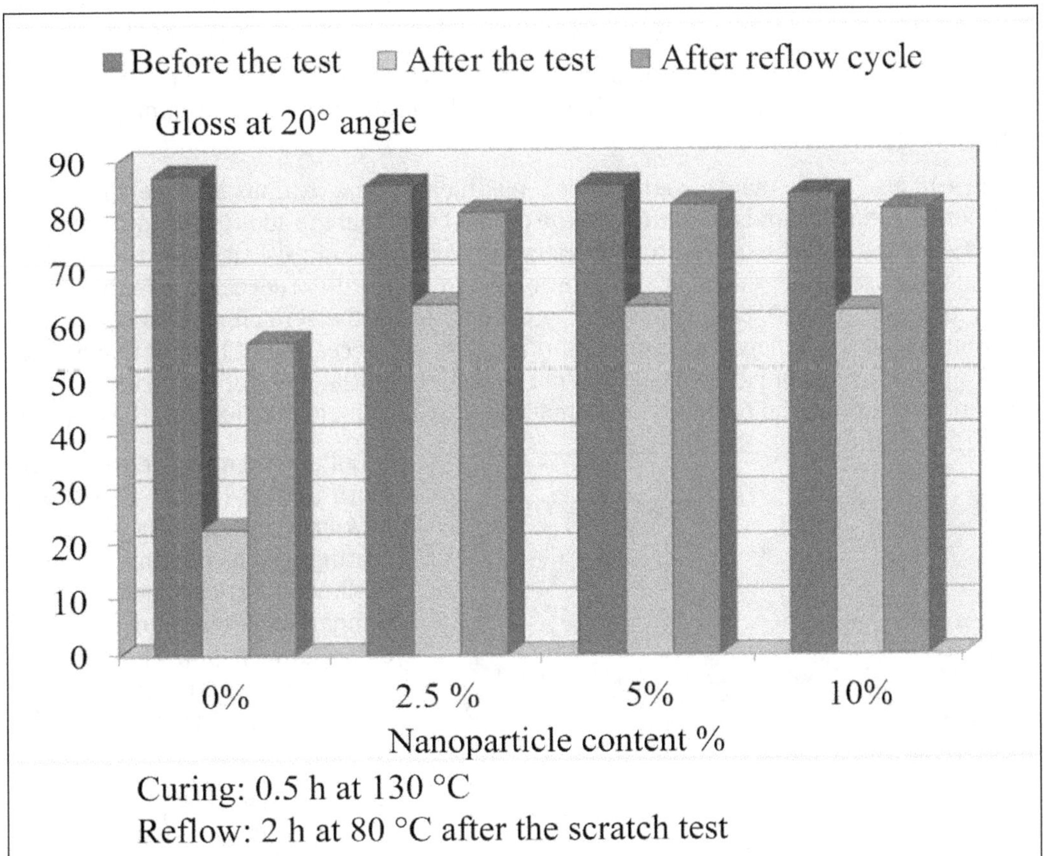

Figure 4.18: Improvement of the scratch resistance of a 2K PU clear coat (Crockmeter test; Daimler Norm PBODC390) by core-shell nanoparticles [74]*. Please note: The y-axis shows the absolute gloss value and not the gloss retention*

Figure 4.19: 2K-PU OEM series automotive clear coat with 5 % and without core/shell nanoparticles after the crockmeter-test. Sample preparation: Coating on polycarbonate/ carbon black shapes via spray coating. Curing: 130 °C/30'[74]

Source: Merck KGaA

rable to the well-known hammer/ steel wool test, but are much more reproducible and accurate.

A piece of precision sanding paper with exactly 9 μm roughness is moved forwards and backwards 10 times under a load of 9 N. The difference in gloss levels between a pristine and a scratched surface at 20° viewing angle is inversely proportional to the scratch resistance.

After scratching, the sample is kept for two hours at 80 °C to test the reflow capability of the coating. At this temperature, the coating softens and some of the smaller scratches can be healed by the coating material. Hard coatings typically show a bad reflow behavior. Figure 4.18 shows the results for a typical car OEM car clear coat, containing different amounts of core/shell nanoparticles.

It can be seen, that the coating without nanoparticles was heavily damaged by the sanding paper. Even a small amount of core/shell nanoparticles is able to reduce the damage significantly (Figure 3.25, page 66).

Figure 4.18 also shows that the nanoparticle-modified coating exhibits an excellent reflow behavior. This result stands in contrast to the earlier belief that the addition of nanoparticles always leads to a higher hardness and brittleness of coatings. Figure 4.18 shows the absolute value of the gloss measurement at an angle of 20° and not the difference between the values before and after the test. So it is possible to not only judge the scratch resistance but also the influence of the nanoparticles on the initial gloss of the coating. It is remarkable, that even higher loadings of nanoparticles do not reduce the initial gloss of the coating, a fact which points out that the particles are homogeneously distributed throughout the coating.

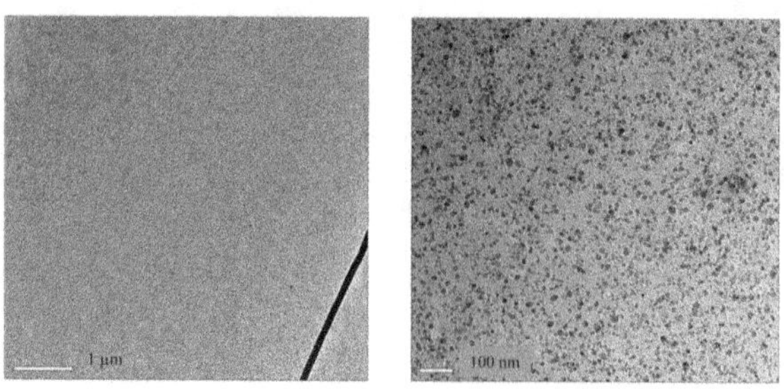

Figure 4.20: Electron microscopy picture of perfectly dispersed core/shell particles in a cured 2K PU clear coat at 2 different magnifications. 40 weight% core/shell particles. An agglomerate-free dispersion is the prerequisite for clear coatings[74]

Source: Merck KGaA

Numbers are only one part of the truth and so Figure 4.19 shows how an unmodified resin and a coating which contains 5 % core/ shell nanoparticles look in direct comparison after being scratched by sanding paper. The coatings were applied on carbon black colored PC and cured at 130 °C for 30 minutes..

An increased scratch resistance does not necessarily mean a higher hardness. Nanoscaled corundum does not work better than nanoscaled SiO_2, despite the fact that the corundum is significantly harder than the silicon dioxide. A by far more decisive factor is the improvement of the viscoelastic properties of the coating. If nanoparticles are covalently bonded to the resin matrix, the storage modulus of the coating is increased over a broad temperature spectrum [133-135].

Due to the surface modification, the particles are stabilized in a very effective way so that they can be dispersed without shear forces down to primary particle size (compare “Sterical stabilization”, page 46).

5 Coating resins made by nanotechnology – the sol-gel process

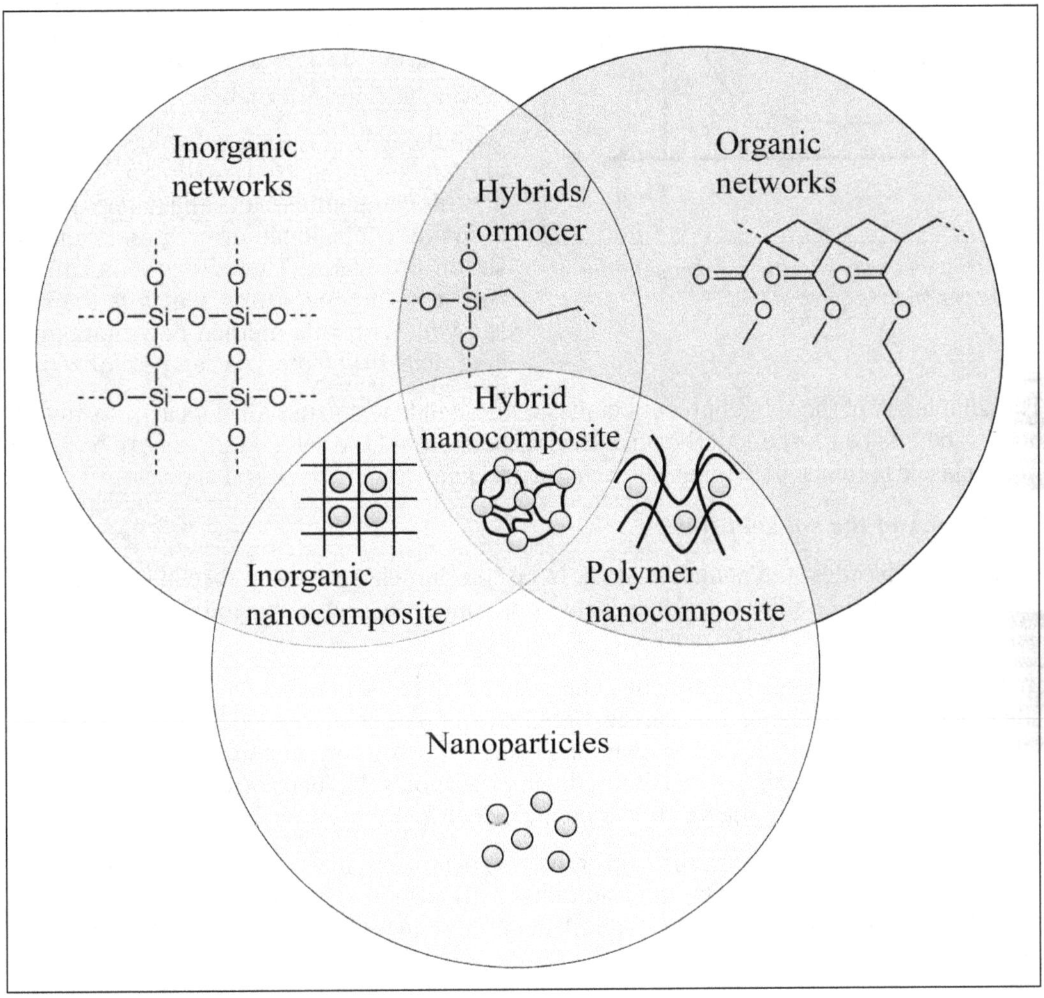

Figure 5.1: Inorganic-organic composites [136]

The sol-gel process describes the synthesis and application of nanoparticles and inorganic/organic structures not only in, but also as coating resins. Its focus is the synthesis of inorganic/organic composites which are known as "ormocer", "ormosil", or "nanomer". Figure 5.1 illustrates these terms.

What is the nature of sols and gels?

An "official" definition does not exist for the sol-gel process, so we can help ourselves with these descriptions:

Gerhard Jonschker: Sol-Gel-Technology in Praxis

ISBN: 978-3-86630-886-2

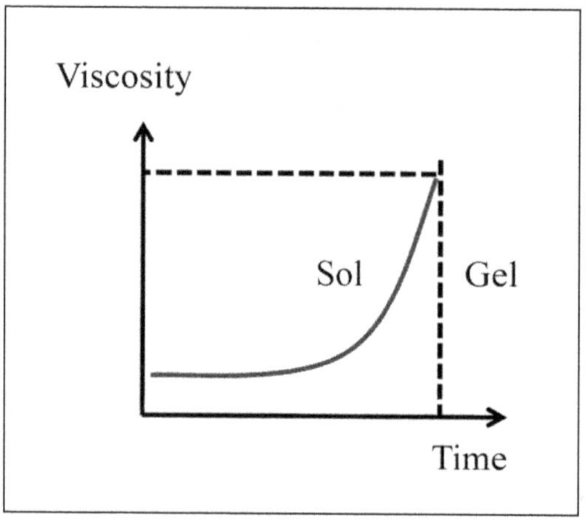

Figure 5.2: Viscosity changes dramatically during the transition from sol to gel

The sol-gel process is a generic term which describes the synthesis of inorganic and hybrid materials at low temperatures by wet chemistry methods, starting from soluble precursors.

A sol is a colloidal dispersion of particles smaller than 100 nm and a gel is a viscoelastic fluid. As soon as mechanical properties can be measured, a sol becomes a gel.

With the help of the sol-gel process, materials can be synthesized starting from metal-organic compounds, salts and nanoparticles by using controllable growth and condensation processes. These reactions can be run at low temperatures, which is the reason why with this method new inorganic/organic hybrid materials can be realized.

The complexity of the sol-gel process makes it impossible to discuss all of its aspects in this book. A detailed survey about the physics and chemistry of the sol-gel process can be found in the classic textbook of Brinker and Scherer and many other books and articles [5, 137-141, 150].

The history of the sol-gel process

Since many decades, the sol-gel process is fascinating chemists and materials scientists alike due to the possibility to synthesize hybrids containing inorganic and organic building blocks starting from a molecular level [8, 87].

In 1845 Ebelmann already described the possibility to synthesize SiO_2 by hydrolysis of silicic acid esters. By using dip-coating, Geffcken and Berger succeeded in coating surfaces with inorganic coatings by using solutions containing the corresponding soluble organometallic precursors. In the United States during the 1960's the focus of sol-gel research and development was mainly the solidification of nuclear fuel.

Soon it became clear, that with the help of the sol-gel process, multicomponent compositions which could not be produced by the traditional methods could be synthesized and that the optical properties of fibers, coatings or bulk materials can be influenced to a so far unknown extent [142].

The first publications mainly dealt with glass- or ceramic-like materials. Then the sol-gel process experienced a huge impulse in popularity, when the inorganic/organic hybrids were discovered. Organofunctional silanes made it possible to synthesize those hybrids well below the decomposition temperature of organic compounds and their applicability as liquid sols opened up a variety of technical possibilities like e.g. for functional coatings [87].

It is difficult to fix the starting point for this scientific breakthrough. Often, the first international conference on coatings on glass in Padova 1981 is being named as starting point for the rapidly growing number of scientific publications [143]. In the following years, the topic grew more and more and today influences almost all aspects of material research. Figure 5.3 shows the reaction scheme of a typical sol-gel synthesis.

First, the precursors are hydrolyzed and thereby generate oligomers via condensation reactions. Then, the resulting colloidal dispersions are processed by manifold ways to yield coatings, fibers, bulk materials, etc.

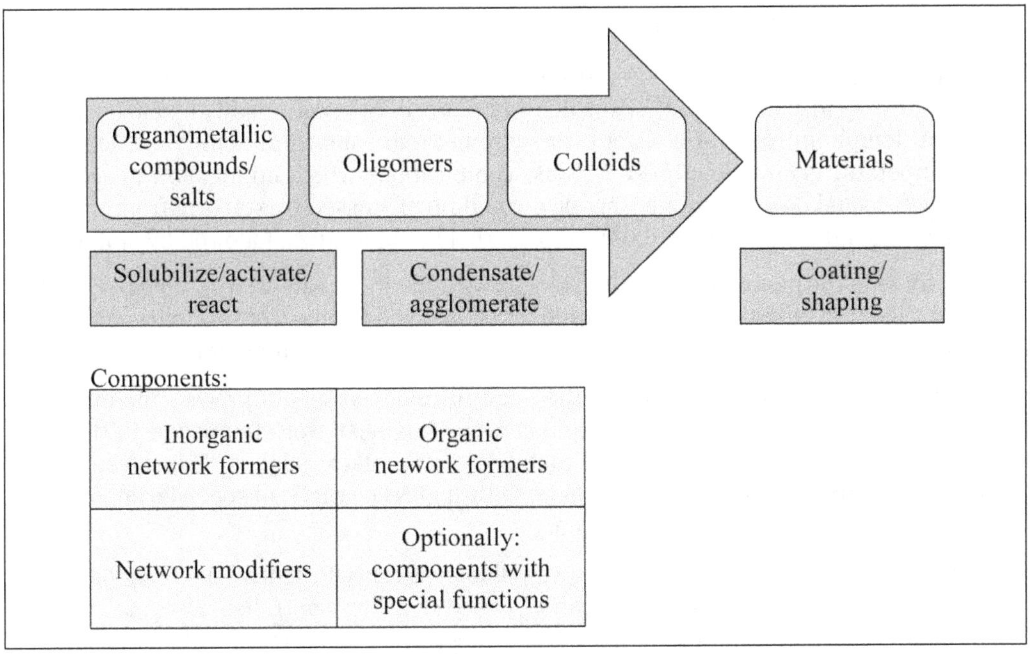

Figure 5.3: Typical components and reaction scheme of the sol-gel process

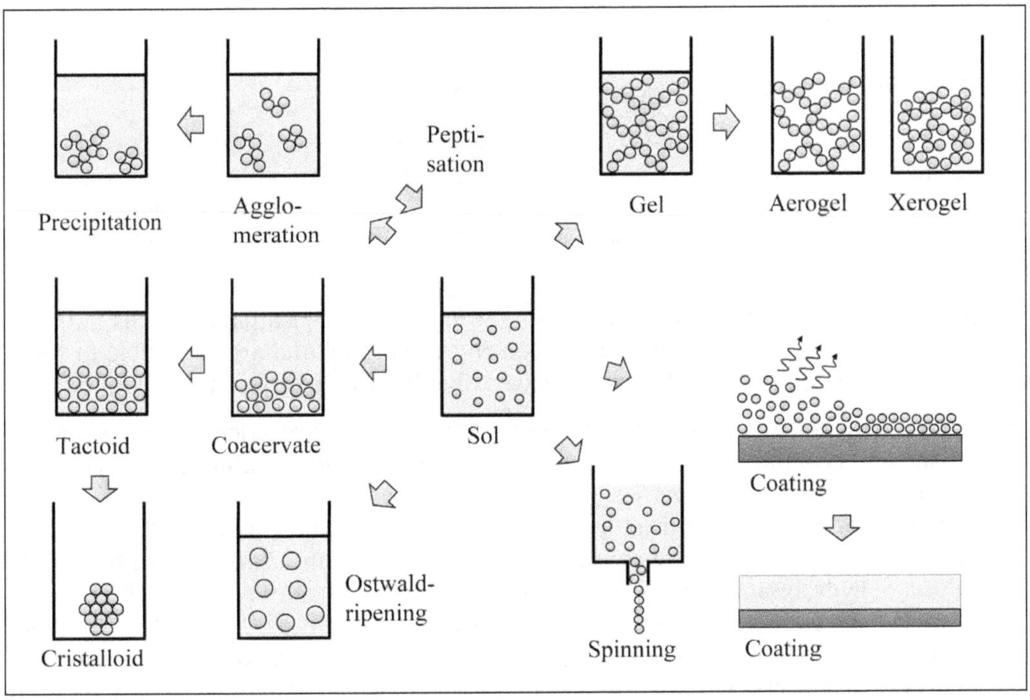

Figure 5.4: The sol-gel process as a tool to produce fibers, coatings or bulk materials

The scope of possible raw materials is wide, ranging from metal-organic compounds, via salts to organic monomers. Figure 5.4 illustrates some possibilities of sol-gel processing, this book however focuses on sol-gel coatings.

In the following chapters, typical components of sol-gel formulations will be discussed, starting with materials which form inorganic networks.

5.1 Inorganic networks

In the early days of the sol-gel process, alternative ways of making glasses and ceramic compositions were in focus. Compositions which when processed traditionally had to be melted, sintered at temperatures >1000 °C, or even were not realizable at all, now were accessible at low temperatures far below 1000 °C. Glass compositions which might show phase-separation when cooled down after melting, now could be processed by starting from the "cold end", involving mixing and processing steps at room temperature, followed by sintering.

The reactivity of the nanoscaled building blocks with their extremely large surface area and stored surface free energy allowed the production of almost dense ceramic parts or glasses at temperatures well below the melting point or traditional firing temperatures.

So it can easily be understood, that in the beginning of the sol-gel process, mainly companies which were active in the glass and ceramic business were interested in this new technology. Typical raw materials were metal salts and alkoxysilanes. The solvent-based formulations allowed a coating of surfaces with thin glass coatings of special compositions, which was considered as a real innovation.

The use of the alkoxysilanes opened up a valuable possibility to control the reaction speed. Water as the primary reaction partner can be added in a very controllable way, so that precipitation due to a too high reactivity can be avoided. In a waterborne solution, like it is the case of metal salts, this is hardly possible. In addition to that, alkoxides release only alcohol as a byproduct of the reaction which then becomes part of the solvent mixture. So very pure materials are accessible by the sol-gel process.

5.1.1 Hydrolysis and condensation

The material synthesis by the sol-gel process starts with precursors, which in the course of controllable chemical reactions generate oligomers and colloids of different morphologies. Typical precursors are the alkoxides of elements which form oxides, like tetraethoxysilane for SiO_2 colloids or zirconium propoxide for ZrO_2 colloids. With the help of suitable catalysts, they react with water in a variety of solvents.

Reactive monomers are formed by hydrolysis, which, via condensation reactions build up a three-dimensional network. These reactions run in parallel and are reversible to some extent, so that an equilibrium of species evolves like that which is depicted in Figure 5.5.

Even after a long reaction time and with an excess of water as a reaction partner, sol-gel formulations still contain many Si-OH and Si-OR groups. The educts almost never condensate to 100 %.

The amount of water which is used for hydrolysis influences the reaction speed and the structure of the resulting polycondensate. Therefore the amount of water used relative to the amount of hydrolysable groups is a valuable steering parameter of the reaction. In the early days of the sol-gel process, most experiments were done with TEOS (tetraethoxysilane) and the Rw-value indicated the amount of water relative to the amount of TEOS. Now, the R_{OR} value is used, because it takes into account that many organo-modified silanes are used which contain less than four alkoxy groups.

In principle H_2O can be replaced by other reactive species like e.g. HF. Then, metal fluorides are generated in this "fluorolysis", which can be used e.g. for low index coatings (compare "Tailoring of the refractive index of sol-gel coatings", page 186) [317-319].

By variation of the reaction conditions, the speed of hydrolysis and condensation can be influenced and thereby the structure of the resulting materials can be changed signifi-

a) Substitution of alkoxy- by hydroxygroups:

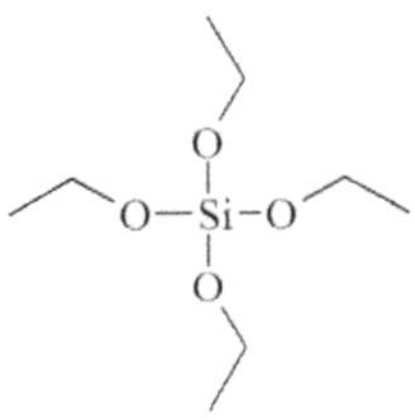

$$Si(OR)_4 + H_2O \rightleftarrows \mathbf{HO}\text{-}Si(OR)_3 + ROH$$

$$HO\text{-}Si(OR)_3 + H_2O \rightleftarrows \mathbf{(HO)_2}Si(OR)_2 + ROH$$

$$(HO)_2Si(OR)_2 + H_2O \rightleftarrows \mathbf{(HO)_3}SiOR + ROH$$

$$(HO)_3Si\text{-}OR + H_2O \rightleftarrows Si\mathbf{(OH)_4} + ROH$$

b) Condensation reactions:

$$n\ Si(OH)_4 \rightleftarrows (SiO_2)_n + 2n\ H_2O$$

$$\equiv Si\text{-}OR + HO\text{-}Si\equiv \rightleftarrows Si\text{-}O\text{-}Si + ROH$$

Figure 5.5: Hydrolysis- and condensation reactions during the sol-gel process of tetraethoxysilane

$$R_w = \frac{M(TEOS)}{M(water)}$$

The R_w-value is an outdated indicator, because for historical reasons it is only valid for tetraethoxysilan

$$R_{OR} = \frac{M(OR - groups)}{M(water)}$$

The R_{OR}-value is the quotient of the number of hydrolyzable groups versus the number of water molecules. In a system with an $R_{OR} > 0.5$ theoretically enough water is present for a complete condensation

Figure 5.6: The R_{OR}-value is an important indicator for the water content of a sol-gel formulation

cantly. Since the number of variables like temperature, catalyst, solid content and additives is high and they influence each other, a precise prediction of the outcome of a process variation is hardly possible. However, a set of well described process-structure relationships in model systems like pure TEOS is available as a guide.

So, the sol-gel process presents itself as a powerful, yet complex tool whose breadth of possible variations grant the developer a wide playground but also demand some effort in the investigation of cause-effect relationships.

5.1.2 Catalysts for hydrolysis and condensation

A change of the pH influences hydrolysis and condensation in a very different way. Figure 5.7 shows that the reaction speed of the hydrolysis (Si-OR → Si-OH) and condensation (Si-OH → Si-O-Si) is a function of the pH value. Instead of a curve, a bandwith is depicted in order to take account for the fact that by variation of alkoxides, solvents or counterion of the catalyst, the reactivity can be changed remarkably.

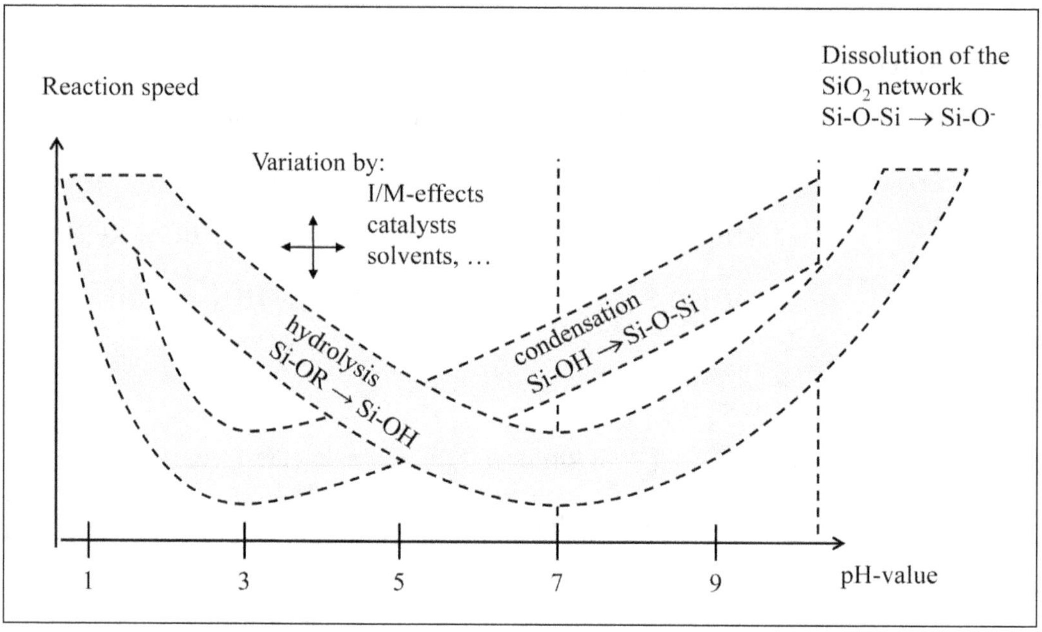

Figure 5.7: Dependency of the speed of hydrolysis- and condensation from the pH value in the SiO_2 sol-gel process

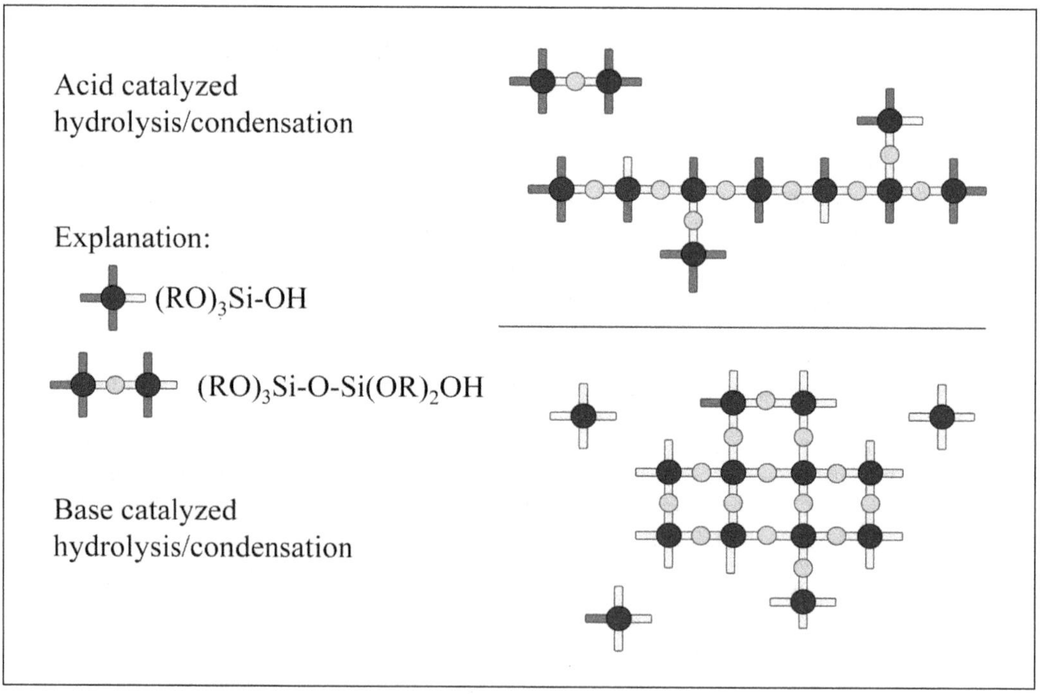

Figure 5.8: Depending on the pH value, hydrolysis and condensation run at different speed. As a consequence different structures result

At mildly acidic pH values, hydrolysis is fast and the slow condensation determines the overall reaction speed. In a first step, small oligomers are generated which carry many Si-OH groups. Reactions preferably take place at the chain ends, so chains with only few branches are generated, which then entangle and react to a polymer-like network.

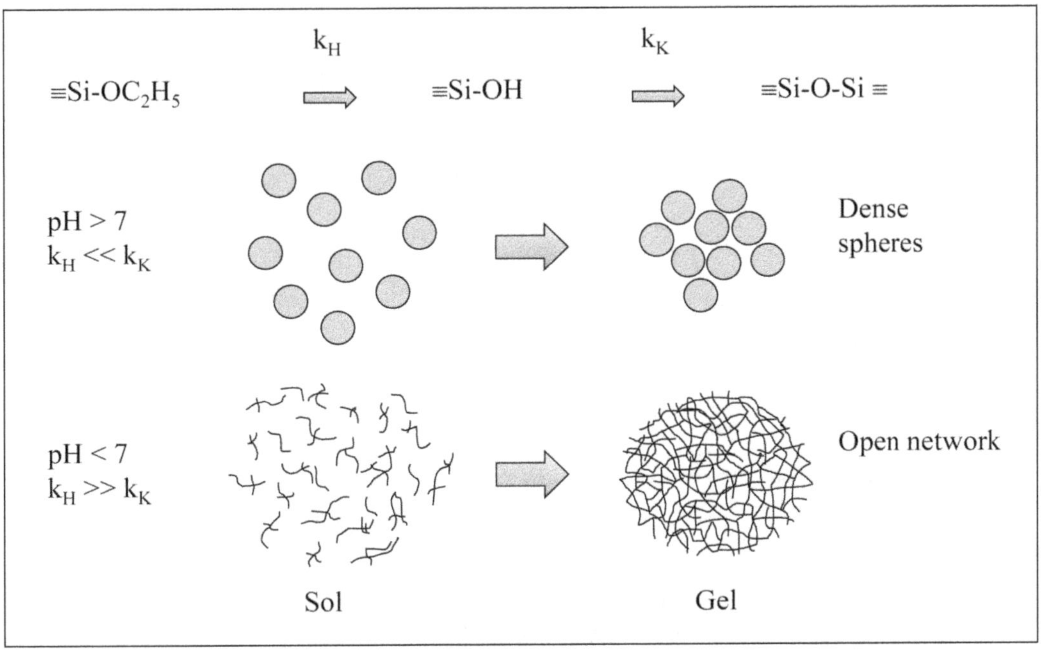

Figure 5.9: Different reaction speeds of hydrolysis and polycondensation lead to different morphologies of the sol particles and the products made from them

At alkaline pH values, condensation is fast and a comparably slow hydrolysis determines the overall reaction speed. Hydrolysis preferably takes place at the silicon atoms which are located in the mid of the chain and at cross-linking points. The resulting structures show a particle-like morphology.

At very high pH values > 10 the Si-O-Si bond can be cleaved again and SiO_2 becomes partly water-soluble. An equilibrium forms, which is determined by the relative amount of alkali ions to SiO_2 and is called “water glass”.

pH value influences the structure

The differences in the structure of the particles also influence the properties of the resulting products. Acid-catalyzed reactions yield open, porous structures which resemble the tumbleweeds known from many Wild West movies. Sols which were synthesized under alkaline conditions consist of rather dense, spherically shaped particles, which can arrange themselves in the following process steps like coating and drying to denser structures (compare “High temperature resistant coatings”, page 72 and Figure 6.19, page 146).

Reactivity

The reactivity of silanes is determined to a large extent by their alkoxide groups. Alkoxygroups with longer alkyl chains react far more slowly than alkoxygroups with shorter chains. Tetramethoxysilane approximately reacts 5 to 10 times faster compared to tetraethoxysilane. The reasons for this behavior are the sterical hindrance of the longer alkyl chains and the smaller +I-effect of the methyl group compared to the ethyl group.

The mechanisms of hydrolysis and condensation are described in detail in the book of Brinker and Scherer [137]. Under acidic conditions, a proton attacks the alkoxy oxygen atom and an alkoxonium ion is generated. Substituents and the alkoxy chain which cause a +I-effect increase the electron density at the silicon and oxygen atoms, stabilize the positively charged transition state and thereby increase the reaction speed.

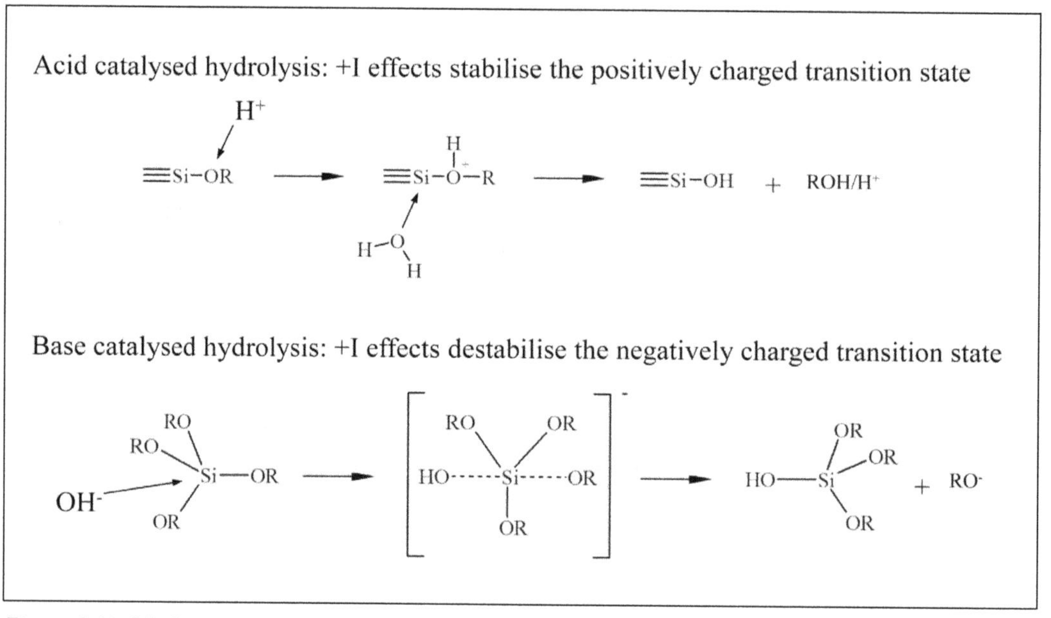

Figure 5.10: Mechanisms of acid and base catalyzed hydrolysis of alkoxysilanes

Under alkaline conditions, an OH^- ion attacks the silicon atom. Substituents and alkoxy groups with +I-effect decrease the positive partial charge of the silicon atom and thereby slow down the reaction speed. So, methyltriethoxysilane under acidic conditions reacts faster than tetraethoxysilane, whereas under basic conditions the order is reversed and methyltriethoxysilane reacts slower than tetraethoxysilane.

Besides their acid or base strength, catalysts are characterized by the profile of their anions or cations respectively. Desired or undesired side reactions and their behavior during further processing steps are decisive criteria for choosing the appropriate catalyst. Table 5.1 lists some examples of typical catalysts and aspects of their use.

Table 5.1: Some frequently used catalysts in the sol-gel process and their characteristics

Catalyst	Properties/characteristics
NH_3, HCl, CO_2	Volatile, usually removed during coating/curing
HNO_3	Oxidizing, metal-passivating, stabilizes metal oxide nanoparticles. Concentrates during coating and can attack the substrate. Salt formation can lead to osmotic pressure.
H_2SO_4, H_3PO_4	Concentrates during coating and can attack the substrate. Salt formation can lead to osmotic pressure. Catalytic ring-opening of epoxides. Phosphate can passivate metal surfaces
Carbonic acids (acetic acid, formic acid)	Reaction with alcolhols to the corresponding esters and water. pH shifts with time. "Hidden" source of water. Reaction with alkoxides yiels esters and polymers.
Trifluoroacetic acid	Carbonic acid with a high acid strength. Source for fluoride ions when decomposed.
NH_4F, HF	Acid catalysts which foster condensation due to the fluoride ions.
Ion exchanger	Immobilized catalyst, removable

Selection of the catalyst

As a general rule, residual traces of catalysts in cured coatings can cause corrosion or negative side effects by salt formation. Salts deteriorate the water resistance of coatings and lead to an osmotic pressure which can easily destroy the coating or cause delamination.

Anions and cations of the catalyst can also deliberately be used in the sol-gel process to synthesize phosphates, fluorides or sulfates. One possibility to synthesize metal fluorides is to use trifluoroacetic acid as a catalyst. During the following thermal treatment, the trifluoroacetic acid decomposes and serves as a fluoride source for the metals which are present in the coating (compare "Tailoring of the refractive index of sol-gel coatings", page 186).

The anions and cations can also show undesired side effects. Sulfuric acid or phosphoric acid are excellent catalysts for the ring-opening of epoxides to the corresponding diols. If epoxy-functional silanes are used in sol-gel synthesis, it is therefore recommended to use HCl or CO_2 as a catalyst to preserve the epoxy group for further reactions. On the other hand, if the hydrophilic diol is the target of the synthesis, the ring opening action of the sulfate or phosphate ion can be used effectively (compare "Hydrophilic coatings", page 179).

An elegant way to catalyze sol-gel reactions is to use compounds which can be removed completely from the formulation when the reaction is finished. Soluble gases like CO_2, ammonia, or solid ion exchangers in the H^+/OH^- form are excellent for this purpose to increase the shelf life of the formulation.

Organic acids react with the alcohol which is released as a byproduct of the hydrolysis and form the corresponding esters. This is the reason, why in the course of sol-gel reactions which are catalyzed by organic acids, the pH value raises slowly over time.

Carbonic acids therefore can be considered as a hidden source of water. The esterification releases water as a reaction product, which reacts with the alkoxygroups. This effect is especially important in sol-gel systems with low R_{OR} values (R_{OR} <1) and can even be used to formulate completely water-free sol-gel systems (compare "Water-free sol-gel techniques", page 100).

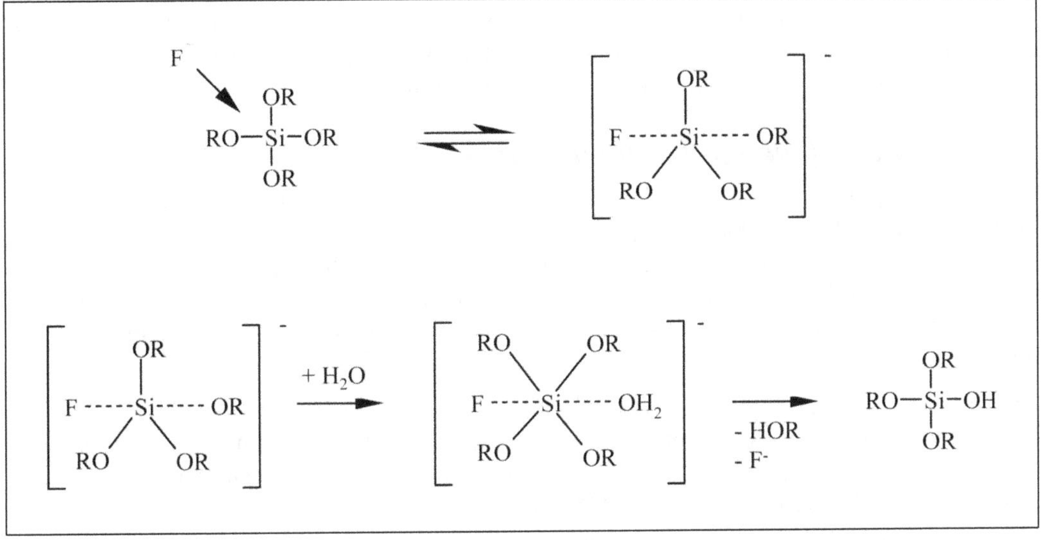

Figure 5.11: Fluoride-anions have a strong affinity to silicium. Fluoride almost has the same size as OH- and can increase the coordination number of Si from 4 to 5 or 6, which increases the condensation speed significantly.

Metal ions can have a very significant influence on the condensation speed of sol-gel systems. Especially Pb, Sn or Ti compounds increase the condensation speed in the sol state dramatically. Alkali ions like Na^+ or K^+ act accordingly during the thermal densification of xerogels. They speed up the sintering and the reduction of porosity during the thermal treatment, even at low temperatures (compare Figure 6.20, page 147). It should be noted, that sols which contain metal ions in general exhibit a decreased content of Si-OH groups, which might be necessary to promote adhesion or to undergo further reactions.

If fluoride ions are present, the condensation speed is increased. This, even under acidic conditions leads to a reduced shelf life. The effect can be measured even in the ppm-range and is caused by the formation of a fivefold coordinated transition state of silicon [137]. Thereby, in the presence of fluoride ions, even under acidic conditions relatively dense sol structures can be realized.

Contaminations have to be avoided

By reviewing the last chapter, it becomes clear that the choice of the catalyst is important and that contaminations can have significant effects on sol-gel reactions. Especially when they shift the pH of the formulation over time or if they contain metal ions, the final material properties can be affected. This is one reason, why sol-gel reactions should not be run in metallic reactors.

5.1.3 Sources of water in sol-gel reactions

Besides the amount of water which is added to sol-gel formulations, water which comes from the humidity of the air or from side reactions has to be taken into account. If carbonic acids are used as catalysts, at least a part of the added amount will react with alcohols which are present in the formulation to esters, thereby liberating water. If such reactions are deliberately used in the sol-gel process, this technique is called CCC "chemically controlled condensation" (compare Example 43, page 166) [137].

The humidity of the surrounding air can play a role during the coating process. Often it causes coating defects or generates so-called "breathe patterns", which are caused by water droplets which condense on the coatings due to the cooling effect of the evaporating solvents. On the other hand, humidity can also be used as a source of water especially for very thin coatings, like easy-to-clean coatings (compare Example 42, page 164). If no water has to be added, formulations with very long shelf life can be designed.

5.1.4 Water-free sol-gel techniques

The common understanding of a sol-gel process is the reaction of alkoxy silanes with water in the presence of a suitable solvent and catalyst. The use of water as a reaction partner however for many reasons is disadvantageous. A water/alcohol mixture cannot dissolve all educts and products of possible sol-gel reactions. High water content can lead to coating defects, if the volatile solvents evaporate first and water with its high surface tension remains in the coating.

Another aspect is that hydrolysis of silanes liberates alcohols, which are undesired either from a toxicological (methanol) or a workers safety (flammability, flashpoint) point of view. Further reasons to avoid water can come from the need for better reaction control or the absence of Si-OH groups in the final product.

Due to these reasons, many efforts have been taken to develop sol-gel processes which work without the addition of water. Two types of processes can be differentiated; processes which generate water by chemical reactions and processes which work completely without water.

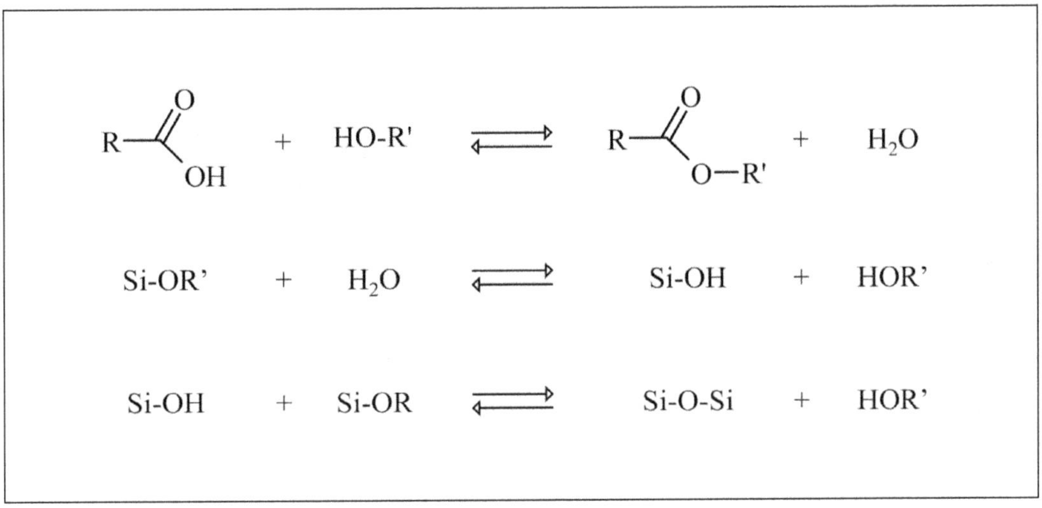

Figure 5.12: "Chemically Controlled Condensation" in the sol-gel process. A separate reaction serves as a source for water and allows a slow, controllable hydrolysis and condensation.

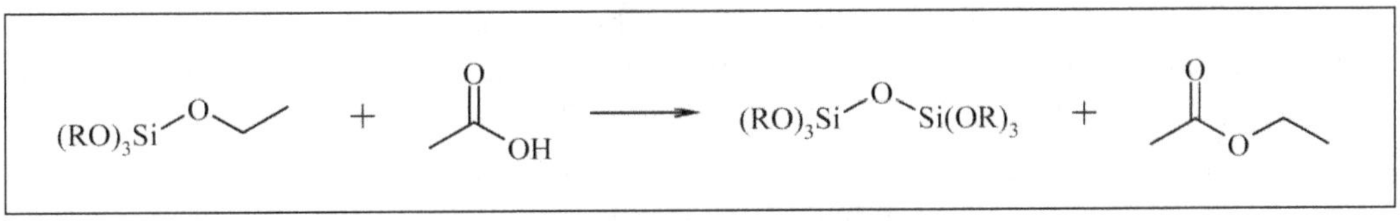

Figure 5.13: Carbonic acids act like water in the sol-gel process and lead to a condensation of alkoxysilanes. They can be regarded as a hidden source of water

Like already described in the previous paragraph, a side reaction which produces water slower than it is consumed by hydrolysis can be used to control the sol-gel reaction. This technique is known as chemically controlled condensation "CCC".

A mixture of carbonic acid, alcohol and silane is a sol-gel system, in which the water generating esterification controls the speed of the hydrolysis and condensation of the silane. Every generated water molecule is consumed by the silane and so the ester formation equilibrium is drawn to the side of the products.

Carbonic acids therefore are hidden sources of water and have to be taken into account when calculating the R_{OR} of the formulation (compare Figure 5.4, page 93). The possibility to formulate sol-gel reactions without water or alcohol opens up an interesting alternative to dissolve even unpolar additives or reaction partners in the formulation.

If no alcohol is added to a formulation which consists only of carbonic acid and silane, then the carbonic acid directly reacts with the alkoxygroup of the silane. In such a formulation, the carbonic acid is consumed as reaction partner of the silane and the corresponding ester is liberated. The product is yielded as a dispersion of a polysiloxane in an ester.

If tetraethoxysilane is reacted with acetic acid, a SiO_2 dispersion in ethylacetate is generated. Since ethylacetate is an excellent solvent for unpolar materials, polymers like e.g. PMMA can be dissolved in the final product without having to take further measures. If the TEOS would have been reacted traditionally, using a water/alcohol mixture, the solvent would have to be replaced subsequently before PMMA could be added.

If carbonic acids which bear further functions are used, like e.g. methacrylic acid, then a SiO_2 dispersion in MMA (methacrylic acid methylester) results, when tetramethoxysilane

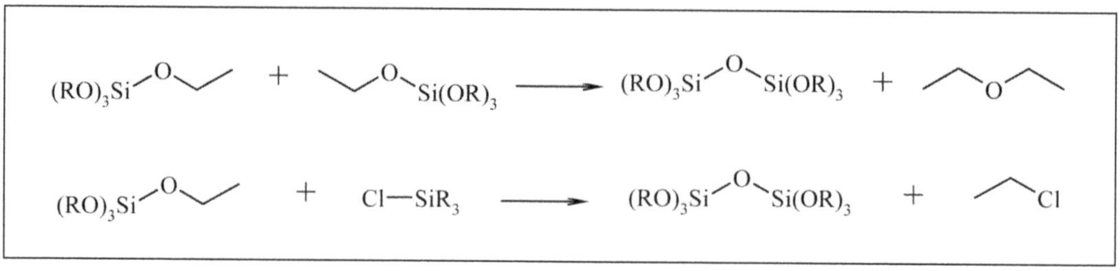

Figure 5.14: Two possible reaction pathways of anhydrous sol-gel condensation processes

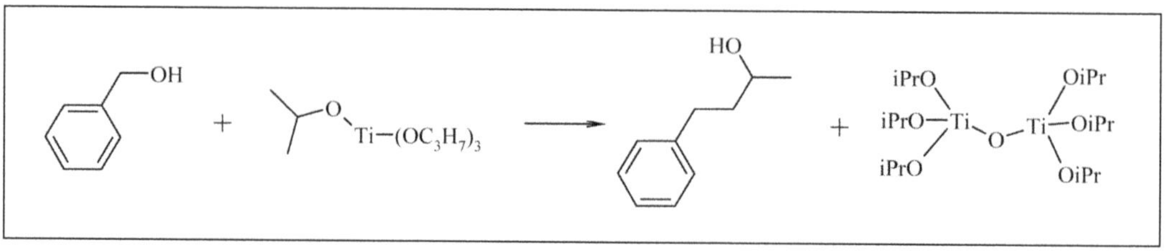

Figure 5.15: Nanoparticle synthesis with benzylalcohol [148]

is used as a starting material. The dispersion can be polymerized directly to yield a SiO_2/PMMA nanocomposite [144]. The possible variations and applications of this principle to design new inorganic/organic composites are manifold.

Besides these special sol-gel reactions, further strategies exist to foster condensation without having to perform the classical hydrolysis step first.

At higher temperatures and in the presence of appropriate catalysts, alkoxysilanes can react directly to the siloxane thereby liberating the corresponding ether. This reaction is used to cure silane-modified coating resins e.g. for scratch resistant clear coats. Metal complexes, like e.g. $Al(acac)_3$ or phosphonates can serve as catalysts for these reactions (compare "Alkoxysilane-functionalized coating resins", page 118) [145].

Equimolar mixtures of chloro- and alkoxysilanes liberate the corresponding chloroalkane at temperatures higher than 100 °C, thereby forming Si-O-Si bonds. This reaction can be used to synthesize precursors of a defined chemical structure like Si-O-metal single source precursors (compare "Metal compounds in the sol-gel process", page 103).

In principle these reactions are not only possible between silanes, but can also be used with metal chlorides and -alkoxides. A variety of other anhydrous synthesis pathways exist, like e.g. the benzylalcohol synthesis or the Peccini method which cannot be described in detail without exceeding the scope of this book. They and other methods can be found in the excellent reviews of Vioux and Niederberger [146, 147].

5.1.5 Solvents

Besides solubilizing the educts and the product, the role and influence of solvents in the sol-gel process is much more complex. As a general rule, alcohols are used to solubilize the educts. Alkoxides however, especially in the presence of metal ions, tends to exchange part of their alkoxy groups against the alcohol of the solvent, a process which is called transesterification [137].

This exchange changes the reactivity of the educts. A titanium-IV-methoxide, dissolved in n-butanol quickly reacts to a mixed titanium-IV-methoxide/-butoxide which is far less reactive than the original titanium-IV-methoxide.

$$(RO)_3\text{—Si—OR} \quad + \quad \text{HO-R'} \quad \rightleftharpoons \quad (RO)_3\text{—Si—OR'} \quad + \quad \text{HO-R}$$

Figure 5.16: Transesterification can happen if alkoxides are dissolved in a protic solvent

A sol-gel reaction of tetramethoxysilane is faster in methanol than in ethanol or in propanol. This effect is caused in part by the transesterification and in part by the formation of alkoxonium ions. An identical amount of acid does not lead to the same acid strength in methanol or in propanol, because the corresponding alkoxoniumion of propanol is more stable than the alkoxoniumion of methanol due to the +I-effect. So, propanol effectively lowers the amount of available H^+ ions compared to methanol.

Ether alcohols

Ether alcohols are popular solvents due to their solvating power, their high flash point and their compatibility with the environment. They tend to form complexes with metal ions and therefore are used like the β-diketones or ketoesters to synthesize metal-oxide/SiO_2 composites.

With their alcohol functionality, they too take part in transesterification reactions with the educts. Since reactivity of alkoxides in the sol-gel process is inversely correlated to their chain length, once they have formed the corresponding alkoxide, they mostly remain attached to the silicon or metal atom and are almost not affected by the hydrolysis reaction. They remain present in the end product and act as softeners, which can be a desired action, but in most cases is unwanted.

Selection criteria for solvents

When choosing solvents for sol-gel coating formulations, their compatibility with the substrate has to be taken into account. Many polymers form cracks upon contact with unsuitable solvents. Tables with suitable solvents are available from the polymer manufactures homepages.

Different coating processes require different solvents. Highly volatile solvents like isopropanol may evaporate almost completely on their way from the nozzle to the substrate during spray coating. Solvents with higher boiling points are a good choice to help generating a homogeneous coating. They have to be chosen in a way that they do not lead to a destabilization, like e.g. a phase separation, when the lower boiling solvents evaporate first during drying.

Ether alcohols like butyl- or iso-propylene glycol in many cases are a good choice. They can also improve wetting and leveling of the coating by retarding the gelation of the sol.

Small (1 to 3 %) amounts of aprotic, higher boiling glycol ethers like e.g. diethylen glycoldiethyl ether can improve the coating result significantly without remaining in the final product.

5.1.6 Metal compounds in the sol-gel process

Besides silicon alkoxides, also metal compounds can serve as building blocks for the inorganic network. Metal compounds are used in the sol-gel process for many different reasons. Zirconium dioxide as an example improves the resistance of SiO_2 against alkaline media [149], or acts as a catalyst for the polymerization of epoxides. Metal compounds can improve adhesion like the known Al/Zr mixed complexes, increase the refractive index or can act as anti-corrosion additive (ZrF_xO_y).

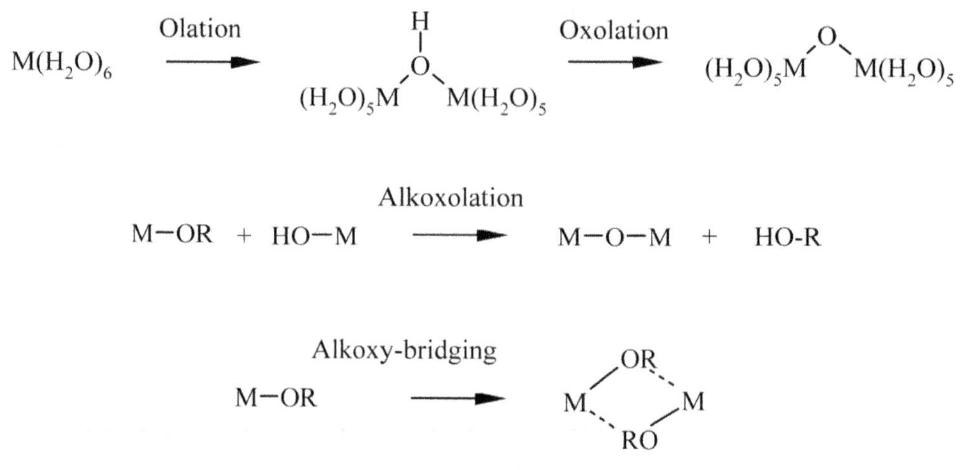

Figure 5.17: Olation, oxolation, alkoxolation and alkoxy-bridging influence the structure of oligomers which form in solutions of organo-metal compounds

During hydrolysis and condensation reactions, complex polyoxometallates are generated, whose structures are dominated by the metal ions and their counterions which might be present [150, 151]. The oligomerization of the metal compounds is governed by the relative speed of five reactions:

- hydrolysis,
- oxolation,
- alkoxolation,
- alkoxy-bridging and
- olation.

These reactions are influenced by the educts, solvent, catalyst, temperature and water content of the formulation.

The oxolation and alkoxolation correspond to the two possible condensation reactions of the SiO_2 sol-gel process, which liberate either water or alcohol (compare Figure 5.5, page 95). The olation describes the bridging of two hydrated, respectively solvated metal ions

Table 5.2: Some typical educts for the production of metal oxides in the sol-gel process

TiO_2	ZrO_2	B_2O_3	Al_2O_3
$Ti(O\text{-}nPr)_4$ Titanium-n-propylate	$Zr(O\text{-}nBu)_4$ Zirkon-n-butylate 80 % in BuOH	$B(OMe)_3$ Boric acid trimethylester	$Al(O\text{-}sBu)_3$ Aluminum-sec-butylate
$Ti(O\text{-}iPr)_4$ Titanium isopropylate	$Zr(O\text{-}nPr)_4$ Zirkon-n-propylat 70%	$B(OEt)_3$ Boric acid triethylester	$AlCl_3$
$TiOCl_2$ Titanylchloride	$ZrOCl_2$	H_3BO_3 Boric acid	$Na[Al(OH)_4]$
$Ti(acac)_3$ Titanium acetylacetonate	$ZrO(NO_3)_2$		
$TiO(NO_3)_2$			

by a hydroxy-group which is the first step of a condensation reaction. This olation also takes place in aqueous solutions of metal salts, which can serve as raw materials in the sol-gel process as well.

Depending on the pH, Al^{3+} either exists as $[Al(H_2O)_6]^{3+}$, $[Al(H_2O)_{6-x}(OH)x]^{3-x}$ up to $[Al(OH)_4]^-$. Ti- or Zr-salt solutions behave in a similar way. So it seems logical, that by a "simple" dilution of a salt solution, the pH and the structure of the solvated species can be changed dramatically.

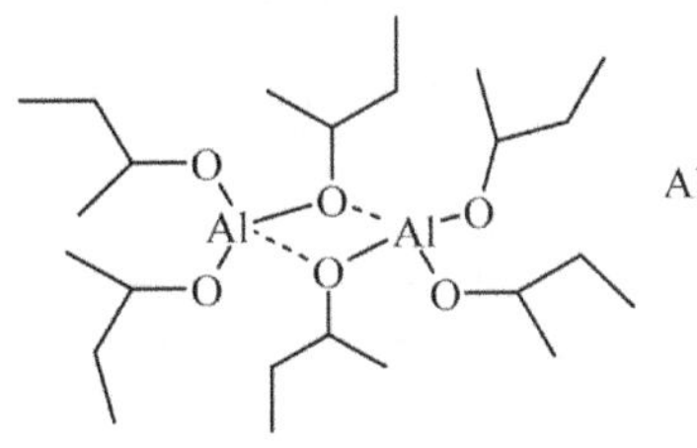

Figure 5.18: Dimer of aluminium-sec-butylate. Alkoxy-bridges lead to oligomerization (compare Figure 5.17, page 104)

By diluting a salt solution, polyisocations or polyisoanions are generated by oxolation. This is the reason, why a precipitate quickly forms after diluting a $TiOSO_4$, $ZrOCl_2$, or $FeCl_3$ solution. If the precipitation shall be avoided, then the shift of the equilibrium has to be countered by addition of H_2SO_4 or HCl respectively.

It is very difficult to predict the exact structure of a salt solution in practice. But what is really important, is to take into account that even "simple" operations like diluting a salt solution often have a tremendous effect on the final product by the invisible formation of oligomers. What makes this consideration even more complex is that these equilibria take their time to recalibrate after a dilution step which further adds the factor "time" to the equation. Quality control is not easy to establish when dealing with these complex systems.

Educts can be metal salts or metal-organic compounds. Alkoxides are available from almost all metals of the periodic system [152]. Under a commercial aspect, especially the alkoxides of B, Ti, Zr, Al, Sn and Na are used in sol-gel formulations.

Depending on their alkoxy-group, the metal alkoxides are either soluble or not in the commonly used solvents like alcohols. They usually are not dissolved as monomers, but rather form dimers or oligomers, which explains their high viscosity [137].

Degree of oligomerization

The degree of oligomerization is determined by the alkoxygroup of the metal alkoxides and the solvent. Short-chain alkoxides and solvents foster polymeric structures, long-chain and especially branched alkoxy groups and solvents rather lead to dimers.

The alkoxy group can be exchanged easily for most metal alkoxides. The transesterification reaction usually is very quick if another alcohol is offered as a solvent.

Selection of the solvents

The selection of the solvent has a remarkable effect on the sol-gel process of metal alkoxides. If formulations tend to gelate or show flocculation, precipitation or coating defects, often changing the solvent can avoid the problem. One example is the precipitate which can be formed if zirconium propoxide or butoxide is dissolved in methanol. If alcohols with longer alkyl chain are used, the formulation is much more stable. Higher alcohols or ether alcohols can cleave the oligomeric alkoxides and form dimers which are better soluble.

If very high boiling solvents are used, they might remain to some extent in the final product. This can contribute to the stability of the formulation or the flexibility of a coating, but in most cases these residues should be avoided (compare Example 14, page 58).

➤ *Example 24: Preparation of a concentrated dispersion of TiO_2 in dimethylacetamide* [153]

0.5 g Dodecyl trimethyl ammoniumbromid, dissolved in 30 g dimethylacetamide (DMAc) are mixed with 9 g HClc. and 9 g water. After the solution has cooled down to room temperature, 110 g tetrapropyl orthotitanate, dispersed in 50 g DMAc is added within 2 minutes. A hazy dispersion results, which turns clear after 45 minutes of stirring. TEM analysis shows TiO_2 particles with a diameter between 4 to 13 nm.

Immediately after its preparation, the TiO_2 dispersion is mixed with 7 g 3-methacryloxypropyl trimethoxysilane and is subsequently stirred for 30 minutes at 70 °C. The low viscous and colorless dispersion is stable for months. If ethanol is used instead of dimethylacetamide, the tetrapropyl orthotitanate immediately reacts to TiO_2, which precipitates and does not yield a stable dispersion.

Highly concentrated titanium dioxide dispersion

The example can be considered as extraordinary, since nanoparticle synthesis usually is carried out only in highly diluted solutions to avoid precipitation. The described synthesis however leads to a 15 weight% dispersion without the formation of agglomerates.

In the previous section, the role of the solvent in the sol-gel process as stabilizer of educts and products has been emphasized. In this example, now an aprotic, highly polar solvent like dimethylacetamide is used to synthesize a concentrated nanoparticle dispersion via the hydrolysis of a titanium alkoxide. The reference experiment with the non-complexing solvent ethanol shows, that it is the ability to form complexes with the titanium compound, which allows for the exceptional reaction control by the dimethylacetamide.

When selecting a complexing solvent, it is important to choose a suitable stability of the formed metal-solvent complex. If the stability of the complex is too high, then the complex does not take part in the sol-gel reaction and remains in the end product. If it is too weak, then agglomeration or precipitation can occur, when water is added.

The amount of water which is used is very small and just enough to hydrolyze the alcoholate. By using small amounts of water, residual alkoxy groups remain in the product and contribute to the sterical stabilization of the formed nanoparticles. Further stabilizing additives in this example are the cationic emulgator and the chloride ions of the hydrochloric acid catalyst.

The reaction speed of the titanium alcoholate already is very high and no catalyst is necessary for a fast and complete reaction. The task of hydrochloric acid in this example mainly is to stabilize the final product [150]. As an alternative, often nitric or citric acid is used, because also nitrate or citrate ions can stabilize metal oxide nanoparticle dispersions well.

A common attribute of metal-organic compounds is their high reactivity with water. Even without a catalyst, hydrolysis and condensation run at high speed. Simply opening the bottle leads to immediate polymerization, which can be recognized by the formation of a white powder or a gel at the spout. If reproducible results are expected, then handling of metal alkoxides under dry inert gas is mandatory.

Catalysts are necessary to adjust a suitable zeta potential for the stabilization of the metal oxide/hydroxide. Hydrolysis of organo-metal compounds without a catalyst yields the metal hydroxide at its IEP (isoelectric point) which is, in contrast to SiO_2, the point of lowest sta-

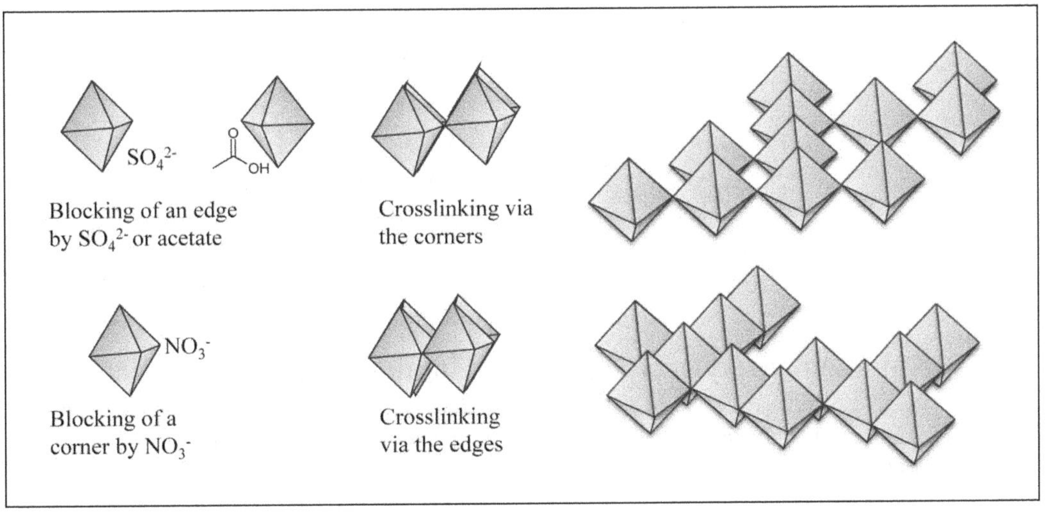

Figure 5.19: According to the partial charge model[150], the choice of the anion has a decisive influence on the structure of polyoxometallates which are generated in the sol-gel process of metal ions

bility (compare "Electrostatical stabilization", page 41). Formation of stable dispersions without addition of stabilizing anions is hardly possible.

The choice of the catalyst determines the anion and thereby the stability of the metal hydroxide/oxide dispersion (e.g. NO_3, citrate, chloride, ...).

Besides alkoxides, also the metal salts can be used as educts in the sol-gel process. The structure of the polyoxometallates in the salt solutions is complex. According to the partial charge model of Livage, an octaeder is the central building block [150]. The metal ion is sitting in its center and the corners symbolize Me-O-Me connections. Anions occupy, depending on their valence, edges or corners and thereby influence the structure of the polymeric metal oxides/hydroxides.

Anions influence the structure

A precise prediction of the change in structure when the anion is varied however is difficult. What should be considered is that if problems like gelation or precipitation occur during-formulation, the variation of the anion is an effective measure to solve them. The anions influence the structure of the smallest building blocks; therefore they also have significant influence on macroscopic properties like porosity, density or crystallinity.

The high reactivity of the alkoxides can be moderated by the solvent (transesterification) and/or by forming complexes. Beta-diketones, ether alcohols or ethanolamines are suitable materials for this purpose.

When choosing the chelate-forming agent, it should be considered that if the stability of the complex is too high, the complexing agent remains in the final product. The same consideration is valid for the choice of a salt, a solvent or a catalyst. Some anions can cause problems in the final coating material like oxidation, corrosion or osmotic pressure.

5.1.6.1 Silane/metal-oxide mixed systems

The sol-gel process becomes exceedingly complex, if metal compounds are combined with silane alkoxides. The organo-metal compounds hydrolyze by far faster than the silicon alkoxide, therefore the synthesis of homogeneous SiO_2/metal oxide hybrid materials is a challenge.

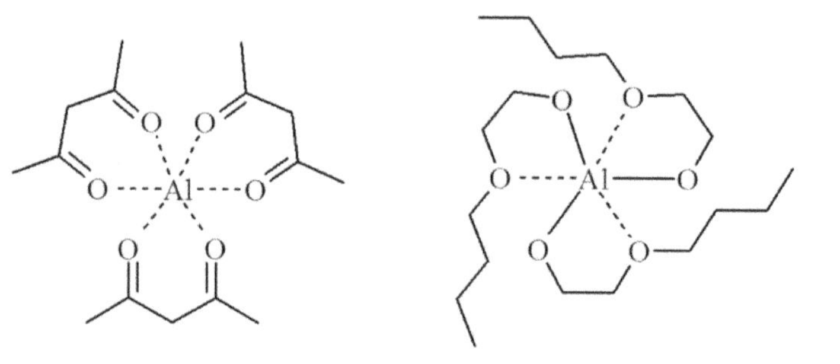

Figure 5.20: Moderation of the reaction speed of aluminum alkoxides by complexation with coordinating organic ligands

If in such mixed systems, the reactivities of metal compounds and silicon alkoxides are not managed properly, the metal hydroxides precipitate (e.g. $Ti(OH)_4 \rightarrow TiO_2 \cdot H_2O + H_2O$) before silanol groups (≡Si-OH) as a reaction partner can be generated.

The consequences are inhomogeneous products, which in certain cases might be acceptable or even desirable, but in the majority of cases homogeneous products will be the objective of the formulation. The following concepts have been developed to realize homogeneous multicomponent material formulations:

- The faster reacting component is slowed down: Chelate complexation of the reactive metal compound with chelate formers (ß-diketones, ethanolamines, butoxyethanol).
- A precursor is synthesized which combines the slower and the faster reacting compound on a molecular level (single source precursor).
- The slower compound gets a head start: Pre-hydrolysis of the silane with a part of the reaction water and then addition of the faster reacting compound.

The concept of the moderation of highly reactive metal compounds already has been discussed. By the choice of the alkoxy group, the solvent, or a complexing agent, the reactivity of metal-organic compounds can be slowed down to such an extent, so that silanes and metal-organic compounds can be hydrolyzed together. It should be considered though, that if strongly binding ligands are used, they might remain in the final product and will not be removed by hydrolysis and condensation reactions.

Acetylacetone is a stronger complexing agent then butoxyethanol. Therefore acetylacetone slows down the reactivity more effectively than butoxyethanol, but also remains in the final product and hinders a homogeneous co-condensation with the other components of the formulation.

As an alternative to the complexation of alkoxides, a metal salt solution (e.g. $TiO(NO_3)_2$ or $Pb(OOC\text{-}CH_3)_2$) can be used in the sol-gel process to synthesize SiO_2-MeOx hybrids. The disadvantage of metal salts is the high amount of anions which is introduced to the formulation. Acetates, formiates and other carbonic acid derivatives are able to react with alcohol and are partially removed as esters from the reaction mixture.

A further possibility to realize homogeneous SiO_2-MeOx hybrids is the synthesis of a "single source precursor", which already contains all components of the final hybrid product in one molecule. If e.g. $TiCl_4$ is reacted with $Si(OR)_4$, RCl is liberated and a precursor like $Ti\text{-}[OSi(OR)_3]_4$ (idealized) is formed (compare "Water-free sol-gel techniques", page 100).

This precursor then is reacted in the following step to the final SiO_2/TiO_2 hybrid material. This concept is particularly suitable for mixed metal multicomponent compositions when extreme homogeneity is crucial for the final product properties (e.g. ITO, YBaCuOx).

Single source precursors

Single source precursors can also be synthesized by heating a mixture of metal acetates and alkoxysilanes. In analogy to anhydrous condensation reactions (compare "Water-free sol-gel techniques", page 100), oligomers are synthesized via ester formation. This procedure is recommended if there is no suitable metal alkoxide available.

The biggest advantage of the single source precursor concept is the highest possible degree of homogenization of the educts prior to the reaction. But hydrolysis is only the first step in the process chain. In subsequent reaction steps, the components can separate again, especially if the thermal treatment program is not tailored to the crystallization behavior of the components.

Prehydrolysis

A simple method to adjust the reactivity of educts is the prehydrolysis of the slower reacting compound, which in most cases is the silane (compare Example 55, page 189). The silane is reacted with a part of the amount of water to yield precursors which are rich in Si-OH groups prior to the addition of the faster reacting metal compound. At this point in time, the amount of free water in the formulation should be very low, so that the metal compound is forced to react with the Si-OH group and does not form Me-O-Me bonds. After the metal compound has reacted and bonded to the silane, the residual amount of water can be added without causing precipitation.

5.2 Network modifiers

Besides pure inorganic networks, via the sol-gel process it is also possible to synthesize organically modified inorganic networks under mild reaction conditions. The inorganic network thereby is modified by the various available organic network modifiers and changes its properties accordingly.

The terms "network former" and "network modifier" originally stem from the traditional glass chemistry and describe inorganic additives like alkali ions to the glass melt, which interrupt the SiO_2 network and lower the viscosity and melting temperature [197]. Polyvalent ions like Al^{3+}, Ca^{2+} or B^{3+} are incorporated into the SiO_2 network and change its acid/base properties and the chemical stability.

In the sol-gel process, due to its low process temperatures, besides inorganic also organic network modifiers are possible (compare Example 33, page 147). They are inert groups which do not form an own network and modify the thermal, chemical and mechanical properties of the inorganic sol-gel formulation. They

- impede the 3-dimensional cross-linking of SiO_2
- diminish brittleness, thereby improving flexibility
- decompose during a thermal treatment at temperatures exceeding 250 °C and
- influence the speed of hydrolysis and condensation by I- and M-effects

Using the example of the smallest possible network modifier, the methyl group, Figure 5.21 illustrates the modification of the SiO_2 building block.

Whereas SiO_2 is very brittle and cannot form crack-free coatings >1 µm (compare "Coatings with sol-gel materials", page 135), methyl modified formulations (polymethylsilsesquiox-

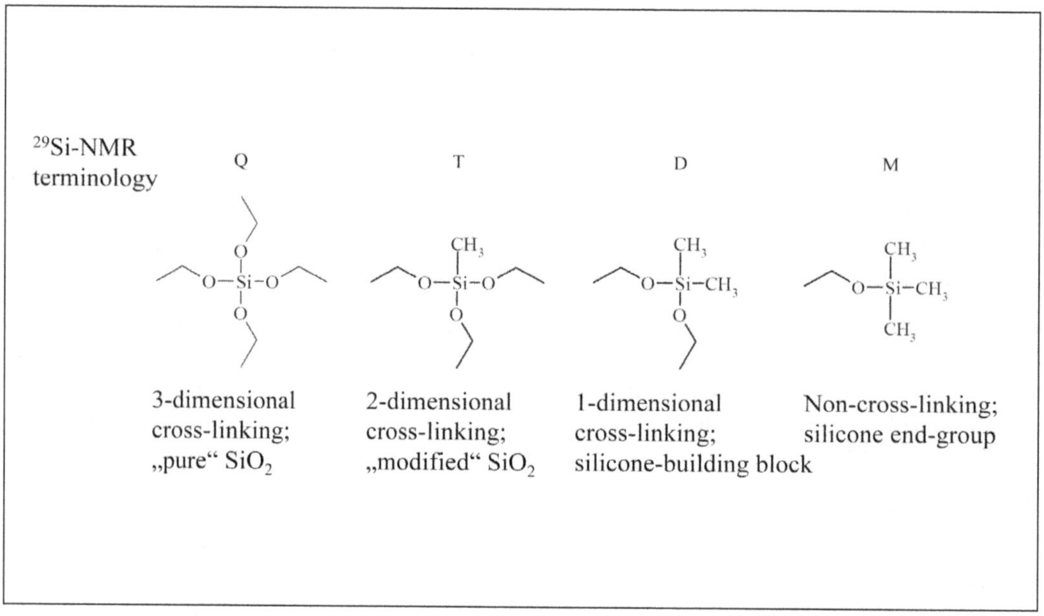

Figure 5.21: The methyl group acts as a network modifier in the sol-gel process

anes) are used as dielectrica (spin on glasses) in the wafer production or as scratch resistant coatings (compare Example 27, page 121) [154]. If one additional methyl group is added to the silicon atom, the typical "-$(CH_3)_2$Si-O-" silicone building block results.

CH_3 group as network modifier

Via the modification with CH_3 groups, the high melting, brittle SiO_2 is turned into a flexible coating material or even a low viscous hydrophobic oil. Besides the methyl group, many other network modifiers exist (linear or branched, aliphatic or aromatic), which are suitable to modify the SiO_2 building block to fit the requirement profile of the product.

Besides the improved flexibility, the organic groups add their own property profile to the resulting hybrid material. Methyl groups are unpolar and hydrophobic. As can be seen in Figure 5.21, the hydrophilic SiO_2 can be turned into a water repelling silicone.

Long chain network modifiers

Alkyl groups with longer chain lengths like butyl- or hexadecyl- and especially partially fluorinated alkyl groups like the tridecafluorooctyl triethoxysilane (compare Example 42, page 164) show even stronger hydrophobicity and more distinct water and dirt repellent properties and therefore are used to formulate easy-to-clean coatings. Due to their network-modifying action, they also weaken the mechanical properties of the coating. This is the reason, why these coatings often contain nanoparticulate fillers or cross-linkers like tetra-ethoxysilane to work against this negative weakening effect.

Temperature sensitivity

When a temperature of approximately > 250 °C is exceeded, the organic network modifiers start to decompose and leave a cavity in the coating network. If the amount of network modifier is not high and the thermal treatment is slow, this is a possible way to synthesize porous inorganic structures. If the amount of organic components is too high or the thermal treatment is too fast, the inorganic network is going to be destroyed.

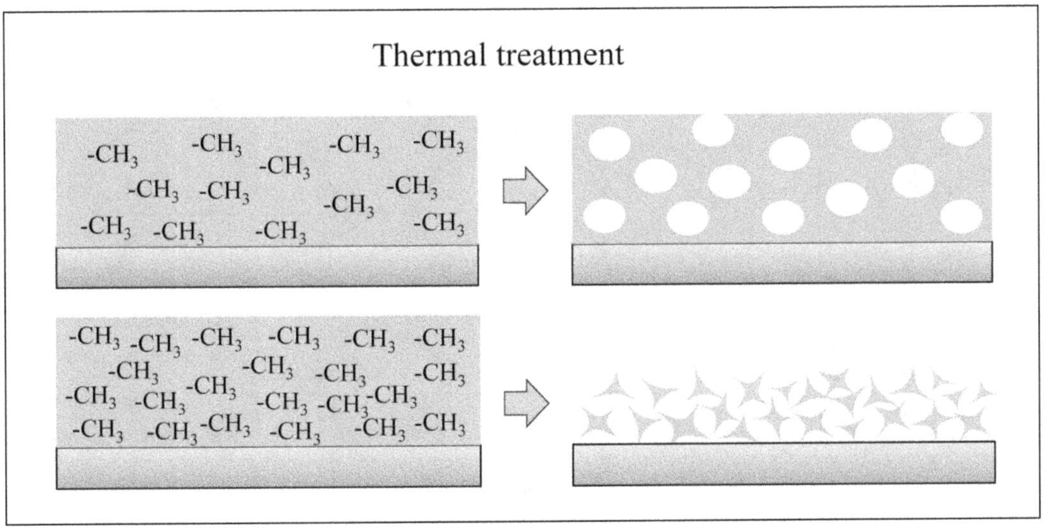

Figure 5.22: Organic components of sol-gel formulations generate porosity during the thermal treatment. If the content of organic components is too high, the coating however may be destroyed

Organoalkoxysilanes

If organoalkoxysilanes are used in the sol-gel process, the organic substituents change the reactivity of the silane substantially [155]. The organic group acts as a sterical barrier for hydrolysis and condensation reactions. Furthermore, it acts via electronic effects on the silicon central atom.

This can be a +I-effect in case of an alkyl group or a +M-effect in case of a phenyl group, whereby in both cases the electron density on the silicon atom is increased. Depending on the pH and as a result the different mechanisms of hydrolysis and condensation, these substituents either act accelerating or decelerating on the reaction speed (compare Figure 5.10, page 98).

Alkaline or acidic catalysis

Under acidic conditions, a methyl group accelerates the hydrolysis whereas under alkaline conditions a methyl group decelerates the hydrolysis compared to the unsubstituted tetraethoxysilane (TEOS). This has the following consequences:

- under acidic conditions MTOS reacts faster than TEOS
- under alkaline conditions TEOS reacts faster than MTOS

If the alkyl/aryl groups carry functional groups like NH_2, COOH, SH, … which can show catalytic effects on their own, the prediction of their reactivity becomes difficult and has to be determined empirically.

5.2.1 *Polysilsesquioxanes/POSS*

Bulky organic substituents force polysilsesquioxanes $(RSiO_{1,5})_n$ to form a band-, layer-, or even cage-like structure (POSS - polyhedralorganosilsesquioxane).

These organically modified silicon oxide polymers are, depending on their composition, soluble in organic solvents and find application as spin-on glasses in the semiconductor industry [154]. A forecast, whether depending on their substituents rather band-, layer- or cage structures are formed, is only possible for few examples in practice.

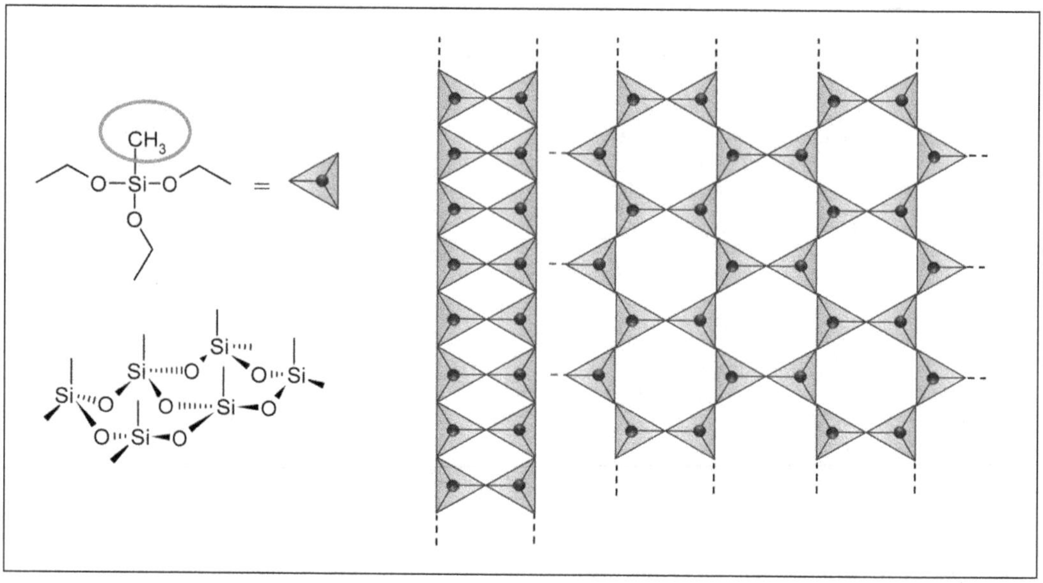

Figure 5.23: Building principles of silsesquioxanes (idealized). Rings, chains and band structures are generated in dependence of the reaction conditions and the space requirements of the substituents

As a simple rule of thumb, small substituents lead to the formation of rings and chains, whereas sterically more demanding groups force the polymer into a three-dimensional folding or cage structure.

For simplicity reasons, the illustrations have been drawn with the methyl group facing upwards. In practice rather an alternating or statistical arrangement and a folded structure can be expected. If strongly alkaline catalysts are used, especially with sterically demanding network modifiers like isobutyl-, cage structures are formed, the so-called POSS (polyhedralorganosilsesquioxanes) [156].

They can be isolated and analyzed as defined compounds. Modified with reactive groups like an acrylic function, they are used as polymer additives to improve hardness and toughness of polymers.

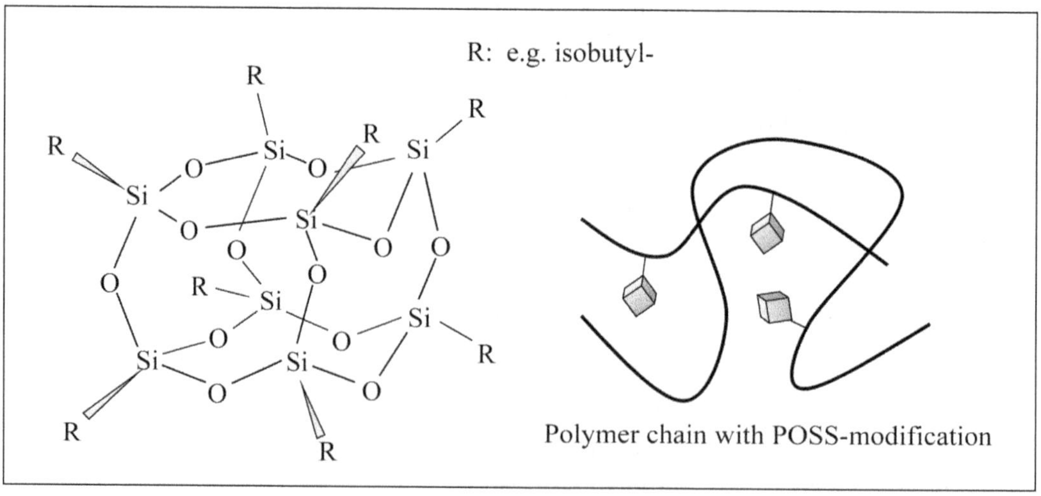

Figure 5.24: Polyhedralorganosilsesquioxanes POSS are defined cage structures, which are used e.g. as polymer modifiers to improve mechanical properties

➤ *Example 25: Preparation of a reactive POSS-polyhedralorganosilsesquioxane [157] - synthesis of (isobutyl)$_8$Si$_8$O$_{12}$ from isobutyltrimethoxysilane*

1st Step: To a solution of 446 g (2.5 mol) of isobutyltrimethoxysilane ("Dynasylan IBTMO", Degussa AG) in 4,300 ml of acetone, a solution of 6.4 g (0.11 mol) of KOH in 200 ml of H_2O is added with stirring. The reaction mixture is subsequently stirred at 30 °C for 3 days. The precipitate formed is filtered off and dried under reduced pressure at 70 °C. The product (isobutyl)$_8$Si$_8$O$_{12}$ is obtained in a yield of 262 g (96 %).

2nd Step: Synthesis of (isobutyl)$_7$Si$_7$O$_9$(OH)$_3$ from (isobutyl)$_8$Si$_8$O$_{12}$ (opening of the cage): At a temperature of 55 °C, 55 g (63 mmol) of (isobutyl)$_8$Si$_8$O$_{12}$ are introduced into 500 ml of an acetone/methanol mixture (volume ratio 84:16) which contains 5.0 ml (278 mmol) of H_2O and 10.0 g (437 mmol) of LiOH. The reaction mixture is subsequently stirred at 55 °C for 18 h and then introduced into 500 ml of 1 N hydrochloric acid. After stirring for 5 minutes, the solid obtained is filtered off and washed with 100 ml of CH_3OH. Drying in air gives 54.8 g (96 %) of (isobutyl)$_7$Si$_7$O$_9$(OH)$_3$.

3rd Step: Synthesis of (3-methacryloyloxypropyl)(isobutyl)$_7$Si$_8$O$_{12}$ from (isobutyl)$_7$Si$_7$O$_9$(OH)$_3$ and 3-methacryloxypropyl trimethoxysilane (modification of the cage): To a solution of 50 g (63 mmol) of (isobutyl)$_7$Si$_7$O$_9$(OH)$_3$ in 50 ml of THF there are added at 20 °C 16 g (64.4 mmol) of 3-methacryloxypropyl trimethoxysilane ("Dynasylan MEMO", Degussa AG). Following addition of 2.5 ml of an aqueous tetraethylammonium hydroxide solution (35 % by weight) the mixture is stirred overnight. Removal of about 15 ml of THF results in a white suspension. By adding 250 ml of methanol, the product is precipitated further. After filtration, the solid which remains is washed with methanol. Drying gives 38 g of (3-methacryloxypropyl) (isobutyl)$_7$Si$_8$O$_{12}$ (70 % yield) as a white powder which can be used as a reactive nanofiller.

Reactive polyhedralorganosilsesquioxanes (POSS)

Sterically demanding organic substituents like isobutyl- are a prerequisite for the formation of a cage structure. The hydrolysis and polycondensation of the corresponding silane is carried out with a strong base as a catalyst to ensure a high condensation speed and a complete reaction. Depending on the size and structure of the organic side chain, also cage structures with more or fewer corners are possible.

To modify the inert cage with an organofunctional silane, at least one corner (silane) has to be removed and replaced. LiOH selectively opens one corner and thereby opens up a variety of possible modifications. The further processing in acidic media stabilizes the Si-OH groups against condensation. After the functional silane is offered to the open cage, the corner is closed again by changing to alkaline conditions with a high condensation speed.

Silicones in the sol-gel process

Synthesis, processing and use of dimethyl silicon compounds, the silicones, has already been described elsewhere [158]. Some silicones, intermediates and reactive silicone compounds can also be used in the sol-gel process. The borderline between polysiloxanes and silicones are not rigid. Due to their chemical similarity, they are widely compatible with each other [159].

Si-OH functionalized silicone oils with various chain lengths are commercially available and can serve for example as anti-fingerprint additive for easy-to-clean coatings (com-

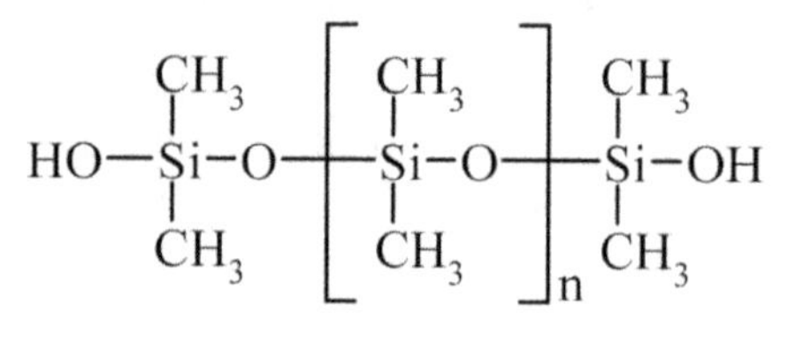

Figure 5.25: Si-OH functionalized silicone oils can be used as raw materials in the sol-gel process to introduce e.g. flexibility or hydrophobicity

pare Example 49, page 177). Their existence and stability is a further proof of the inertness of Si-OH groups in the absence of a suitable catalyst (compare "Electrostatical stabilization", page 41ff).

Via the allyl-Si-H addition, monofunctional alkoxysilyl-modified silicones can be synthesized (compare Example 23, page 85). They can serve as modifiers for nanoparticles too.

Last but not least, short chain volatile silicones are a suitable solvent for many applications. Not always they are as inert as one might think at a first glance.

D4, a silicone ring made of four $Si(CH_3)_2O$ units has been banned due to its toxic potential and, under acidic conditions, hexamethyldisiloxane acts as reactive modifier for Si-OH groups and introduces the inert, hydrophobic $(CH_3)_3Si$-group as a surface modifier [160].

Based on this behavior, a process to produce aerogels at ambient conditions has been developed, in which a colloidal silica dispersion is added drop-wise to boiling hexamethyldisiloxane. The water evaporates, the SiO_2 sol-gelates and is simultaneously surface-modified by the hexamethyldisiloxane. Due to the very low capillary forces between the hexamethyldisiloxane and the $(CH_3)_3Si$-modified silica, the solvent can be removed without causing pore collapse or cracks [160].

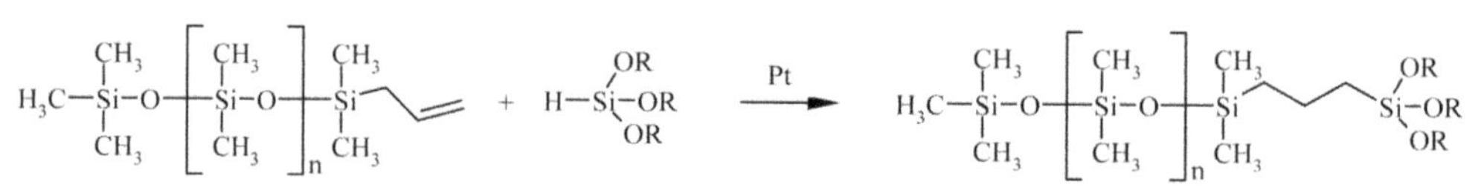

Figure 5.26: Alkoxysilyl-functionalized silicones can be used as raw materials in the sol-gel process or for surface modification of nanoparticles

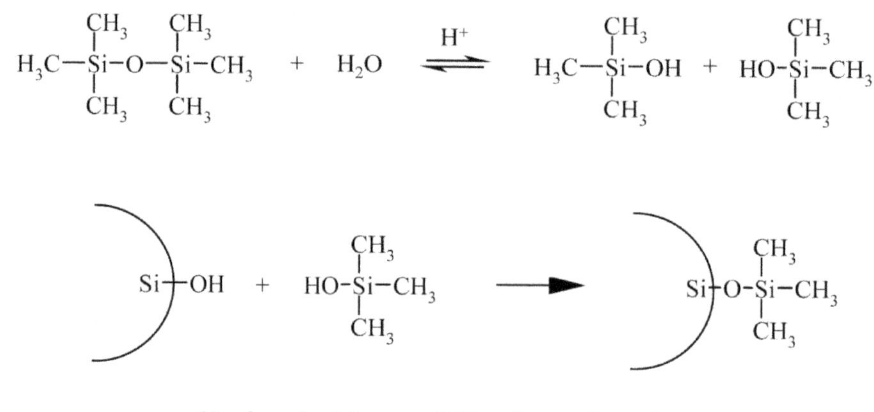

Figure 5.27: Surface modification by hydrolysis of hexamethyldisiloxane

5.3 Organic network formers

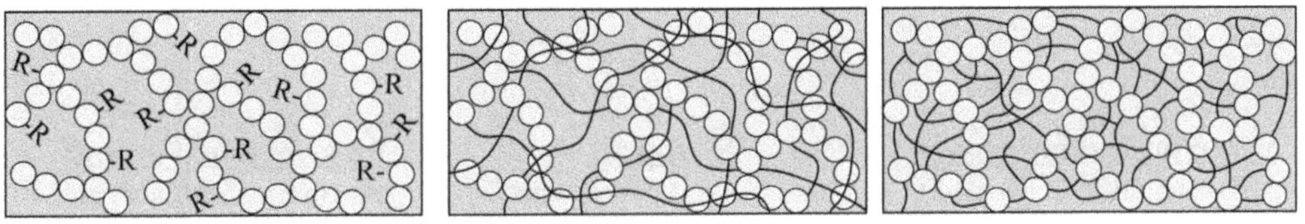

Figure 5.28: Some examples how inorganic/organic hybrids can be generated by modification of inorganic networks

If network modifiers carry functional groups which are capable to polymerize, then besides the inorganic network, an organic network can be synthesized. By this process, hybrid materials, consisting of organic polymers and glass- or ceramic-like structures are accessible, the so-called nanocomposites [161].

Inorganic/organic networks are different compared to the already discussed organically modified inorganic networks (compare "Network modifiers", page 109) and the interpenetrating networks (compare "Nanoparticles in coating resin", page 59).

Inorganic/organic networks are a new class of materials besides metals, glasses and polymers and offer a unique property profile which cannot be realized with any other material. They combine the flexibility of polymers with the hardness and abrasion resistance of glass and ceramic materials.

Nanocomposites are synthesized starting from organofunctional silanes which are used since a long time as adhesion promoters. Due to the presence of inorganic and organic reactive groups they adhere well to many surfaces.

Inorganic and organic building blocks make nanocomposites versatile materials for a variety of applications. The discovery of the nanocomposites and their broad scope of applications contributed decisively to the success of the sol-gel process [162]. In the following chapters the synthesis and some chosen applications of inorganic/organic nanocomposites are discussed.

5.3.1 Organofunctional silanes

The synthesis of nanocomposites often starts with organofunctional silanes, which are known as adhesion promoters. Most silanes are available in a commercial scale [163]. Their general structure follows the scheme from Figure 5.29.

Hydrolysable group, spacer and functional group influence the reactivity of the silanes (compare "Hydrolysis and condensation", page 94), so that a prediction of their reaction speed is hardly possible. As a general rule, α-silanes with only one CH_2 spacer group between silicon and functional group react much faster than γ-silanes with a -$(CH_2)_3$-propyl-spacer [164].

Organofunctional silanes often carry three hydrolyzable groups, but there are silanes available in which one or two hydrolyzable groups have been replaced by a methyl group. These silanes can only cross-link to linear chains and lead to more flexible structures.

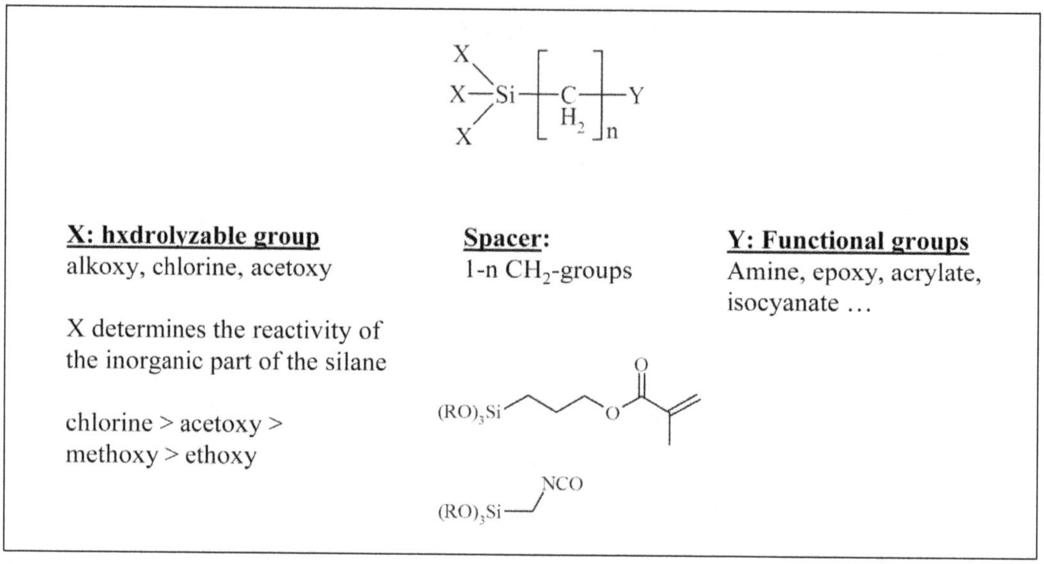

Figure 5.29: Building blocks of organofunctional silanes [163]

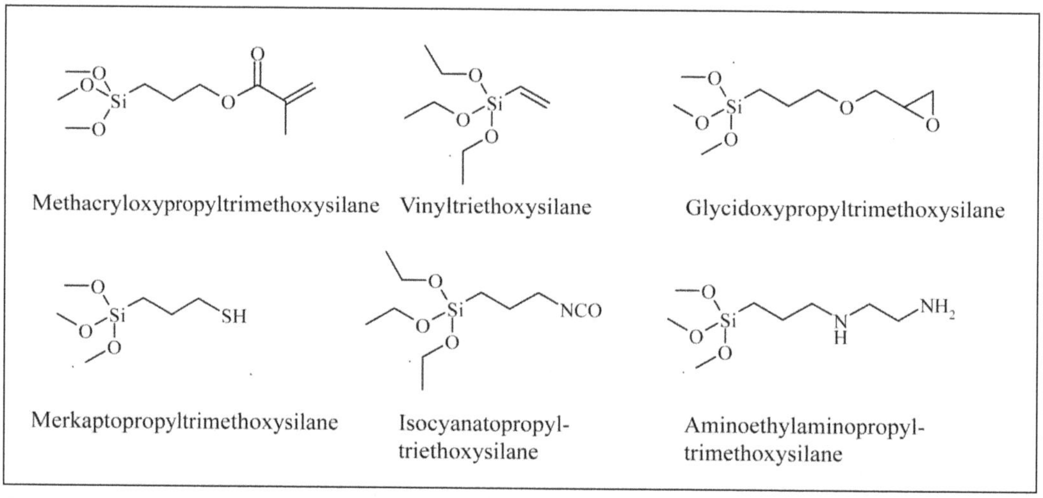

Figure 5.30: Examples of organofunctional silanes with polymerisable groups

Via the functional organic group, almost the whole spectrum of organic synthesis reactions is possible with silanes. So silanes which are not yet commercially available can be synthesized to fulfill special requirement profiles.

Non-aqueous reactions are preferred to avoid hydrolysis of the alkoxy-groups. Some examples of targeted synthesis reactions are depicted in Figure 5.31.

Addition reactions are favored, because a liberation of byproducts like water is avoided. The Pt^0-catalyzed addition of $(RO)_3Si$-H to unsaturated compounds is an important tool to synthesize tailor-made silanes. Especially allyl-functionalized compounds easily react with the Si-H group (compare Figure 5.26, page 114) [165, 166].

When organofunctional silanes are hydrolyzed, possible side reactions of the organic functional group are to be considered. Under acidic and alkaline conditions, e.g. epoxy groups tend to react with water under ring opening to the diol, amino groups react basic and easily form intramolecular complexes with the silicon atom (compare Figure 5.39, page 124).

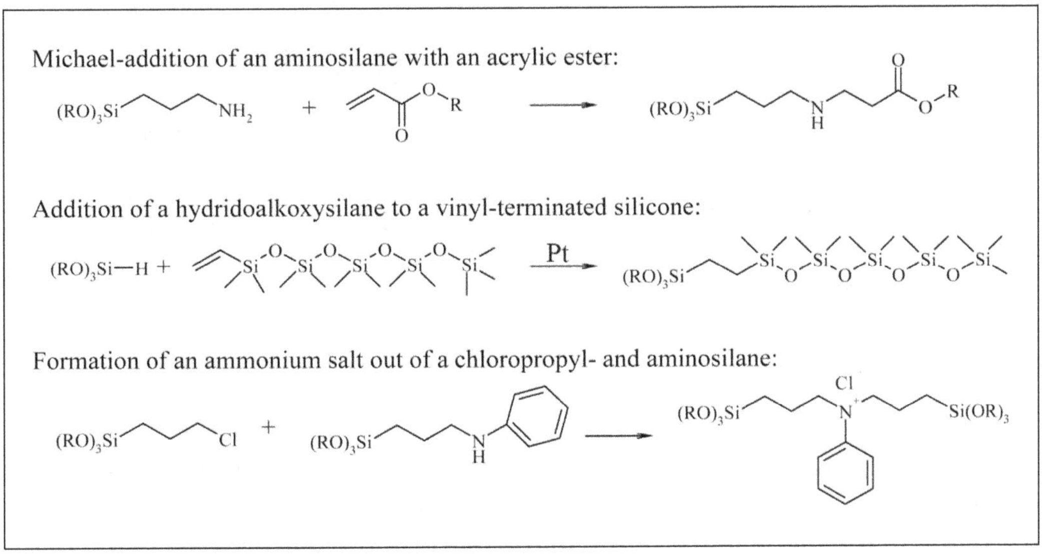

Figure 5.31: Three examples of problem-specific silane synthesis

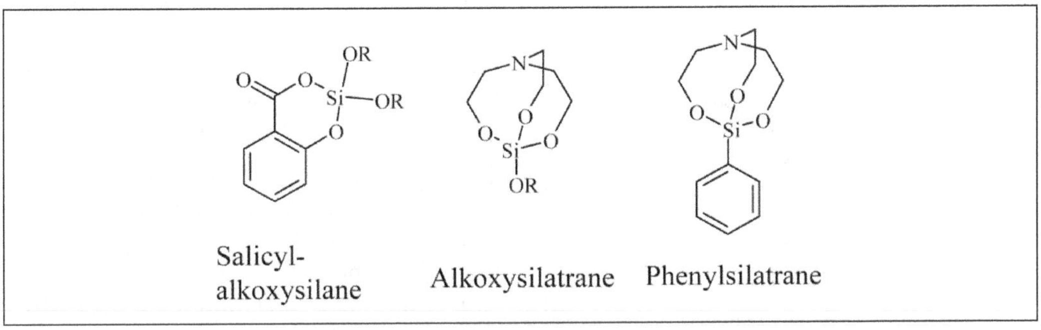

Figure 5.32: Ethoxysilane salicylate, alkoxysilatrane and phenylsilatrane as examples for reaction products of silanes with chelating groups

The alkoxy-groups of the silanes can also undergo transesterification reactions. If ethoxy- groups are exchanged by glycols, water-soluble silanes can be synthesized (compare "Waterborne formulation", page 124). Also, chelates can be formed with compounds like salicylic acid or triethanolamine.

Triethanolamine coordinates via the N-atom to the Si-atom and forms so-called silatranes, which have been proposed for medical applications like the treatment of a weakness of the connective tissue or alopecia (loss of hair). The example of phenyl silatrane shows, that caution is always recommended when handling new compounds.

Whereas ethoxy silatrane is non-toxic, phenyl silatrane exhibits an exceptionally high toxicity (LD_{50} appr. 0.2 mg/kg). In the body, the odorless and flavorless silatrane is decomposed quickly to non-toxic, hard to identify substances, which even led to an appraisal as "the perfect poison" in a movie [167].

Figure 5.33 illustrates the synthesis of an inorganic/organic network using the example of methacryloxypropyl trimethoxysilane. In a first step, the inorganic Si-O-Si network is built up and in the following step the organic network is synthesized by a radical polymerization reaction. Both steps can be processed individually, a fact which has been used to produce deep drawable coatings (compare Example 40, page 158).

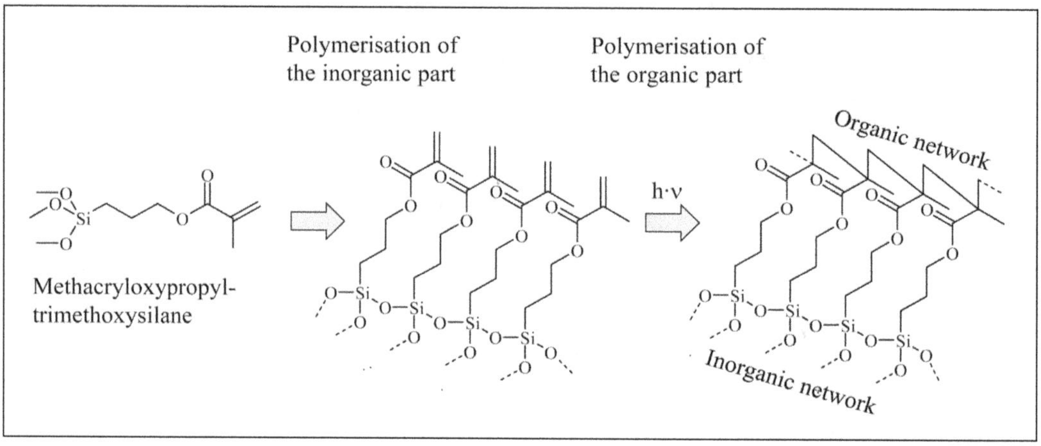

Figure 5.33: Dual cure inorganic-organic polymerisation of an organosiloxane hybrid material based on methacryloxypropyl trimethoxysilane

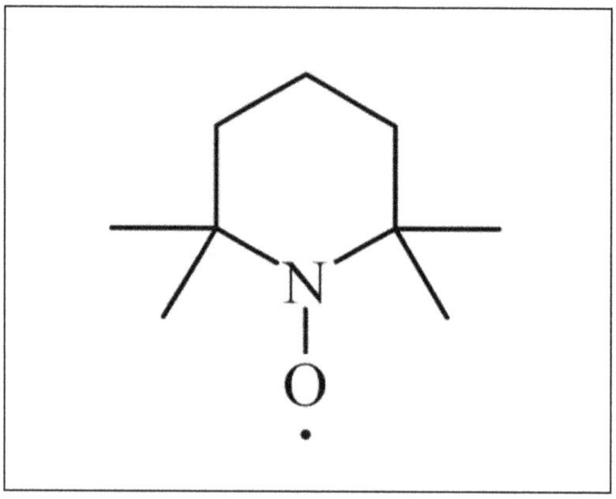

Figure 5.34: The stable radical (2,2,6,6-tetramethylpiperidin-1-yl)oxyl, "TEMPO" is an excellent stabilizer for unsaturated compounds, especially for handling and storing under an inert atmosphere [168]

Methacryloxypropyltrimethoxysilane contains stabilizers to prevent an unintended polymerization. They can be removed prior to a formulation by adsorption on aluminum oxide. Stabilizer-free silanes with (meth) acrylic function should be handled with care, just like other acrylates.

Hefty uncontrollable polymerization reactions are possible in unstabilized formulations. The same can easily happen under anaerobic conditions (e.g. in a rotary evaporator) when common stabilizers like hydrochinon are no longer effective. Alternative stabilizers like TEMPO (2,2,6,6-Tetramethylpiperidin-1-yl)oxyl) which are active even without oxygen being present are available [168].

In the following chapters, now various principles for the formulation and synthesis of nanocomposites are discussed and several examples are given.

5.3.2 Alkoxysilane-functionalized coating resins

In the previous chapter, the synthesis of tailor-made organofunctional silanes has already been described. If the amount of the organic functionalization is now increased further and further, the organosilane at some point can also be regarded as silanized polymer. Via radical copolymerization of methacrylic-group functionalized silanes together with organic monomers, binders with pendant alkoxysilyl groups can be generated.

After curing, polymers with a high network density are created, which show advantages in their chemical and mechanical resistance (compare Figure 5.13, page 101). They find application as latex dispersion paints with increased rub- and washing resistance [163], or as scratch resistant automotive clear coats [169]. Figure 5.35 illustrates the structure of a silanized resin as an example for a possible future UV-curable automotive clear coat.

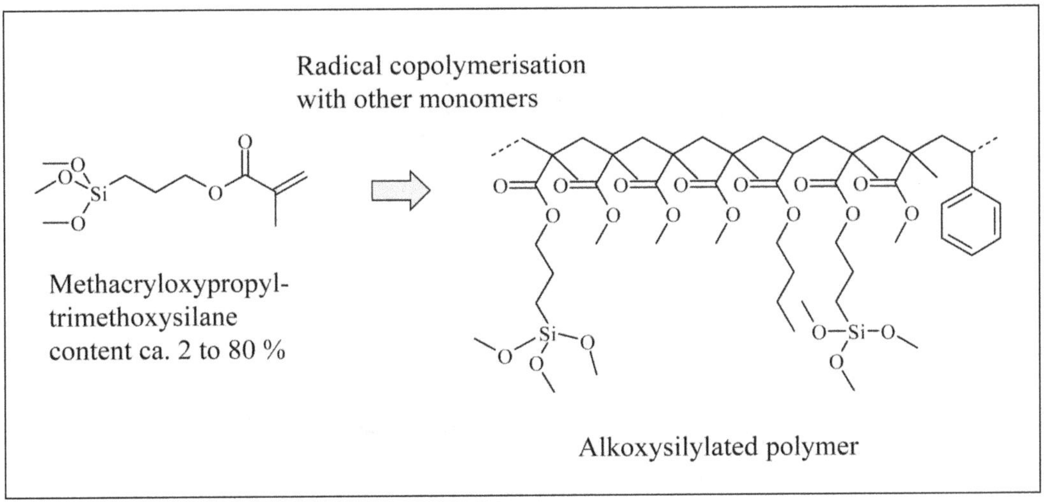

Figure 5.35: Alkoxysilane-modified methacrylate copolymer [170–172]

➤ *Example 26: Preparation of a silane-modified polymer resin for clear coats* [173]

Preparation of a methacrylate copolymer:

A three-necked flask was charged with 669.5 parts of ethoxypropanol. The initial charge was heated with stirring to 130 °C under a nitrogen atmosphere. Subsequently the first feed stream, consisting of 377 parts of glycidyl methacrylate, 658.5 parts of methacryloxypropyl trimethoxysilane ("Dynasilan MEMO") and 48.25 parts of 1,1'-diphenylethylene, and the second feed stream, consisting of 11.75 parts of tert-butyl peroxy-2-ethylhexanoate and 134 parts of ethoxypropanol, were subsequently metered slowly into the initial charge, beginning simultaneously, with stirring. While the first feed stream was metered in over two hours, the second feed stream was metered in over 2.5 hours. The resulting reaction mixture was postpolymerized with stirring at 130 °C for five hours.

Preparation of a clear coat:

A round-bottomed flask was charged with 1,502.7 parts by weight of the methacrylate copolymer, 1,2693.7 parts by weight of isopropanol and 365.4 parts by weight of 0.1 N aceticacid. This initial charge was heated at 70 °C for three hours. Subsequently 1,394.4 parts by weight of "Solventnaphtha" were added, after which the resulting reaction mixture was stirred at 70 °C for five minutes. Thereafter, the low-boiling constituents, especially water, were distilled off under reduced pressure at 70 °C (3,737.9 parts by weight), to yield 2,218.3 parts by weight of the condensate. The residual water (0.2 % by weight) and isopropanol (2.5 % by weight) contents were determined by gas chromatography. The condensate was stable on storage at 40 °C for more than four weeks. It was outstandingly suitable for producing clear coats.

Silanized resins

Silanes with an unsaturated double bonds can be used as co-monomers in polymerizations like any other organic monomers. Silane-functionalized polymers result, which then can cross-link further by hydrolysis and Si-O-Si condensation of the alkoxysilane groups.

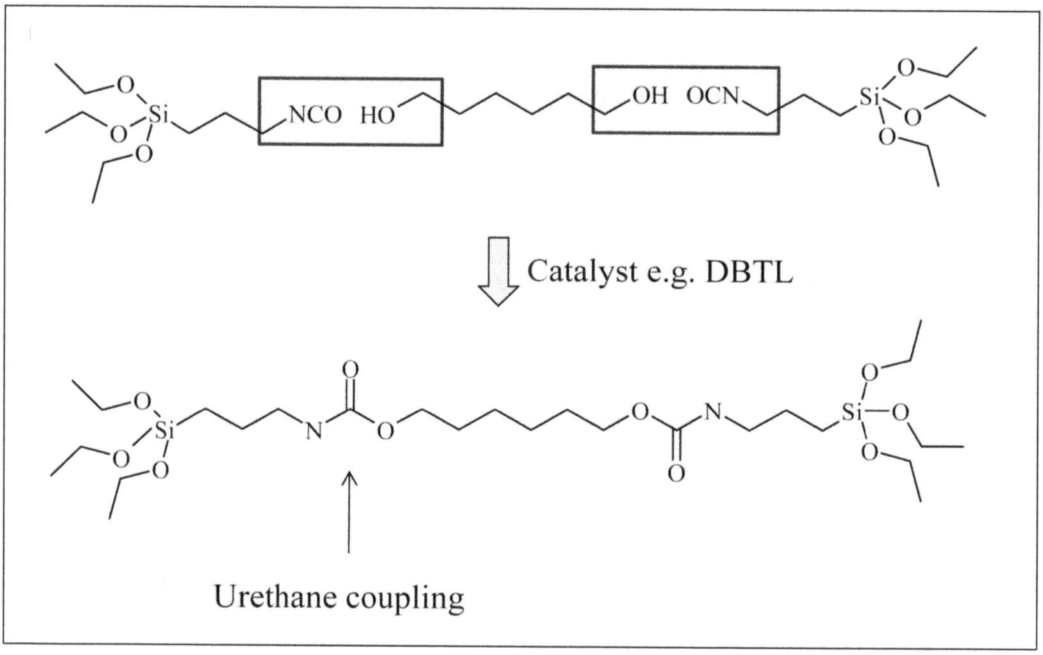

Figure 5.36: Alkoxysilane-modified resins like e.g. "Silixanes" are used for coatings with exeptional scratch and chemical resistance[145]. DBTL: Di-n-butyltindilaurate

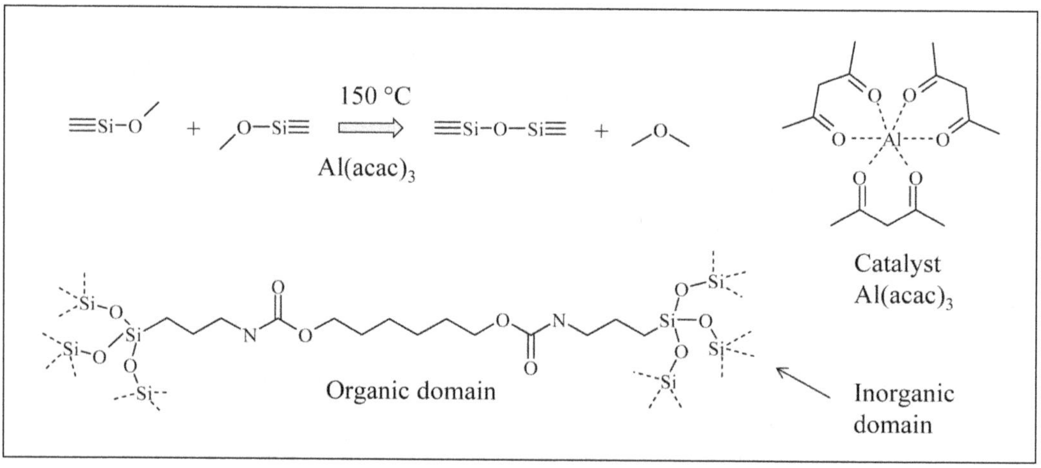

Figure 5.37: Curing of alkoxysilane-modified resins by "direct condensation" with liberation of ethers[145] (compare "Water-free sol-gel techniques", page 100)

In this example, a combination of alkoxysilane and epoxy groups is used for curing the resin. The alkoxysilane functionality can be cured without using water or like in this example by conventional hydrolysis and condensation. The slightly acidic conditions are suitable to generate many Si-OH groups which do only condensate during the thermally induced polymerization of the epoxy groups. The curing parameters are orientated at the typical curing scheme of automotive clear coats (130 °C, 20 minutes). The silane and epoxy groups form inorganic and organic networks with a high scratch resistance (compare "Scratch resistant coatings", page 150).

Alkoxysilane-functionalized resins do not have to be almost organic polymers. If the volume fraction of the silane equals or exceeds that of the organic cross-linker, then the silane-functionalized polymer rather becomes an organically cross-linked polysilsesquioxane [174]. Again, the borders between the two material classes are not precisely defined. Figure 5.36 as an example shows the reaction of 1,6-hexanediol with isocyanatopropyl triethoxysilane which form a silane cross-linked urethane.

Alkoxysilyl-functionalized binders do not necessarily need water for cross-linking. Via a thermal direct condensation, the corresponding ether can be liberated and a Si-O-Si bond is built. This mechanism ensures that curing takes place even in thicker layers of the coating. Typical catalysts for the direct condensation are phosphonates and/or metal complexes like $Al(acac)_3$.

➤ ***Example 27: Preparation of a scratch-resistant, thermally curable nanocomposite*** [175]

49.5 g ICTES (isocyanatopropyl triethoxysilane) are added to 11.8 g hexanediole under continous stirring. While stirring, the reaction mixture is heated up to 50 °C and 0.1 g dibutyltin dilaurate are added. Stirring continues for 30 min at 50 °C, followed by cooling down to room temperature.

5 g of the reaction product and 0.2 g aluminium acetylacetonate are dissolved in 10 g 1-methoxy-2-propanol. After application (e.g. by flooding) onto a polycarbonate panel, hardening is performed for 50 min at 120 °C in a circulating air oven. The resulting coating exhibits excellent scratch-resistance.

Scratch resistant nanocomposites

In a first step, the silane-terminated polyurethane precursor is synthesized by addition of an isocyanate-functionalized silane to an aliphatic diol. The cross-linking of the PU resin is done via the non-aqueous Si-O-Si condensation and not, similar to conventional resins, via the isocyanate-OH addition.

As a catalyst for this reaction, the metal complex $Al(acac)_3$ (aluminum acetylacetonate) is used. A variety of silane-terminated urethanes and ureas are meanwhile available under the brand name "silixane".

The synthesized nanocomposite excels in scratch and chemical resistance which is caused by the combination of nanoscaled hard inorganic and elastic organic domains.

A variation of this building principle uses the urea- instead of the urethane coupling mechanism by reacting the well-known curing agent HDI (hexamethylene diisocyanate) isocyanurate with a bipodal amino silane [176]. The result is a resin binder with high scratch- and chemical resistance for automotive clear coat applications, which is cured by the non-aqueous condensation with phosphonates used as catalyst.

The excellent coating properties can be explained beside other reasons by the increased network density (compare Figure 4.14, page 83). The number of possible cross-linking points has increased from three (HDI trimer) to 18 in its silanized derivative. Even if it is clear that due to sterical reasons not every 18 SiOR groups will react, the cross-linking density in the final coating will be much higher than in the conventional PU resin.

The change in the curing principle from NCO/OH addition to Si-O-Si condensation also brings with it a changed curing behavior. In isocyanate-cured resins, after a short period of time no residual isocyanate groups can be detected after curing is completed. So, there is only a minor post-curing effect observable.

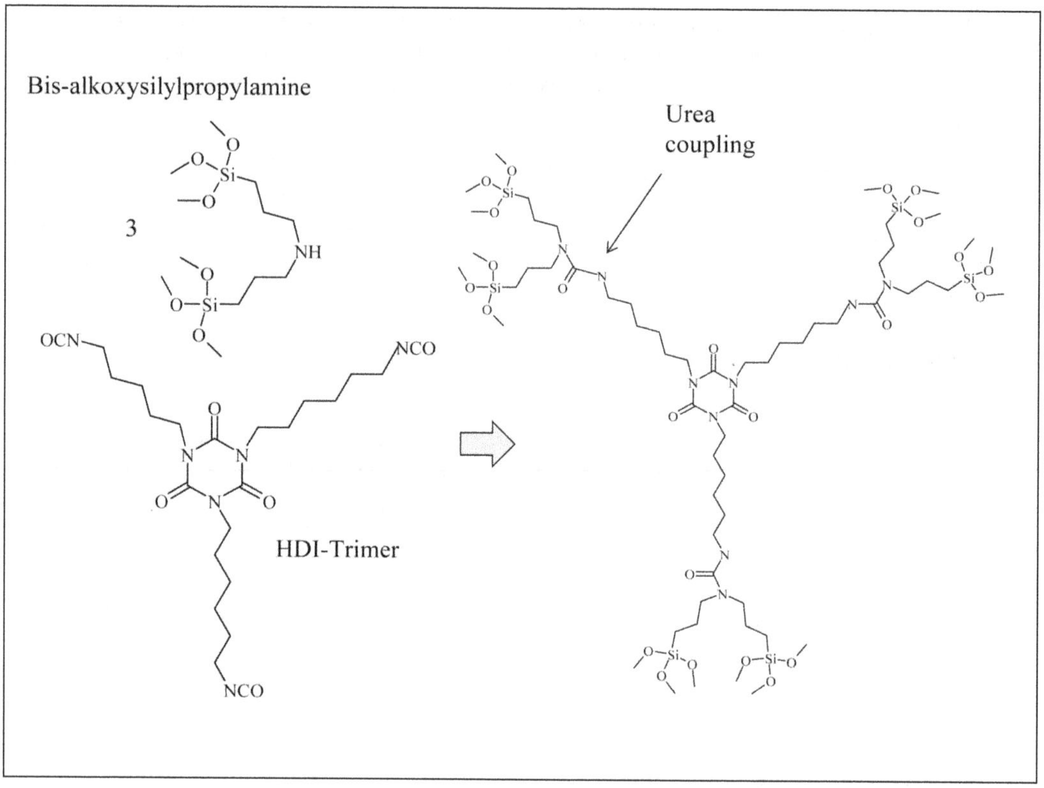

Figure 5.38: Silylated isocyanurates can be used as highly scratch- and chemical resistant resins [176]

The silane cross-linked resins are different in that the cross-linking agent Si-OH remains active over the whole lifetime of the coating. Even months or years after the application, Si-OH groups can and will lead to further cross-linking. If this effect plays a role concerning the long-term property profile of the coatings has to be investigated individually.

5.4 Formulation of sol-gel coatings

5.4.1 Solvent-based formulations

Solvents are used to ensure a homogenous mixture of the silanes, other educts, catalysts and the water which is needed for hydrolysis. The alcohols which are released during the hydrolysis become part of the solvent, a fact which can turn out to be problematic for safety reasons.

Most sol-gel formulations contain ethanol and/or methanol. The reason is the hydrolysis of the commonly used ethoxy- and methoxy silanes. These alcohols evaporate very fast and can lead to problems with film formation during the coating step, especially when the formulation is being sprayed on. Another important aspect is their low boiling point and high flammability which makes special safety measures necessary during transport, handling and application. In addition to this flammability hazard, methanol is highly toxic and has to be removed or avoided in most coating formulations by law.

In some cases the low boiling alcohols can simply be removed by distillation and replaced by higher boiling alternatives with lower toxicity. An elegant method is the phase separation technique, which can be performed with unpolar formulations. By addition of water to the hydrolyzed and partially condensated reaction mixture, the polarity of the solvent is raised so high, that the polysiloxane separates as an oily phase. After the polysiloxane has been separated and if necessary washed, it can be dissolved in another solvent, processed as 100 % system, or dispersed in water as an emulsion.

➤ ***Example 28: Preparation of an easy-to-clean additive for industrial cleaners*** [177]

25.5 g Tridecafluoroctyl triethoxysilane are mixed with 13,8 g octyltriethoxysilane. To this mixture are added 15 g isopropanol, 2.5 g water and 0.03 g HClc subsequently. After stirring for 6 h and 12 h, each time 2.5 g of water are added. After stirring for 24 h, a phase separation is induced by adding 50 g of water. The lower silane-phase is separated, dissolved in 50 g ethylacetate and diluted with 1,900 g mineral spirit (boiling point 50 to 80 °C).

An industrial steel surface which is contaminated with fat and oil is first precleaned in a conventional way and then cleaned intensively with a cloth, which had been soaked in the prepared mineral spirit. After cleaning, residual solvent is removed with a dry cloth. A clean water- and dirt-repellent surface results.

Easy-to-clean additive

Catalysts, byproducts, water, salts and low boiling solvents often are undesired in the final product formulation. They are however indispensable for or released during the sol-gel process. By using distillation, filtration or chromatography as purifying methods, most byproducts can be removed, but the increased effort and process complexity is substantial.

One solution can be the deliberate phase separation when the sol-gel formulation consists of mainly hydrophobic components. By the addition of water in excess after the polycondensate has been synthesized, the polarity of the solvent is raised up to a point when the condensates separate as oil or resin. Water, salts, water-miscible solvents or catalysts remain in the water phase and a purified product can be separated.

In analogy to Example 43 (page 166), the example describes a reactive additive for mineral spirit, which after the cleaning procedure stays on the surface thereby creating an easy-to-clean coating. The removal of the catalyst HCl by the phase separation process prevents corrosion of the cleaned metal parts.

The phase separation process also has been described for the synthesis of UV- and thermally curable sol-gel resins where its main use is the removal of toxic methanol from the formulation [178].

When the catalyst is removed, the sol-gel formulations exhibit a pH near the IEP (pH 2) where the stability maximum of SiO_2 is found. This is the reason why the phase-separated Si-OH group rich polycondensates can be stored even many months. The further processing involves either other solvents, the emulsion in water or the use as 100 % solventless system [145, 179].

Suitable solvents

Suitable solvents for sol-gel formulations are polar, protic or aprotic solvents like ether alcohols, ethers or esters. If protic solvents are used, the possibility of a transesterification reaction has to be considered. Higher boiling solvents might remain in the coating, but on the other hand they improve flow and leveling of the coatings. A general rule does not exist; the suitable solvent has to be evaluated for each application individually.

When solvent mixtures are used, attention should be paid that the higher boiling fraction of the solvents is capable of dissolving all components of the formulation. At the end of the coating and drying step, the higher boiling fraction enriches in the coating. Phase separation and coating defects will be the inevitable consequence of an unsuitable solvent mixture.

Water, compared to many organic solvents is high boiling and evaporates slowly. If it enriches in the solvent mixture, its high surface tension can cause wetting problems and high capillary forces in the coating which might lead to cracks or delamination. As a consequence, the water content should be minimized and/or balanced by suitable high boiling solvents like ether alcohols.

5.4.2 *Waterborne formulations*

For safety and environmental reasons, there have been many attempts to realize waterborne sol-gel formulations [180, 181]. The special complexity of these attempts evolves from the fact that water not only is a solvent, but also a reaction partner for hydrolysis. It can be expected, that waterborne silane-based formulations gelate quickly or precipitate.

If sol-gel building blocks shall be dispersed in water, they have to be stabilized like nanoparticles (compare "Stabilization of nanoparticles against agglomeration", page 40). Sterical, electrosterical and electrostatic stabilization, as well as the SiO_2 anomaly, the unusually high stability of Si-OH groups at the IEP can be used.

The primary goal of the stabilization method is to prevent the premature condensation of reactive Si-OH groups. Figure 5.39 illustrates some possibilities.

The ether- and epoxy-group of glycidoxypropyl trimethoxysilane (GPTES) both show a high polarity. This is a reason, why a hydrolysate of GPTES is stable in water over a long period of time. GPTES can also be used to stabilize nanoparticles in water, either in its original form or ring-opened as diol [182, 183].

Co-condensates of GPTES and the extreme water-repellent tridecafluorooctyl triethoxysilane are water dispersible, if the amount of GPTES exceeds 30 %.

Aminosilane-hydrolysates also can be water-soluble. They easily form an intramolecular ammonium salt with the Si-OH group. The salt formation generates a hydrophilic structure,

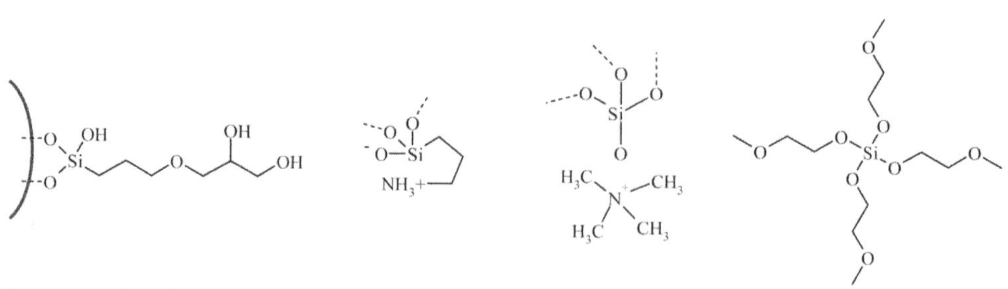

Nanoparticles with a surface modification made of ring-opened glycidoxypropylsilane

Inner salt which forms in aqueous dispersions of aminosilanes

Salt formation with quarternary ammonium-compounds

Watersoluble glycolethersilane

Figure 5.39: Several possibilities to realize waterbased sol-gel coatings

which renders the molecule or the corresponding polycondensate water-soluble. A similar principle is used in Example 44 (page 167). There, a fluorinated silane is rendered water-soluble with tetramethyl ammoniumhydroxide. Again, a hydrophilic salt-like structure is generated which leads to a soluble product.

As a general rule, the solubility of a polysiloxane in water has to be secured by the addition of stabilizers or dispersing aids. These stabilizers have to be removed from the coating after the deposition or they will affect its properties. A thermal curing process can help, if the thermal stability of the dispersing aid is lower than the stability of the coating material.

➤ *Example 29: Preparation of a waterborne scratch-resistant sol-gel coating material* [184]

27.8 g (0.1 mol) of (3-glycidyloxypropyl) triethoxysilane (GLYEO) were admixed with 27.8 g of silica sol (30 % by weight SiO_2, “Levasil 200 S”, Akzo Nobel). The mixture was subsequently stirred at room temperature for 5 hours. Thereafter, the ethanol formed by hydrolysis was removed by distillation (rotary evaporator, maximum bath temperature 40 °C). The resulting waterborne coating material was admixed with 1.11 g (0.0005 mol) of N-(2- aminoethyl)-3-aminopropyltrimethoxysilane (DIAMO) and stirred at room temperature for one hour.

The sol-gel coating material was used to coat polycarbonate and aluminum sheets and CR-39 lenses. The coated polycarbonate and aluminum sheets were stored at room temperature for 30 minutes and then cured at 130 °C for 4 h. The CR-39 lenses were stored at room temperature for 30 minutes and then cured at 90 °C for 4 h.

The example was repeated by using 3.05 g (0.001 mol) of (3-(triethoxysilyl) propyl) succinic anhydride (GF20) instead of DIAMO. Investigation of the abrasion resistance of polycarbonate sheets coated with this composition by the Taber abrasion test (wheel material CS 10F, 1000 cycles, wheel load 500 g), showed an increase of haze by only 7 %.

Waterborne abrasion resistant sol-gel coating

In this example, an organoalkoxysilane which carries a hydrophilic group (glycidoxypropyl) is used to synthesize water-dispersible polysilsesquioxane structures. A cationic dispersion of colloidal silica serves as mildly acidic hydrolysis- and condensation catalyst and reaction partner.

Colloidal silica sols can be stabilized with a cationic charge (compare “Stabilization of nanoparticles against agglomeration”, page 40) by adsorption of aluminum chloride. Via the reaction with Si-OH groups of the hydrolyzed silane, the sol is further stabilized sterically, which prevents it from gelation or precipitation during the sol-gel reaction (compare Example 31, page 140).

The byproduct of the hydrolysis is ethanol, which lowers the dielectric constant of the solvent (water) and weakens the electrostatical stability by limiting the range of the repulsive forces (compare Figure 3.11, page 42).

The cross-linking of the epoxy groups can be started by a base-catalyzed polymerization (formation of polyethers), addition of amines or reaction with carbonic acid anhydrides (the formation of polyesters). Polyethers or amine adducts are more hydrophilic than polyesters, which is the reason why they most often are used in indoor applications.

If aminofunctional silanes or amines are used as curing agent for epoxy-functionalized polycondensates, the pH shift should be considered. A part of the aminosilane will form a salt with the acidic catalyst and is no longer available for a reaction with the epoxide. The pH raises and the condensation speed is increased significantly which can have negative implications on the shelf life and the pot time (pot life) of the formulation. A temporary blocking of the amine as carbonic acid salt or ketimine is a suitable measure to avoid these negative side effects (compare Figure 5.41, page 127).

Water is not the most favorable solvent for coatings. It evaporates slowly, therefore has to be force-dried and due to its high surface tension, it does not wet many surfaces well. This is the reason, why waterborne coatings contain many additives like wetting aids and defoamers or biocides. These additives remain in the coating and influence the properties of the final product.

In practice, waterborne sol-gel formulations are often used as anti-corrosion primers for coatings (compare "Anti-corrosion coatings", page 190) and for the coating of mineral building materials [325]. Premium optical coatings like scratch-resistant coatings on polymers are formulated rather with solvents due to the reasons which have been discussed in this chapter.

5.4.3 Storage stability of sol-gel formulations

In contrary to organic polymers whose polymerization process is finished in their form of delivery, sol-gel formulations are living systems. Hydrolysis and polycondensation are equilibria, which are influenced by pH, temperature, catalyst, solvent, transesterification and which change the product consistently during storage time. So old formulations can show haziness or an increase of viscosity and the resulting coatings can crack or delaminate.

The storage stability of sol-gel systems is influenced by a variety of factors and has to be determined individually. The time frame reaches from hours to years and can be optimized by several measures:

- lowering the solid content
- storage at low temperatures
- long-chain alcohols as solvents to foster transesterification
- pH which slows condensation, taking advantage of the SiO_2 anomaly at pH 2 (IEP)
- removal of the catalyst/latent catalysts/2K systems
- incomplete hydrolysis with sub-stoichiometric amounts of water
- reversible blocking of Si-OH groups.

If a smaller amount of water is used than it would be necessary to hydrolyze and condensate all alkoxygroups in the formulation, some Si-OR will groups remain in the polysiloxane. They stabilize the product which cannot cure, until after the addition of water (humidity, process water) the residual alkoxygroups are hydrolyzed.

The reversible blocking of a curing catalyst is an elegant method to extend the storage stability of a formulation. Some of the already known blocking agents of the polymer resin chemistry can be used as shown in Figure 5.41.

The topic "storage stability" also opens up the question, how to determine whether a coating formulation already shows some undesired change of properties or not, which is an essential question of quality control in industrial processes.

The decisive measurement tool is the application test, but the observation of the viscosity as a function of shear force can give valuable hints to identify a negative change of properties.

The viscosity of a sol is correlated with its structure and the interparticulate interaction. With the same solid content, spherical particles exhibit a lower viscosity compared to branched particles because the structures cannot entangle.

The more the condensation reaction advances, the more agglomerates are generated which increase the viscosity of the sol. A higher degree of entanglement and interparticulate interaction in the sol expresses itself in a shear force dependent viscosity with a hysteresis of the viscosity curve. Therefore, the measurement of the viscosity, especially as a function of the shear force, is a sensitive tool for detecting changes in the structure of the coating formulation. IR, NMR or other more sophisticated analytical tools often cannot be correlated in practice with the shelf life or performance of sol-gel coatings.

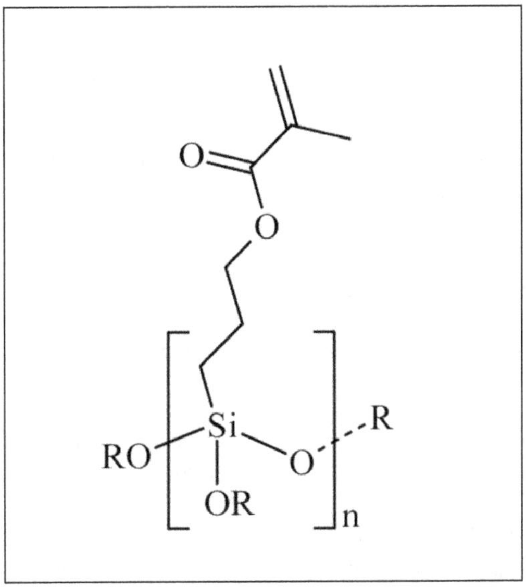

Figure 5.40: Methacryloxypropyltrimethoxysilane-oligomer (MEMO) with unreacted methoxy-groups

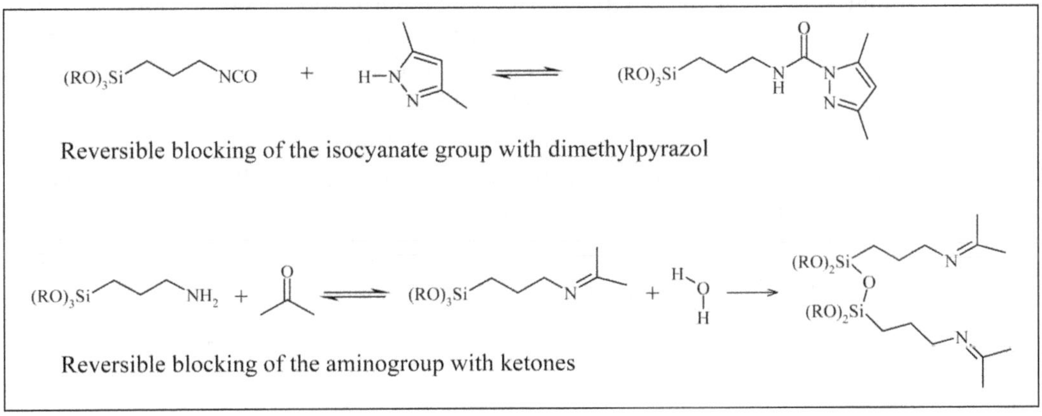

Figure 5.41: Reversible blocking of the isocyanate function with dimethylpyrazol or other other leaving groups like caprolactam [185, 186]. Ketimine-blocking of aminosilanes with ketones. Liberation of water during the blocking reaction leads to oligomerization. Contact with water unblocks the amine again.

5.4.4 Sol-gel powder coatings

Due to environmental reasons, a strong trend developed to avoid or minimize solvents. So powder coatings have found widespread use because they allow for elegant and effective application methods with a possible reuse of powder overspray and the complete avoidance of solvents during their application.

For a long time it seemed impossible to synthesize powder coatings via the sol-gel process, because the building blocks of the sol irreversibly react with each other during drying. Sterically demanding substituents however allowed to dry sol-gel building blocks which are rich in Si-OH groups without causing permanent cross-linking [189].

The main problem in the conceptual design of sol-gel powder coatings is how to prevent the premature cross-linking of Si-OH and Si-OR groups while maintaining a molecular weight

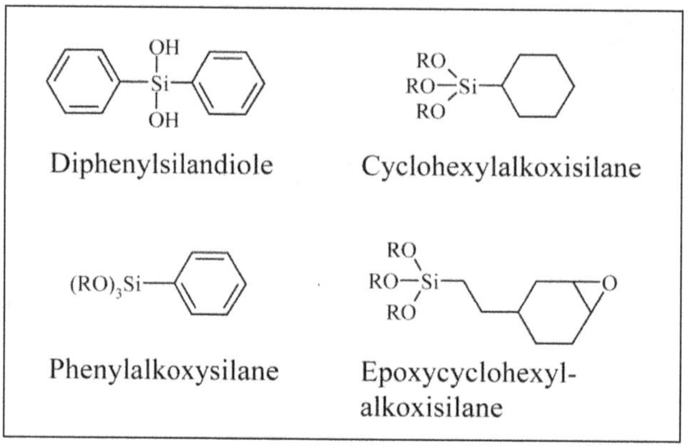

Figure 5.42: Silanes with sterically demanding substituents allow the formulation of sol-gel powder coatings

and cross-linking density of the network, which allows for grinding and milling, but also easy meltability of the final product. As a general rule, Si-OH group bearing oligomers can remain stable for a long period of time, like it can be seen for example in the case of diphenylsilanediol or Si-OH terminated reactive silicone fluids (Figure 5.25, page 114).

In the absence of catalysts (bases, metal ions) and at moderate temperatures these compounds are storage-stable for years. For the synthesis of many stable Si-OH groups, a mildly acidic pH is suitable to support hydrolysis and to slow down condensation (compare "Hydrolysis and condensation", page 94).

The second problem which has to be solved is how to generate structures which are brittle enough so they can be grinded and milled and which easily melt to a film at curing temperatures below 200 °C. Sterically demanding groups like phenyl-, or especially cyclohexyl- are suitable for stabilizing the Si-OH group. Such oligomers form xerogels which can be milled to a fine powder and can cure at elevated temperatures when suitable catalysts are present. An additional organic cross-linking can be realized by the condensation with epoxy-, amino-, carboxy-, isocyanate- or acrylate-functional silanes.

Another possible way to synthesize sol-gel powder coatings is to copolymerize organic monomers with silanes to a solid polymer, which can be grinded and milled and which cures via its silane function (compare "Figure 5.37, page 120) [187, 188].

Figure 5.43 depicts a further concept for a sol-gel powder coating, in which unhydrolyzed Si-OR groups protect the polysiloxane oligomers from premature cross-linking.

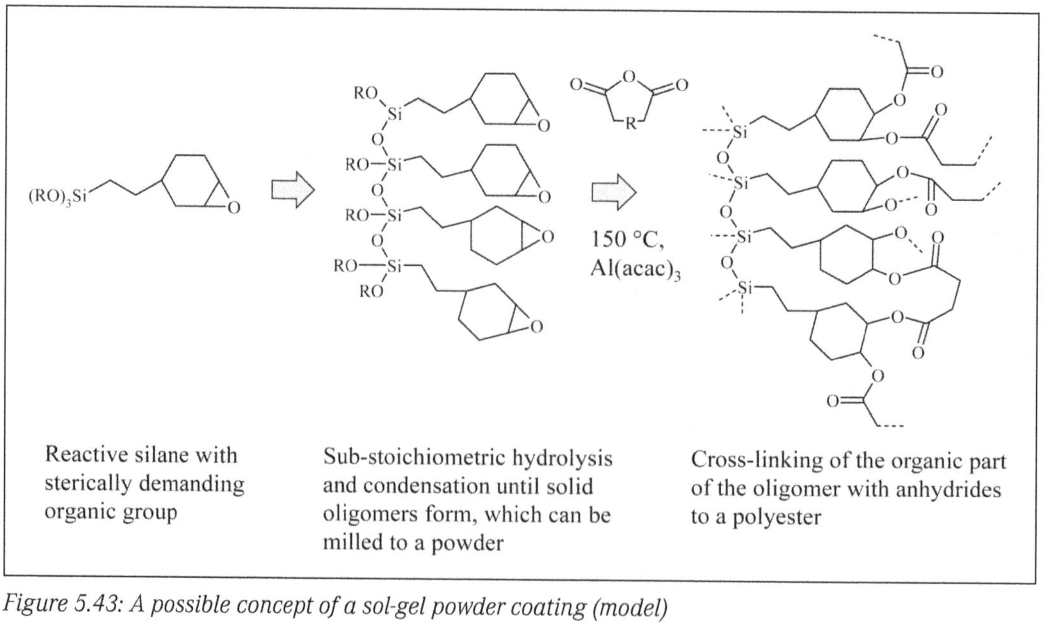

Figure 5.43: A possible concept of a sol-gel powder coating (model)

➤ *Example 30: Preparation of a sol-gel powder coating* [189]

24.63 g (0.1 mol) [ß-(3,4-Epoxycyclohexyl)ethyl] trimethoxysilane are mixed with 12.22 g (0.05 mol) diphenyldimethoxysilane by stirring which results in mixture A. In parallel, a mixture B is prepared by adding 5.8 g (0.05 mol) maleic acid to 10.8 g of a 0.1 N HCl. Under vigorous stirring and ice-cooling, mixture B is added within 30 minutes to mixture A and is subsequently stirred for 4 h.

The reaction product is dried by removing the solvent with a rotary evaporator followed by 5 h storage under vacuum (7 mbar, 45 °C). The solid residue is grinded to a fine powder with a median density of 1.29 g/ cm^3 and a grain size distribution of 20 to 100 µm. The powder is applied electrostatically to aluminum sheets and is cured for 20 min @ 130 °C. The coating exhibits a thickness of 10 to 20 µm and shows excellent adhesion.

Sol-gel powder coating

The general concept of sol-gel powder coatings involves educts with sterically demanding groups to ensure that a polymer which contains many Si-OH groups can be dried and milled without premature cross-linking. In this example, a cyclo-aliphatic epoxysilane is used and combined with a di-phenyl substituted silane.

Acidic catalysts promote hydrolysis and the formation of many Si-OH groups (compare “Hydrolysis and condensation”, page 94). The condensation reaction is hindered by the mildly acidic pH and the bulky substituents. Maleic acid is being used not only as cross-linking agent for the epoxy group, but also as carbonic acid catalyst. Since its acidity alone would be too weak, hydrochloric acid is added to ensure a reasonable hydrolysis speed.

During the thermal curing step, maleic acid reacts with the epoxy silane and forms an unsaturated polyester. The double bond then can react in a second curing step with other unsaturated components. As an alternative to maleic acid, succinic acid could be used to synthesize an inert polyester.

Instead of the diphenyldimethoxysilane, also diphenylsilanediol could be used as an example of a stable silanol. As a general rule, aromatic components should only be used in indoor coatings, since they have a tendency to yellow under UV exposure. The cycloaliphatic components are more UV resistant, but also exhibit a lower index of refraction and as a consequence a lower gloss of the coating. The composition of the formulation can be adapted to the individual requirements of the final product.

5.4.5 Additives for sol-gel formulations

Like it is the case for conventional polymer coatings, additives can help to improve the properties of sol-gel coatings. In general, there is hardly a difference compared to conventional coatings.

As discussed in “Improving the scratch resistance of coatings”, starting with page 82, the scratch resistance of a coating is also a function of its surface. Slippery, plane surfaces are harder to scratch than rough, blunt surfaces. This is also the reason for the use of silicones and waxes as scratch resistance increasing additives. An addition of silicone/polyether-based slip- and scratch resistance enhancing additives can further improve the already high scratch resistance of sol-gel formulations [107, 108, 130].

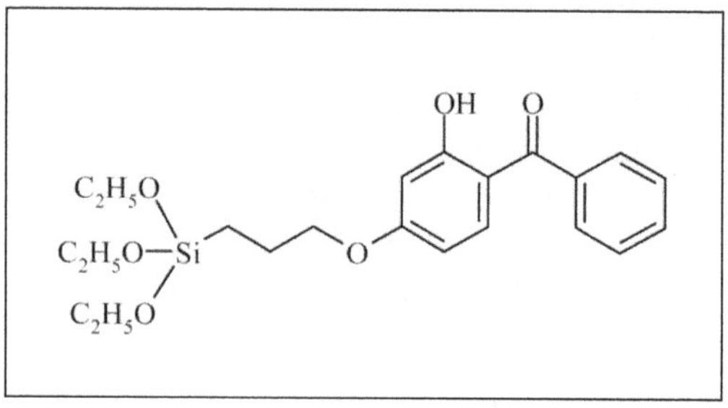

Figure 5.44: Silylated UV absorbers can copolymerize with other silanes and thereby avoid migration [163]

A protection of sol-gel coatings against UV radiation usually is not necessary, because they consist to a large extent of inorganic building blocks. A UV protection for the substrate however can be a reasonable measure, since a deterioration of the substrate surface would affect the optical properties and the adhesion of the sol-gel coating as well.

Typical combinations of UV absorbers and HALS (hindered amine light stabilizers) can be used, as well as nanoscaled inorganic zinc oxide or titanium dioxide particles [190]. To prevent migration, also silane functionalized UV absorbers are available, which can cross-link with the coating.

HALS exhibit a free amine group, which can affect the pH of the coating formulation and can interact with other components of the coating, like e.g. epoxy groups. Since organic UV absorbers weaken the mechanical properties of sol-gel coatings, it is beneficial to use inorganic UV absorbers like nanoscaled cerium oxide, zinc oxide or titanium dioxide (compare "UV absorption", page 78).

Many sol-gel formulations are based on alcohols as solvents. During the coating step alcohols evaporate fast and the coating quickly gels. Only a few percent of a higher boiling solvent like diethylenglycolethers or ether alcohols can improve leveling and gloss significantly.

5.4.6 Compatibility with organic resins

Organic resins can be mixed with sol-gel formulations. In principle, every polymer is suitable which dissolves in the used solvent and does not lead to macroscopic phase separation during drying. This can happen, if the polarity of the sol-gel building blocks and that of the polymer are very different. One measure to suppress a phase separation is to introduce reactive sites in the polymer and the sol-gel building blocks, which lead to covalent bonds between the two.

Silicone resins, especially when they exhibit reactive Si-OH groups, are suitable additives for sol-gel formulations due to their chemical similarity. They can improve the flexibility and alkaline resistance of sol-gel formulations or bring about special wetting properties (compare "Anti-fingerprint coatings", page 176 and Example 49, page 177).

Often, a combination of UV curable resins with methacrylate functional sol-gel formulations has been described. During the UV curing step, both components cross-link with each other and form scratch- and abrasion resistant coatings (compare Example 40 and Example 41) [145, 191, 192].

Polyesters can react with their free -OH groups via transesterification with sol-gel systems. The resulting Si-O-polymer bond is not very stable against hydrolysis, but contributes to the stability of the mixture and prevents a phase separation [193]. This technique reminds of the modification of conventional resins with silicones [194, 204].

Conventional polymers can help to improve the mechanical and chemical stability of the sol-gel components. The same argument however is true vice versa. Sol-gel formulations can introduce a distinct inorganic character to formulations when mixed with conventional polymers. This can be very positive, but can also lead to brittleness and reduced flexibility.

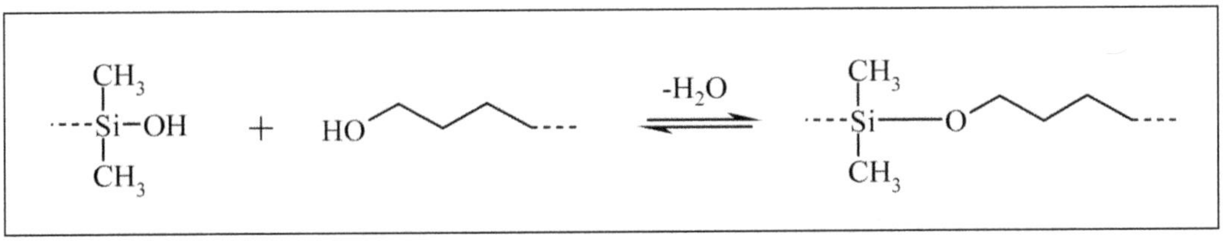

Figure 5.45: Esterification or transesterification reactions can be used to modify organic resins with silicones or siloxanes

One possibility to overcome this problem reminds of the leafing technology, which is used to bring hydrophobic pigments to the surface of a coating. If the sol-gel polycondensate is modified by co-condensation with hydrophobic long chain alkyl- or fluoroalkyl silanes, an intentional phase separation takes place during the coating step and the mechanically more robust polysiloxane orientates itself to the surface, without affecting the bulk properties of the polymer (compare "Improving the scratch resistance of coatings", page 82ff) [195, 196].

6 Application, drying and densification

6.1 Pre-treatment of surfaces

Sol-gel formulations, due to their versatility are suitable for coating a variety of materials. To ensure a good adhesion, the reactive groups of the sol-gel coating material must gain access to the reactive groups at the surface of the substrate. Technical surfaces however are contaminated by dust, oil and dirt, so that a direct coating without a prior pre-treatment hardly ever makes sense.

The pre-treatment methods of glasses, ceramics, metals and polymers differ significantly. From a chemical point of view, glass is an alkali/earth alkali silicate with additions of other oxides. In contact with humidity, alkali hydroxide forms at the surface, which attacks the glass network. CO_2 is absorbed and the resulting carbonates are washed off. A porous colloidal silica gel remains, which in extreme cases like glass corrosion in the dishwasher, can show itself as iridescent or white layer.

Due to its high adsorption capacity, the gel layer readily adsorbs organic contaminations like process fluids (oils, cooling liquids) or fingerprints which can hardly be removed completely [197, 198, 220]. Only the mechanical removal of the gel layer with the help of polishing aids like cerium oxide or silica guarantees that the contaminations are removed and the reactive groups of the coating have unhindered access to the glass surface.

Glasses, ceramics and metals carry -OH groups on their surface which can react with silanes to Si-O-Si bonds. The surface however, even when it had been cleaned and dried, still is not accessible without barriers. Some layers of adsorbed water always block the surface -OH groups and so silanes in a first step only can form hydrogen bridges with the surface. Covalent bonds are only formed during a curing step at elevated temperatures, with very active catalysts or after some time [199-201].

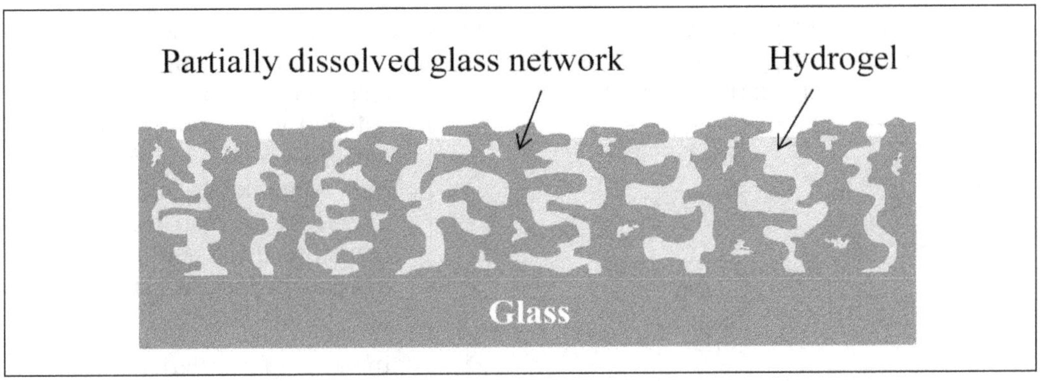

Figure 6.1: Aged glass is covered by a hydrated gel layer which easily adsorbs organic contaminations. In order to achieve good adhesion, this gel layer has to be removed.

Gerhard Jonschker: Sol-Gel-Technology in Praxis

ISBN: 978-3-86630-886-2

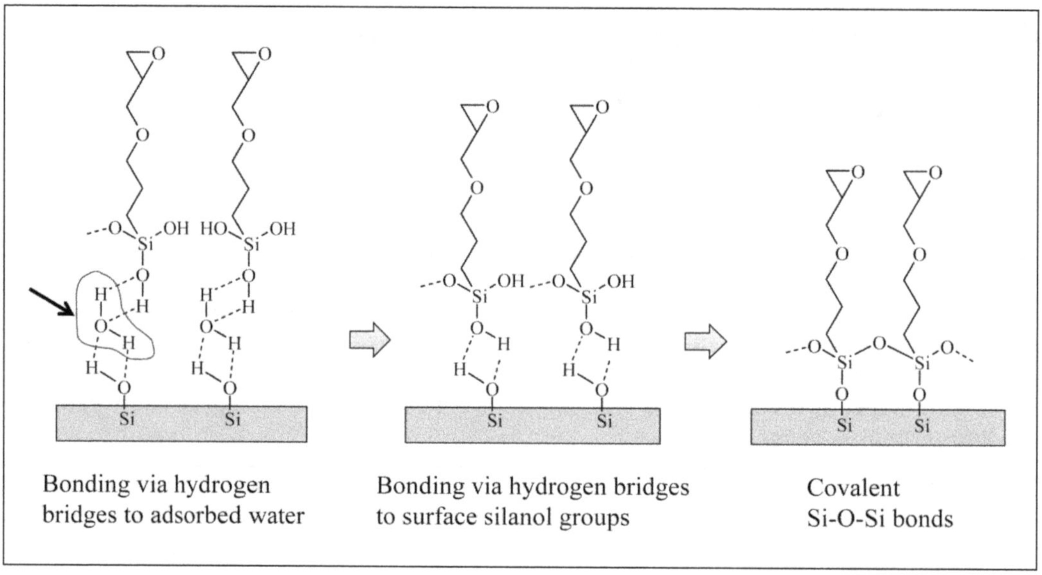

Figure 6.2: Adhesion of sol-gel coatings to inorganic surfaces like glass. Si-O-Si bonds are only formed if the reactive groups are activated by suitable catalysts and get access to the surface.

Polymer surfaces exhibit only a few binding sites on their surface. Silanes in principle can undergo a transesterification reaction with C-OH groups, but the bond is not very stable against hydrolysis. Further options for a covalent bond exist, if the silanes carry reactive groups like -Cl, -NCO, epoxy or amine, which can react with the corresponding functions of the polymer. It is however clear, that the majority of bonds will be hydrogen bridges or other dipole interactions.

The primary target of a polymer pre-treatment is to remove dust and dirt and, if possible, to increase the number of potential reaction sites. Corona discharge or flame treatment are processes, which are often used and which generate polar groups via partial oxidation of the polymer. Another aspect of these treatments is that the polar groups increase the surface energy which eases the wetting of the treated surfaces. Among the generated groups are alcohol-, carbonic acid or peroxide groups, which are capable to react with functional silanes.

The reactive flame treatment combines the aspect of a partial oxidation of the surface with the incorporation of SiO_2 nanoparticles into the polymer surface. These nanoparticles are generated via the combustion of highly volatile silicon compounds (e.g. trimethylmethoxysilane) which are added to the gas stream of the burner. At the high temperatures, the generated nanoparticles melt into the polymer surface and now can serve as inorganic adhesion point for coating materials [202].

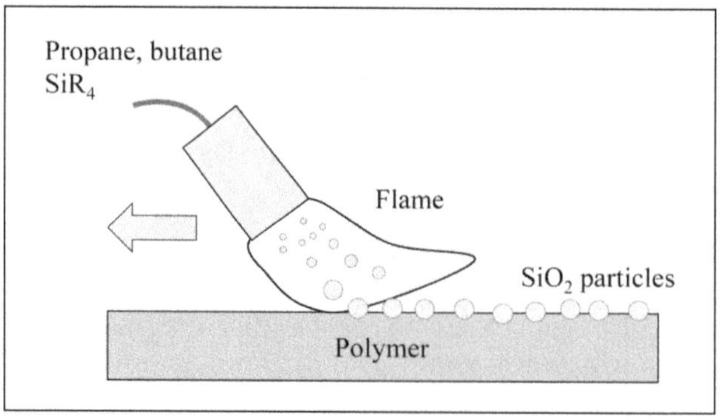

Figure 6.3 Treatment of polymer surfaces with a "reactive flame". Volatile silicon compounds yield SiO_2 nanoparticles which are melted into the activated surface.

In sol-gel formulations, primers often fulfill other functions than in conventional coatings. Sol-gel coatings typically consist to a large extent of organically modified silanes, which are adhesion promoters by themselves.

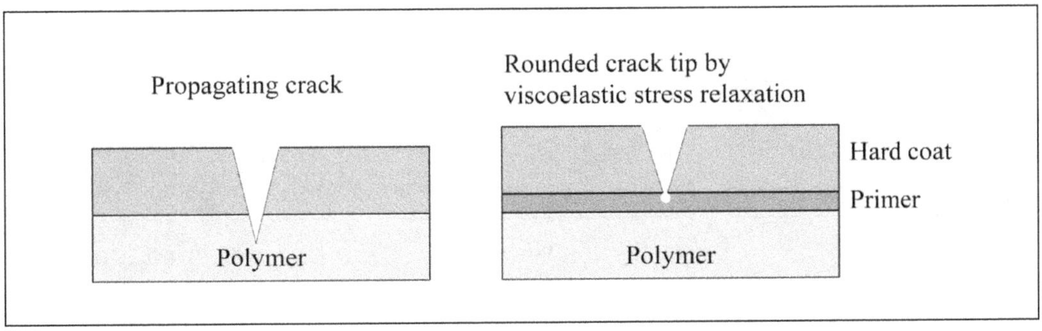

Figure 6.4: Viscoelastic primers can stop propagating cracks by rounding the crack tip and thereby dissipating the energy

Primers are used for sol-gel formulations, if the difference of the mechanical properties between coating and substrate are too big, like it would be the case of a hard, brittle coating on plastic substrates. A thin, flexible primer can help to avoid stress cracks and to pass demanding tests like the stone chipping test [203, 204]. If a hard, but brittle coating has a good adhesion to a plastic substrate, a crack can propagate right through the plastic substrate, if no elastic primer is used.

A further application for primers in sol-gel formulations is to use them as carrier for functions, which cannot be realized in the coating material. Typical sol-gel coatings are very resistant against UV irradiation because of their high content of inorganic building blocks. The substrate however needs protection, but organic UV absorbers would weaken the mechanical properties of the coating material. A solution can be to incorporate UV absorbers into the primer layer.

When metals are coated, primers can enhance the corrosion protection. Silane-based sol-gel formulations can replace chromate- or phosphate-based treatments with equally good performance (compare “Anti-corrosion coatings”, page 190) [205, 206].

Tensions in glasses or plastic substrates can impede the formation of well adhering coatings. They are generated, if the material is cooled too fast in its production and have to be relaxed in order to guarantee a good adhesion of coatings.

Sol-gel formulations with a very high amount of inorganic building blocks show only a weak stress relaxation ability. When they are applied, the relaxation of tensions in the substrate prior to the coating step is highly recommended to avoid stress cracks or delamination. The substrate is heated up to temperatures which exceed its transformation temperature (T_g) and then is slowly cooled down again [220, 207].

6.2 Coatings with sol-gel materials

Sol-gel formulations can be applied with almost all available coating methods like e.g. spraying, rolling, via doctor blade, dipping, flooding or printing.

During the evaporation of the solvent, the interparticulate distance gets smaller and smaller until agglomerates form and the sol gelates. The wet gel dries to a xerogel and eventually is treated at higher temperatures afterwards. The densification temperature is chosen accordingly to either maintain or deliberately destroy the organic components of the coating. If the temperature is increased further, the porous inorganic network sinters to a dense coating.

Very early it became clear, that sol-gel formulations can only be coated up to a critical coating thickness before cracks and delamination where observed. For inorganic sol-gel coat-

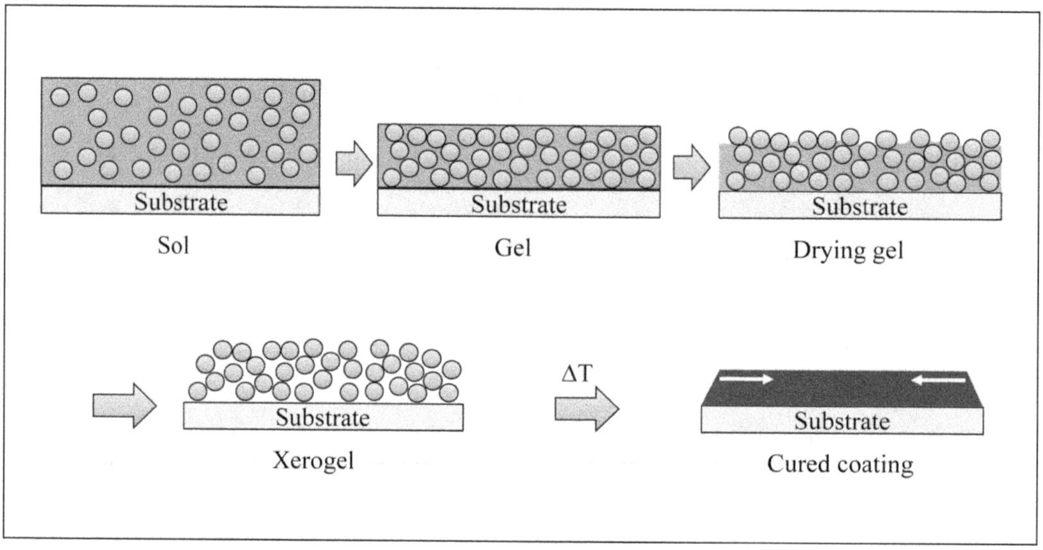

Figure 6.5: Illustration of the drying and curing process of sol-gel coatings. Because the coating is chemically linked to the substrate, tensile stress develops during shrinkage. This is the main reason for the generation of cracks.

ings, the critical thickness is well below 1 μm. The two decisive factors which determine the critical thickness are the degree of shrinking and the ability of the coating material to relax stresses.

The effectiveness of the particle stabilization determines the maximum packing density of a sol until it gelates. Early agglomeration leads to open, irregular structures which have to shrink more during drying and densification and lead to the generation of more stress.

At a certain point during the evaporation of the solvent, mechanical properties can be measured – the sol has gelled. The gel cannot shrink as fast as the liquid level of the solvent sinks, so the liquid level will enter the gel. At this point, a capillary pressure develops, caused by the interaction of the pore wall with the pore liquid. Its extent is dependent on the pore radius, the surface energy of the gel and the surface tension of the pore liquid [208, 209, 210].

Formula 6.1:

$$P_c = -\frac{2\gamma \cdot \cos(\Theta)}{r}$$

The capillary pressure is determined by the pore radius r, the contact angle of the pore liquid Φ and the surface tension γ

If the surface tension of the pore liquid stays low during drying, only small capillary forces develop and the pore structure remains intact. The usual solvents which are used in sol-gel processing however contain alcohols and water which lead to high capillary forces. Especially when water is enriched during the evaporation of the higher volatile alcohols, huge capillary forces develop which exceed by far the mechanical stability of the pore walls. A reduction of the water content and addition of less volatile, water-miscible solvents can be of help to avoid cracks.

At the supercritical point of a solvent, no phase boundaries or surface tension exists anymore. The pore liquid can be removed without causing a deformation of the pore structure, an aerogel has been generated [211, 212].

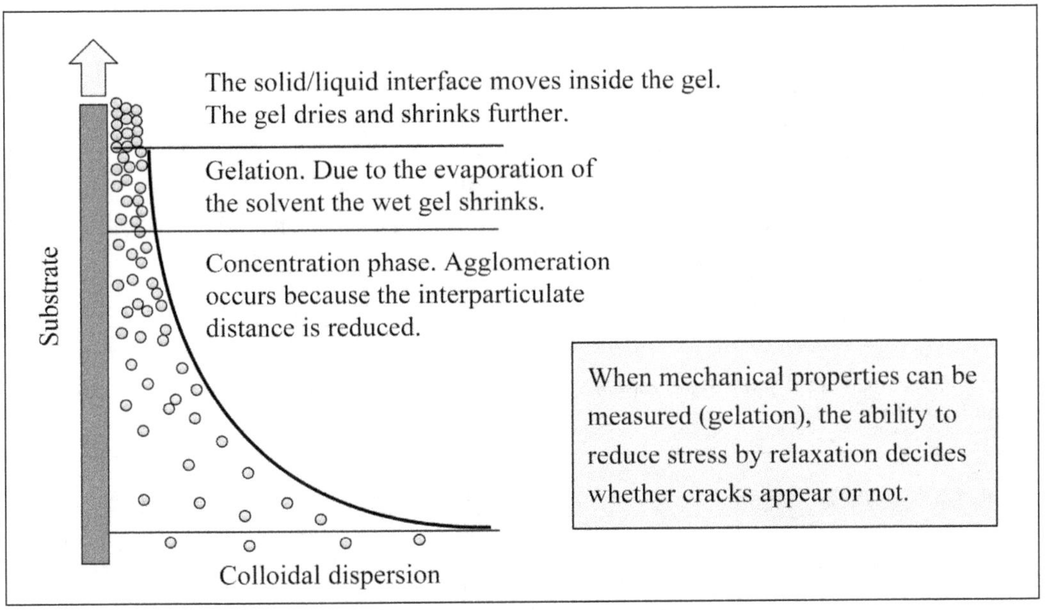

Figure 6.6: Illustration of the dip coating process with colloidal dispersions

Also fluorocarbon solvents or volatile silicones can leave the gel almost without causing capillary forces. They can be used to produce aerogels at ambient pressure [213, 214].

The surface energy of the pore wall is dependent on its chemical composition. Hydrophobic components, like unreacted alkoxy groups (Si-OR), as well as alkyl or aryl groups lower the surface energy, a high content of Si-OH groups or hydrophilic organic substituents increases it.

As a general rule, the higher the surface energy and thus the interaction between pore wall and the pore liquid, the higher are the mechanical forces which act upon the pore wall. A coating of the pore walls with low energy materials like fluorocarbon silanes or silicones can lower

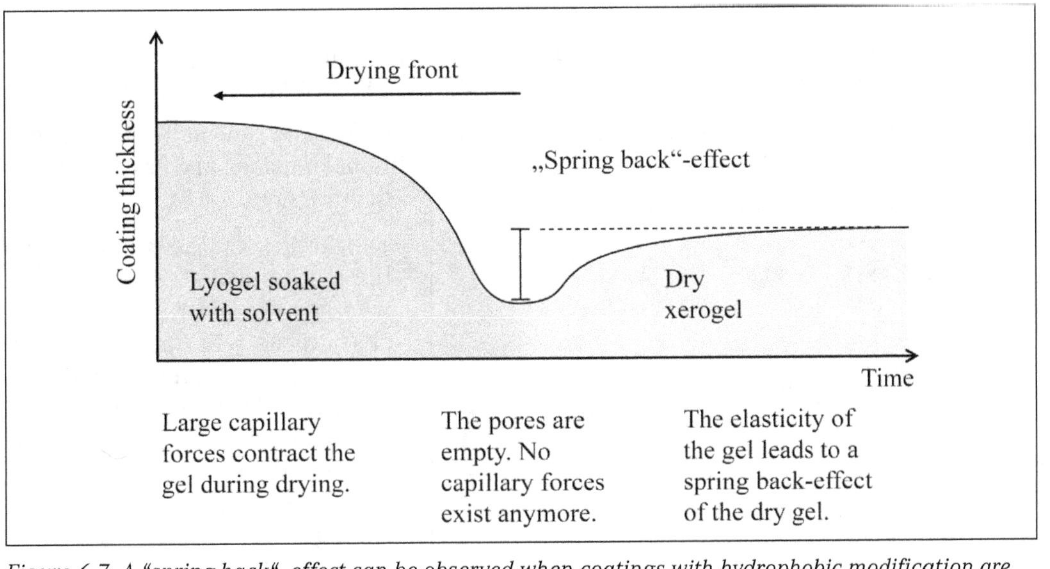

Figure 6.7: A "spring back"- effect can be observed when coatings with hydrophobic modification are dried

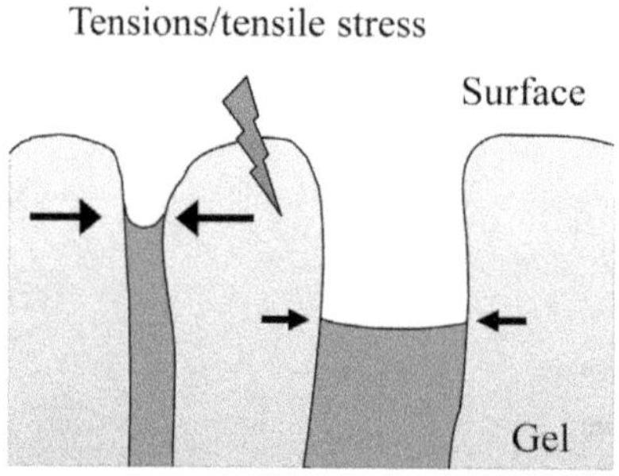

During drying, the gel shrinks and the capillary forces are causing large tensions. Big pores empty faster than small pores which additionally leads to a tension gradient.

Figure 6.8: Tensions develop during drying of sol-gel coatings if pores of different diameter empty at different speed

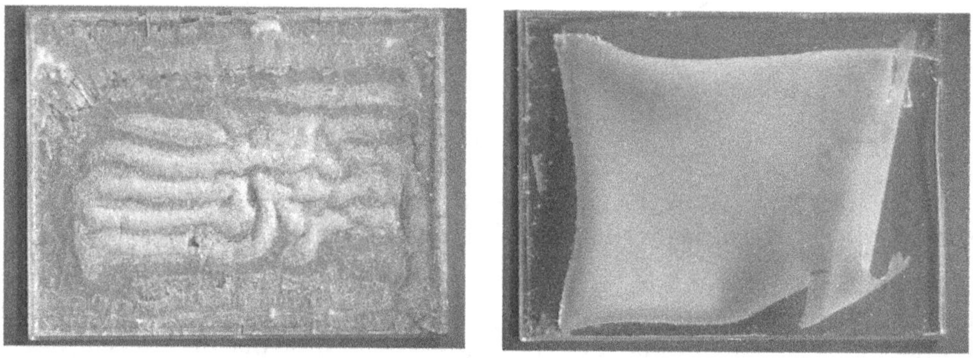

Dried (80 °C) dispersion of 10 nm SiO_2 particles in water. Capillary forces destroyed the film.

Dried (80 °C) film of a dispersion of surface-modified 10 nm SiO_2 particles. The elastic modification prevents cracks.

Figure 6.9: Comparison of dried colloidal silica dispersions with and without an elastic surface modification [220]

the capillary forces significantly and prevent a pore collapse during drying. This can even cause a so-called spring back-effect of the deformed gel, if the solvent leaves the pores [215].

Since bigger pores empty faster than smaller pores, the front of the solvent level in the gel is not even, but rather inhomogeneous, which causes additional tensions and cracks during drying (Figure 6.8).

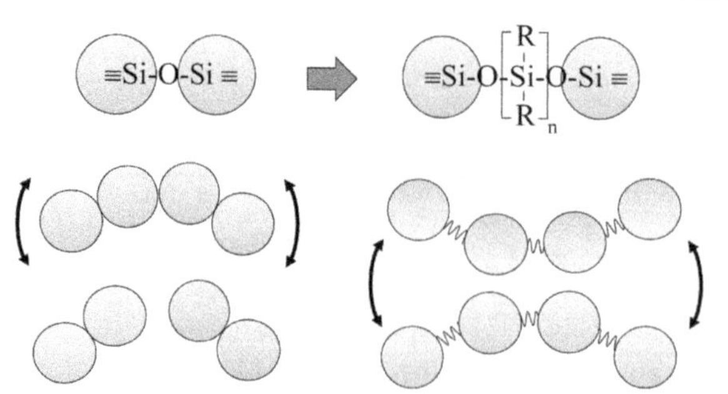

Figure 6.10: Silsesquioxane-bridges increase the flexibility of SiO_2 nanoparticle aggregates [220]

The ability of the gel to relax the mechanical stresses decides whether the gel remains intact or if cracks will appear. This is illustrated in Figure 6.9 and Figure 6.10. A layer of colloidal silica sol dries to a crumbling mass, whereas the same colloidal silica with a flexibilizing methyl polysilsesquioxane coating dries to a flexible film (compare Example 31, page 140).

After the removal of the solvent, usually a thermal densification step follows in order to improve the mechanical properties and the adhesion of a sol-gel coating. The maximum possible temperature is determined by the substrate and the most labile component of the coating material.

For polymer substrates, usually the transformation temperature (T_g) sets the highest possible densification temperature. So especially if scratch resistant sol-gel coatings shall be realized, a polymer substrate should be chosen which exhibits the highest possible thermal resistance.

If the xerogel is heated up, condensation reactions of the remaining free Si-OH groups take place and the gel shrinks further. The mechanical stress in this process step is inversely proportional to the already achieved packing density of the gel during the gelation and drying step.

Open, porous xerogels will have to shrink more than already densely packed xerogels (compare Figure 6.19, page 146). The mechanical stress during the thermal densification of SiO_2 coatings can easily reach 200 to 500 MPa, which exceeds by far the average strength of silica glass (approximately 50 MPa) [137].

These numbers reveal the difficulties in realizing thick, crack free sol-gel coatings. For almost pure inorganic coatings the critical thickness at which cracks appear is well below 1 μm [216]. Techniques how to realize thick crack free inorganic coatings will be discussed in the next chapter "High temperature resistant coatings".

6.3 Sol-gel coatings - examples

6.3.1 High temperature resistant coatings

High temperature resistant coatings are used e.g. as binders for glass wool or for coatings on metals. It is a special advantage of the sol-gel process, that glass-like or ceramic coating compositions can be deposited on various substrates starting from liquid precursors at moderate process temperatures. Thereby the material gap between enamels and silicones can be closed. The temperatures used for the production or during use of these coatings exceed by far the decomposition temperature of common organic compounds.

When the temperature starts to rise, residual water and solvents leave the xerogel and condensation of Si-OH groups starts. The gel is fixed to the substrate and can only shrink freely in a vertical direction [217-219].

The resulting tensions must be relaxed by the gel in order to avoid crack formation. However the mostly inorganic components only possess limited stress-relaxation ability and the acting forces exceed the mechanical resistance of the gel by far.

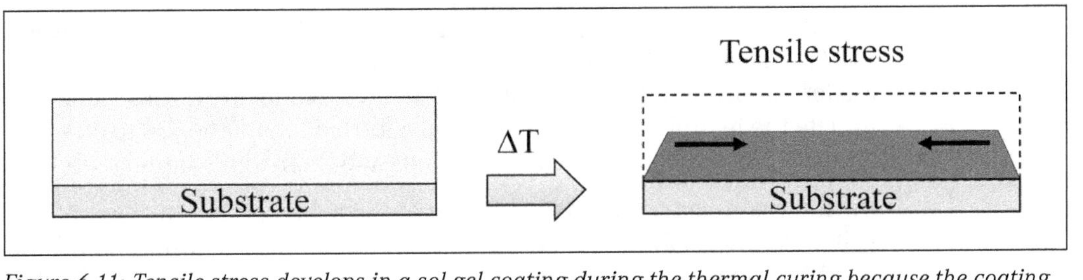

Figure 6.11: Tensile stress develops in a sol-gel coating during the thermal curing because the coating film is attached to the substrate and can only shrink in vertical direction

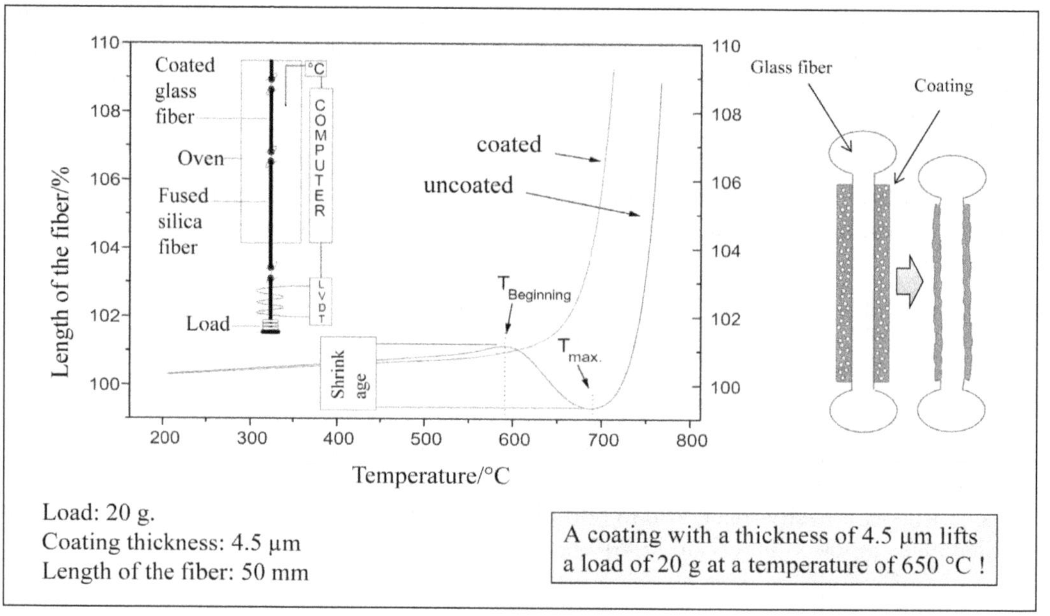

Figure 6.12: Demonstration of the forces which develop during the thermal shrinkage of sol-gel coatings with the help of a fiber elongation viscosimeter[220]

An unusual experiment can be used to visualize the magnitude of these forces during the thermal densification step [220, 221]. A fiber elongation viscosimeter is used to measure the T_g (transformation temperature) of glass. In a furnace, a weight is attached to a suspended glass fiber and its elongation at elevated temperatures is measured with a travel sensor.

The experiment compares uncoated and sol-gel coated fibers (material: Example 31, page 140, coating thickness: 4 to 4.5 μm). Whereas the uncoated glass shows the expected behavior and elongates at temperatures exceeding its T_g, the shrinking coating leads to a shortening of the coated fiber and thereby additionally lifts the weight of 20 g!

During the application and drying of inorganic sol-gel coatings, a nanoscaled porosity and a large inner surface is generated. Via its surface energy, this porosity contributes to the total energy content of the coating material.

During the thermal densification, this energy is released and reduces the necessary densification temperature in comparison to the non-porous bulk material. In the fiber elongation experiment, the glass fiber substrate can be deformed by the huge forces which occur during the shrinking of sol-gel coatings when its T_g is exceeded at approximately 550 °C.

➤ *Example 31: Preparation of thick crack free SiO_2 coatings* [222, 223]

A mixture of 20 ml methyltriethoxysilane and 6 ml of tetraethoxysilane is provided and under vigorous stirring, 15 g of colloidal silica sol ("Levasil 300/30", concentrated to 45 % by weight SiO_2) is added thereto. Following the formation of an emulsion (after about 20 s) 0.3 ml of HClc are added to initiate the hydrolysis. The mixture stays turbid for another 20 to 60 s and thereafter immediately turns first viscous and then highly liquid and clear.

During that reaction, the temperature of the sol increases to about 40 °C. Following the cooling to room temperature (optionally in an ice bath) the resulting sol is filtered through a filter of a pore size of 0.8 μm, preferably by using a prefilter of a pore size of 5 μm. The viscosity of the sol thus prepared may be adjusted to a desired value by addition of e.g. ethanol, propanol or other solvents and it may be used for coating for at least 6 hours.

Window glass is coated with the aid of an dip drawing apparatus with speeds ranging from 3 to 10 mm/s. The freshly drawn layers are dried at 60 °C for 15 minutes and are subsequently densified in a furnace according to the following temperature program: Room temperature to 400 °C with 1 K/min, 400 °C to 500 °C with 0.3 K/min. The thickness of the resulting coatings ranged from 2 to 6.5 µm. If metallic substrates are to be coated, preferably H_3PO_4 should be used as a catalyst instead of HCl to avoid corrosion.

Crack free SiO_2 coating

The synthesis of this coating material is an example of a sol-gel formulation without the addition of solvents. The two silanes which are used as educts are not miscible with water, therefore a two-phasic system is generated after the addition of the aqueous colloidal silica sol. HCl as mineral acid triggers the start of the hydrolysis at the boundary between water and silanes. As a consequence, ethanol is released and finally leads to a one-phasic homogeneous solution.

The advantage of this process is that the colloidal silica sol is not destabilized at the beginning of the reaction by the addition of a solvent. At the time when the solvent is released by hydrolysis, simultaneously reactive silanes with Si-OH groups are generated which can sterically stabilize the silica sol. As an alternative, ion exchanged colloidal silica sol at pH 2 (IEP) can be used, because it shows a higher compatibility with solvents (compare "Electrostatical stabilization", page 41). From an economical point of view however, the use of the silica sol in its delivery form is preferred.

The used silica sol is stabilized by sodium ions at an alkaline pH of about 8.5 to 9. In the first step of the reaction, the pH is rapidly changed to the acidic region and thereby travels through a pH region of minimal stability (approximately pH 4 to 7). This is the reason why the addition of the acid has to be done as quickly as possible under vigorous stirring to avoid gelation or agglomeration.

The acid generates a destabilizing salt together with the sodium ion of the colloidal silica. The formed sodium phosphate or -chloride however is hardly soluble in the reaction mixture, precipitates during the ice cooling step and can be removed by filtering.

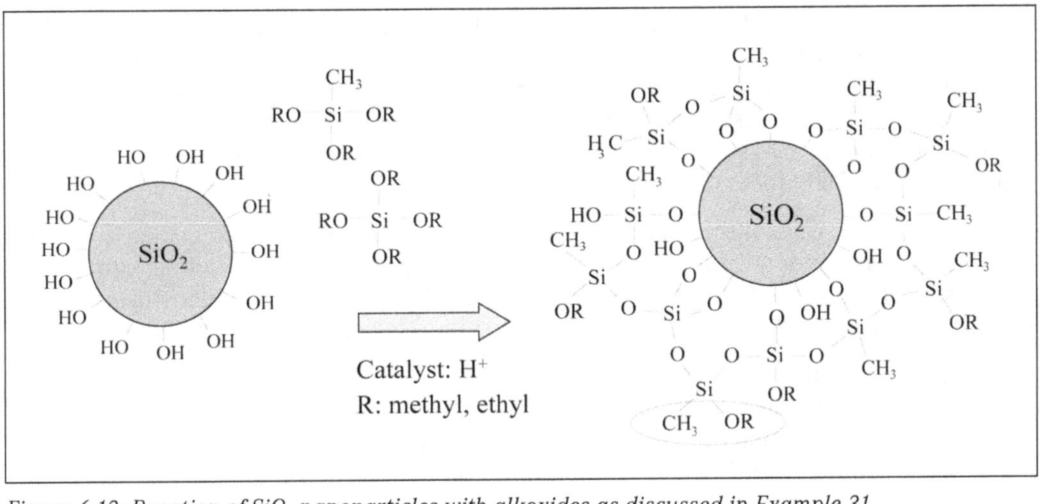

Figure 6.13: Reaction of SiO_2 nanoparticles with alkoxides as discussed in Example 31

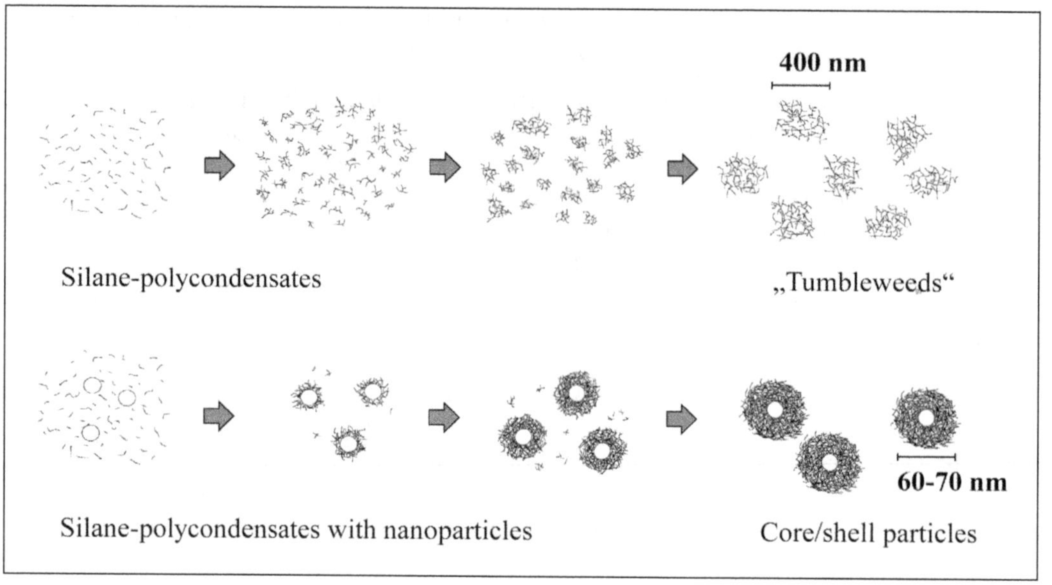

Figure 6.14: The growth of silane polycondensates is affected by the presence of nanoparticles. They lead to a denser packing of the polysiloxanes in proximity to their surface [220].

The acid should be chosen according to the application of the coating. Instead of HCl, e.g. also H_3PO_4 can be used for coatings on metals to help prevent corrosion.

If nanoparticles are present during the hydrolysis and polycondensation of silanes, the intermediately generated silanols are adsorbed on their large surface. The used colloidal silica sol exhibits a surface area of approximately 300 m^2/g and so preferably core/shell structures are generated and only few free siloxane oligomers are observed [220].

From dynamic light scattering measurements [220] it is known, that acid-catalyzed silane formulations form densely packed core/shell structures when nanoparticles are present, whereas in the absence of nanoparticles loosely packed agglomerates are generated.

The sol-gel formulation without nanoparticles shows 400 nm large structures and the identical sol-gel formulation with nanoparticles contains 60 to 70 nm small structures (composition: compare Example 31, page 140).

The synthesis is conducted in an acidic environment. This fosters the generation of branched chain and band structures. Without nanoparticles being present, these structures entangle and form loosely packed agglomerates, whereas in the presence of nanoparticles these structures grow on the large nanoparticle surfaces to much denser structures. So nanoparticles do not only act as a sort of inert filler for sol-gel formulations, but also direct the structure of the final product.

In Example 31, the critical thickness of a SiO_2 coating could be increased to approximately 5 µm by a small amount of organic modification ($-CH_3$) and the addition of nanoscaled particles. That the methyl groups still can act as flexibiliser at the high densification temperatures of up to 500 °C can be attributed to the stabilizing effect of the SiO_2 nanoparticles.

Figure 6.15 depicts the shift of the peak maximum of the methyl group decomposition in a DTA/TG measurement as a function of the SiO_2 nanoparticle content. In the experiment, a series of sol-gel formulations with increasing nanoparticle content was heated up to 1000 °C in a thermogravimetric analyzer. The maximum of the DTA peak which indicates the decomposition of the methyl group then was plotted against the nanoparticle content. Thereby the influence of the SiO_2 nanoparticles on the thermal decomposition of the methyl groups can be illustrated.

It can be seen that the peak maximum is shifted to higher temperatures with increasing nanoparticle content. Via ^{29}Si-NMR, density and PCS (photon correlation spectroscopy), it could be shown, that the siloxane structures grow around the nanoparticles instead of forming a separate particle fraction of loosely packed agglomerates (compare Figure 6.14, page 142) [87, 220].

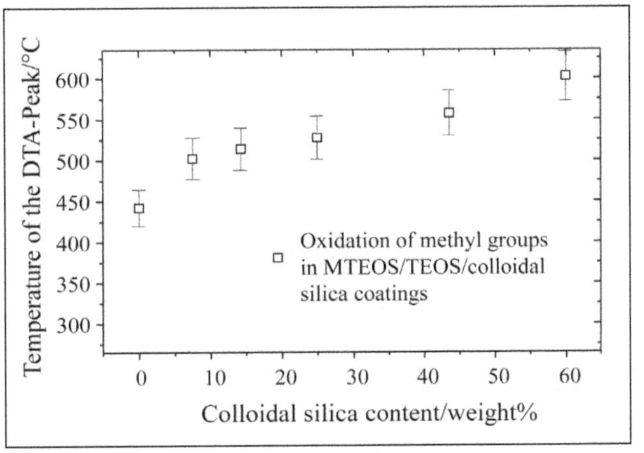

Figure 6.15: DTA/TG measurements prove: SiO_2 nanoparticles increase the thermal decomposition temperature of methyl groups in xerogels made of MTEOS/TEOS (methyltriethoxysilane/tetraethoxysilane) (Example 31, page 140) [87, 220]

The higher packing density of the polysiloxane network can hinder the access of oxygen to the organic components of the xerogel and thereby delay their oxidation. From investigation of the thermal decomposition of methyl-group-containing coatings in an inert gas atmosphere, it is known that the anaerobic decomposition starts at temperatures exceeding 600 °C [224].

These results are consistent with the result depicted in Figure 6.15 and support the hypothesis that it is the hindrance of the free access of oxygen which increases the thermal stability of the organic groups in the presence of nanoparticles.

Methyl groups are not the only organic compounds which can be protected against early decomposition at higher temperatures. In general, it can be observed that organic groups or molecules which are embedded in a sol-gel matrix exhibit a higher temperature resistance than in their free state. These effects can be enhanced by the addition of nanoparticles. It has been reported that e.g. the thermal decomposition temperature of fluoroalkyl groups in anti-adhesive coatings could be increased by over 100 K (compare "Easy-to-clean/anti-adhesive coatings", page 160) [220].

Factors which influence the critical thickness

A survey about the factors which influence the critical thickness of sol-gel coatings can be found in Figure 6.16.

Starting from a critical thickness of <1 μm for inorganic sol-gel coatings like those e.g. derived from TEOS (tetraethoxysilane), the crack formation can be suppressed by a variety of measures.

A formulation consisting of methyltriethoxysilane, tetraethoxysilane and colloidal silica (Example 31, page 140) can be densified at 500 °C on glass up to a coating thickness of approximately 10 μm [220].

The reasons for this extraordinary high critical thickness can be found in Figure 6.16. Three of four major factors which influence the critical thickness are stresses which are generated during the application and densification and which cause cracks and delamination. The avoidance and relaxation of stresses in the different phases of the application of a coating is of utmost importance in order to realize thick and crack free coatings.

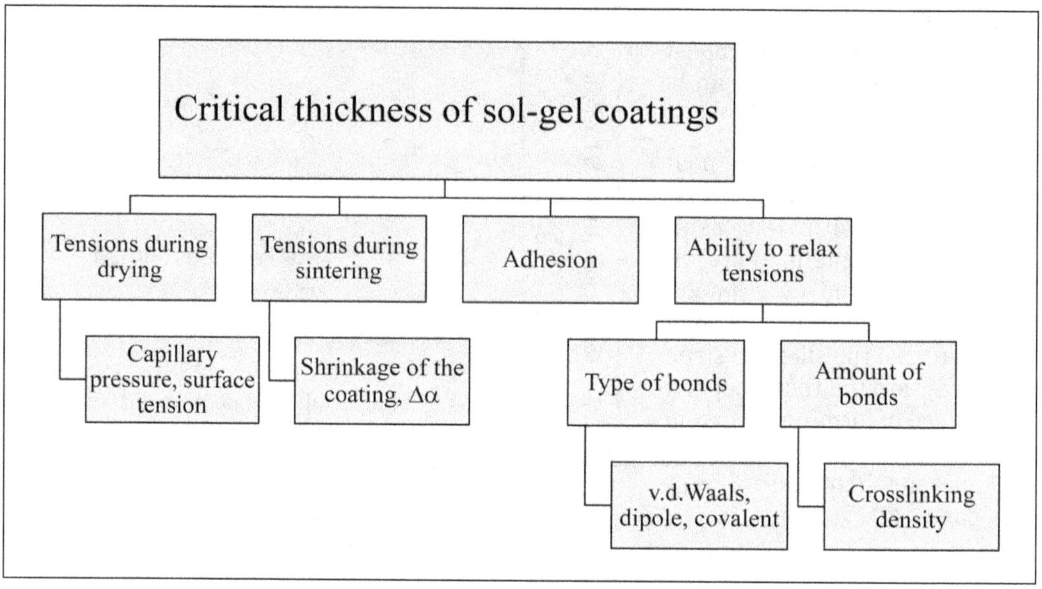

Figure 6.16: Survey of the factors which influence the critical thickness of sol-gel coatings

The stresses develop in sol-gel coatings, starting from the point at which mechanical properties can be measured in a gelling sol film. After this point, the capillary forces and the shrinking of the network determine the extent of the development of stress.

The higher the solid content of the sol during gelation, the lower is the amount of shrinkage which happens during drying and thermal densification. All factors which influence the solid content at which a sol gels, also influence the extent of stress which is generated in the following steps of sol-gel processing. SiO_2 nanoparticles increase the density of the sol-gel particles and thereby have a positive influence on the critical thickness.

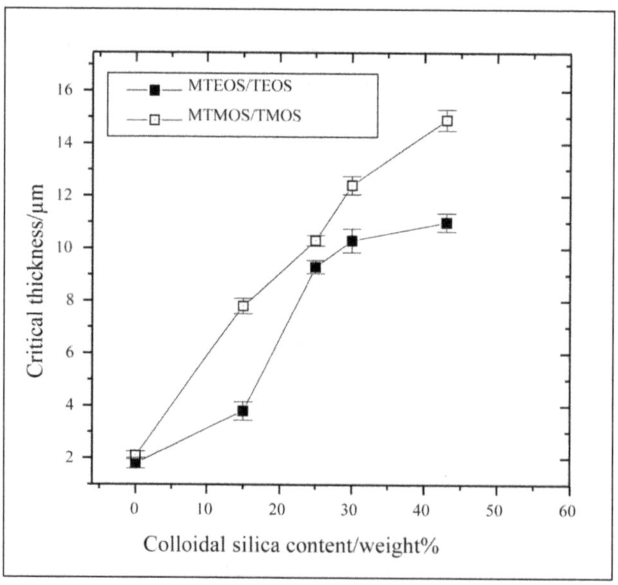

Figure 6.17: Influence of collodial silica on the critical thickness of coatings made from a mixture of 80 mol% of methyltriethoxysilane and 20 mol% tetraethoxysilane. Curing at 500 °C, substrate: glass[220].

The capillary forces in the lyogel (gel soaked with solvent) are determined by the interaction of the pore wall with the liquid. Methyl groups can modify the pore walls to be hydrophobic and in general lead to a reduced surface energy. Thereby they reduce the interaction between pore liquid and pore wall and decrease the mechanical forces which act on the pore walls.

A good adhesion is a prerequisite for achieving a high critical thickness. The pre-treatment of the substrate should be chosen accordingly.

The decisive factor to prevent cracks still is the ability to relax the inevitable stresses during the thermal densification. At higher temperatures, the coating shrinks at a point in time at which many covalent bonds already have been formed and the ability to relax stresses diminishes quickly. Organic groups which help to flexibilize the network decompose more and more, which leads to an increased brittleness.

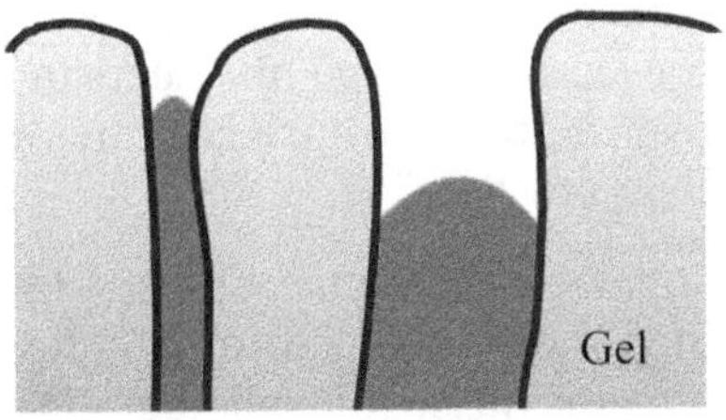

Pore walls with a hydrophobic treatment

Figure 6.18: A hydrophobic coating decreases the capillary forces which act upon the pore walls and helps to prevent crack formation (compare Figure 6.8, page 138)

In Example 31, the methyl group introduces enough flexibility up to temperatures of 500 °C to yield an excellent high critical thickness. This is only possible, because the high content of SiO_2 nanoparticles stabilizes the methyl groups at the high temperatures and prevents their early oxidation.

By using methoxy- instead of ethoxysilanes, even higher critical thicknesses are possible. Due to the smaller methoxy groups, a higher solid content is possible at the gelation point which means less shrinkage is necessary in the following steps.

The higher critical thickness of the methoxy silane formulation brings with it a short pot life. Whereas the formulation consisting of ethoxy silanes can be used for hours, the methoxy silane formulation after only 30 minutes shows a drastic reduction of its critical thickness.

➤ *Example 32: Preparation of a self-supporting SiO_2 glass film* [225]

A mixture of 0.92 mol methyltriethoxysilane, 0.08 mol tetraethoxysilane and 0.25 mol colloidal SiO_2 ("MA-ST", produced by Nissan Chemicals; 30 % SiO_2 in methanol) is diluted with 4.5 mol ethanol and hydrolyzed by adding 4 mol aqueous ammonia solution (0.34 g of 25 % aqueous ammonia in 73 g of water) under magnetic stirring. Stirring then is continued for another 3 minutes.

Upon aging the reaction mixture at 80 °C in a closed vessel for 4 days, a viscous sol with a viscosity of 20 to 25 Pa·s is obtained. That sol is cast onto polystyrene molds. The mold is covered and kept in an oven at temperatures ranging between 40 to 65 °C for 15 h. During this period of time, the viscous sol gels and an intact gel film can be removed. The film is being dried at 20 to 30 °C for 3 h and subsequently at 50 °C for 15 h.

The gel films produced in that manner may be sintered in air at temperatures of up to 400 °C regardless of their thickness, which results in intact transparent glass films. Gel films having a thickness below 50 μm may be sintered in air at up to 1000 °C without crack formation to form transparent glass films. Thicker films may be sintered in nitrogen between 400 to 1250 °C resulting in mechanically and thermally stable black glass films. The thermo-mechanical properties thereof approximately correspond to those of pure silica glass.

If sintered in an ammonia atmosphere at temperatures of up to 1000 °C, glass films of excellent transparency may be obtained which exhibit better chemical and thermal stability and higher breaking strength, microhardness and temperature-resistance than pure silica glass.

Self-supporting SiO_2 sheet

During the processing of coatings and self-supporting sheets, the avoidance and relaxation of stresses is a critical point. Sheets in contrary to coatings are not fixed to a substrate. The drying starts from the surface of the wet film and causes shrinkage which curls the sheet and causes cracks.

The example describes measures to lower the stresses which occur during the drying and shrinking of the coating/film. The formulation consists to a large extent of the relatively flexible methylsilsesquioxane building block and only of a small portion of tetraethoxysilane as a cross-linker. Organosilica sol particles act as inorganic filler and seed for the growth of the polysilsesquioxane network.

By the use of ammonia as a volatile catalyst, dense spherical particles are generated, which show fewer pores than particles which have been synthesized under acidic conditions. Identical formulations which are catalyzed with acids do not dry to an intact foil but rather curl and crack.

During the thermal treatment, the methyl groups begin to decompose at temperatures exceeding 350 °C and a SiO_2 foil is generated. Pigments, dyes or other components like dopants can be added to modify the foil according to the requirement profile.

The shrinkage during drying is dominated by the morphology of the sol components. Figure 6.19 illustrates the difference between a sol-gel formulation consisting of dense spheres and a formulation made of entangled, branched structures (compare "Hydrolysis and condensation", page 94) [226].

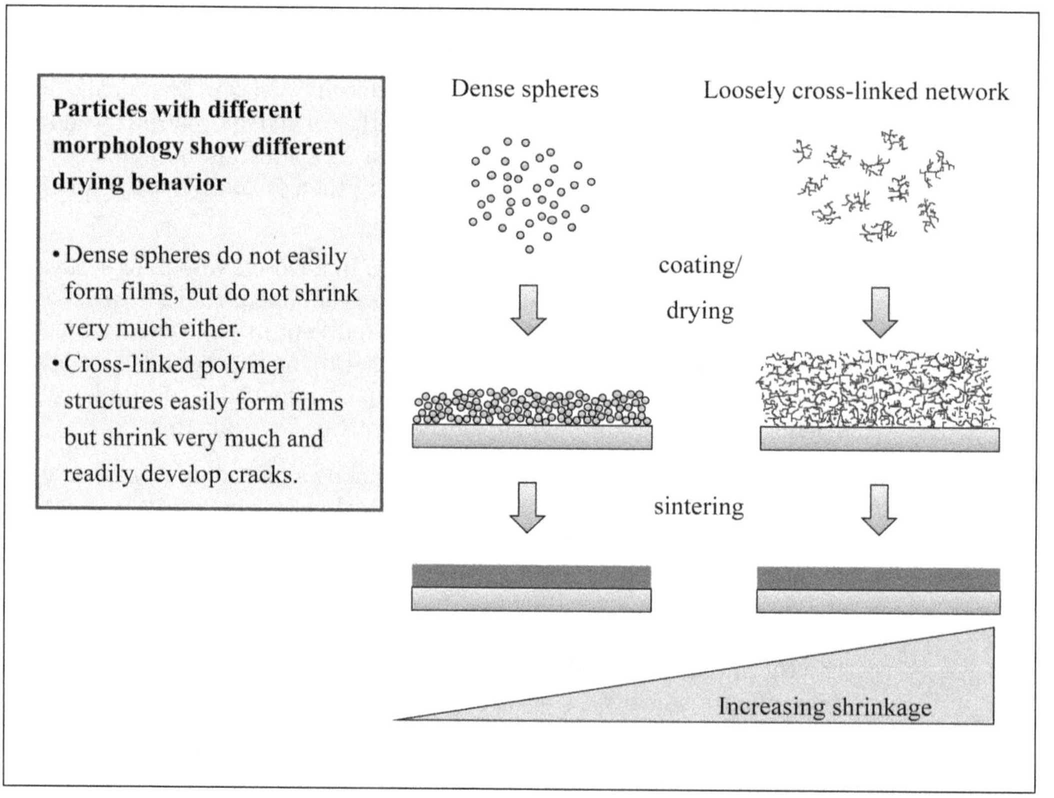

Figure 6.19: Particles with different morphology show different drying behavior. Polymeric structures show a tendency to shrink much and to develop cracks.

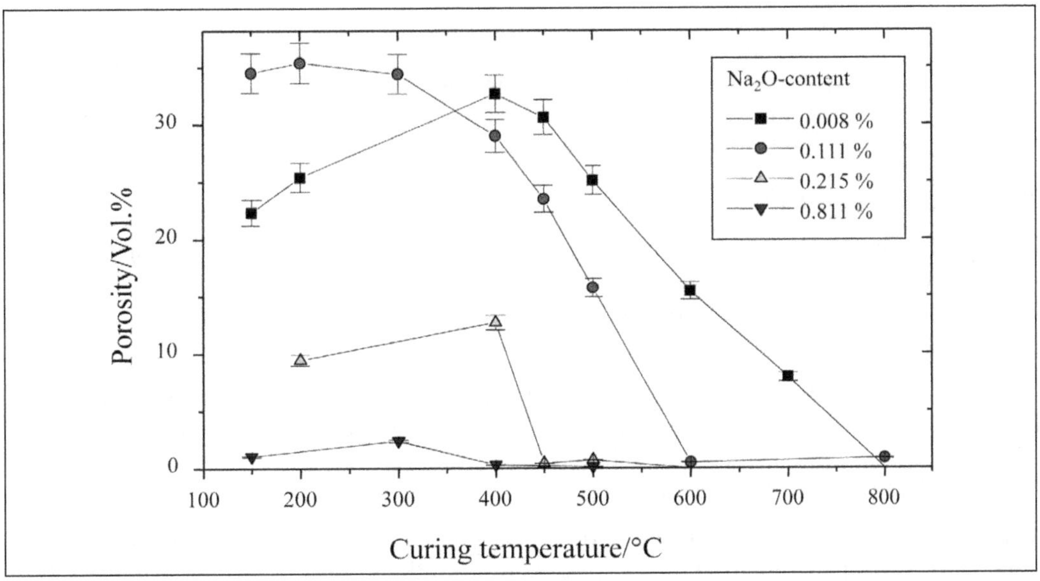

Figure 6.20: Development of the porosity of xerogels made from MTEOS/TEOS as a function of Na_2O content and densification temperature. Sodium-free xerogels still show porosity even after densification at 700 °C, whereas only 0.1 % sodium oxide leads to dense silica after a treatment at 450 °C [220].

When applying the formulation of Example 31 on glass with coating thicknesses larger than 2 µm, it can be seen that adhesion problems occur at densification temperatures between 200 and 350 °C, whereas at higher or lower temperatures the coatings adhere well [220].

Like already discussed, sol-gel coatings bond to the glass surface at lower temperatures by hydrogen bridges and at higher temperatures Si-O-Si bonds are generated (compare Figure 6.2, page 134). In the area are of 200 to 350° the number of polar Si-OH groups and adsorbed water molecules is reduced at a time, when not enough Si-O-Si bonds with the surface are generated, leading to adhesion failure.

In order to achieve good adhesion in this temperature range, another adhesion principle has to be used. From literature it is known, that Na^+ ions form bonds to Si-O groups which show a comparable strength like Si-O-Si bonds [199]. Furthermore, alkali ions act as strong sintering additives at higher temperatures like can be seen in Figure 6.20 [199, 197, 220, 227, 228].

Whereas sodium-free compositions even after densification at 700 °C still show some porosity, even a small amount of sodium ions of only 0.2 % leads to a complete reduction of porosity after a densification temperature of 450 °C.

Sodium ions therefore, even in small concentrations, act as adhesion promoters and catalyst of the Si-O-Si bond formation. With an addition of 0.3 to 0.5 % Na_2O even after densification at 200 to 350 °C well adhering SiO_2 coatings were realized. If the amount of alkali ions is increased further, organo alkali silicates are generated, which can also be used as inorganic binders or coatings.

➤ *Example 33: Preparation of a sodium silicate coating sol* [229, 230]

25 ml (124.8 mmoles) of methyltriethoxysilane (MTEOS) are stirred with 7 ml (31.4 mmoles) of tetraethoxysilane (TEOS) and 0.8 g (20 mmoles) of sodium hydroxide overnight (for at least 12 hours) at room temperature until all of the sodium hydroxide has dissolved and a clear yellow solution is formed.

Subsequently 3.2 ml (177.8 mmoles) of water are slowly added dropwise at room temperature, resulting in a rise of the temperature of the solution. After the addition of water is completed, the clear yellow solution is stirred at room temperature until it is cooled down again and is subsequently filtered through a filter having a pore size of 0.8 µm.

Thick organo-alkali silicate coatings

Example 33 again shows that sol-gel processes can be conducted without adding a solvent. Sodium hydroxide reacts as solid educt with silanes to sodium alkoxysilanoates, a molecular precursor ("single source precursor") of the coating material, which in the second step is reacted with water to the final product.

"Single source precursors" are molecules which contain all components of the final product in one molecule. Their synthesis is especially useful, if the reaction speed of the components or the conditions under which precipitation occurs differ to a large extent and inhomogeneity result during the traditional sol-gel process.

Via a reaction with water in a single reaction step, the targeted product composition is generated. The added amount of water is smaller than theoretically necessary to hydrolyze the alkoxygroups of the formulation. The resulting coating solution therefore is stabilized against gelation by residual Si-OR groups.

During the coating step, humidity serves as a source of water. Residual Si-OR groups in the coating later can cross-link during the thermal densification or via the slow uptake of water from the surrounding air (compare "Water-free sol-gel techniques", page 100).

The very high critical thickness of the coating material (approximately 5 µm) can be explained by the flexibilizing methyl groups and the residual Si-OR groups of the formulation. The high alkali content of the coating speeds up the thermal densification of the SiO_2 network so that the methyl groups are hermetically enclosed from the surrounding air and protected against early oxidation. The alkali silicate coating can be used as well for anti-tarnish coatings on metals since the coating densifies at temperatures far lower than necessary to form the colored oxidation products of the metal surface.

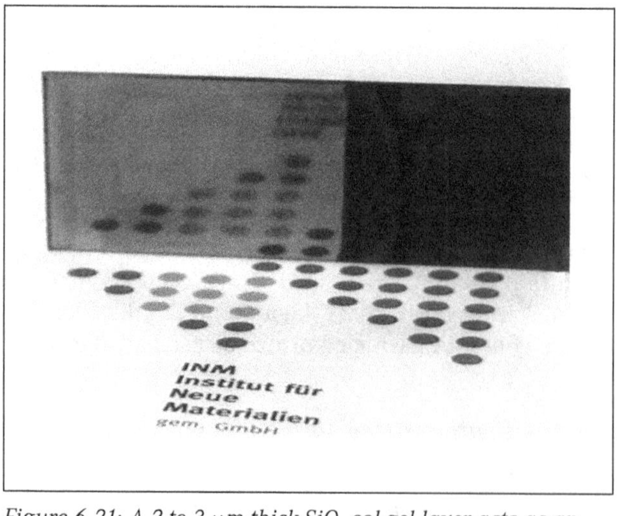

Figure 6.21: A 2 to 3 µm thick SiO_2 sol-gel layer acts as an effective anti-tarnish coating on steel. Even if the steel is heated up to > 1000 °C its surface remains unchanged

Source: INM Institute for New Materials [87, 220]

It is obvious that this type of coating material applied on stainless steel leads to shiny coatings with almost unchanged color of the steel, whereas alkali-free coatings (compare Example 31, page 140) lead to discolored steel parts at densification temperatures exceeding 300 °C. The alkali-silicate coating material can also be used as a primer and anti-tarnish protection for other coating materials, if the solid content is reduced appropriately [87, 220].

The molar Si/Na ratio is 7.75:1 which is significantly higher than in conventional aqueous alkali silicates (2.5 to 4 : 1, "waterglass"). Therefore, the chemical stability of the SiO_2 coating is not affected in a negative way. The formulation of the example can be used as functional coating with additional pigments, nanoparticles or functional silanes, e.g. with anti-adhesive properties (compare "Easy-to-clean/anti-adhesive coatings", page 160).

High temperature resistant coating

Besides its application as binder resin for high-temperature coatings, the formulation of Example 31 can also be used as protective coating for glass.

Figure 6.22 depicts an experiment, in which a coated and a non-coated microscopic glass slide was treated with gas burners. After a few seconds, the uncoated glass melts whereas the only 3 μm thick coating stabilizes the glass to such an extent, that even with two burners after 10 minutes the glass is still stable [220, 231].

By varying the composition of Example 31, a binder for glass wool was synthesized, which can be used as an alternative to phenolic resin. Products have been introduced into the market as formaldehyde-free insulating materials e.g. for kitchenware and trains [87].

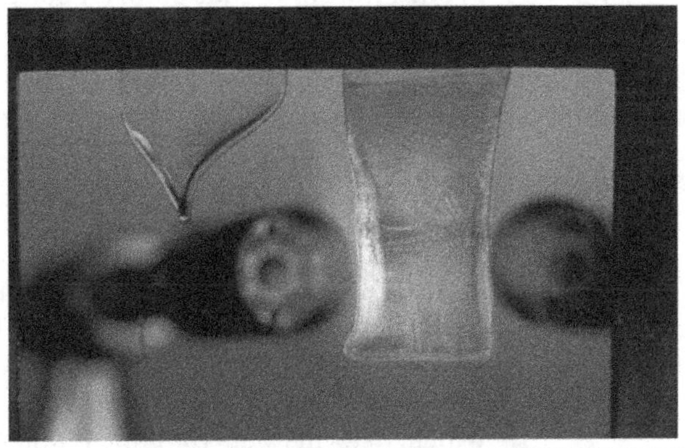

Figure 6.22: Transparent fire resistant coating for glass

Figure 6.23: Glass-wool based insulating boards manufactured with an inorganic sol-gel binder instead of the usual phenol-formaldehyde resin [87]

6.3.2 Colored and pigmented sol-gel coatings

6.3.2.1 Inorganic pigments

Like it has been shown in chapter "Application, drying and densification", starting with page 133, sol-gel coatings cannot be applied in comparably high coating thicknesses like conventional organic coatings.

Especially for formulations which contain a high percentage of inorganic components, typically 1 to 10 µm is the limit to achieve flawless, glossy surfaces. This limits the choice of possible pigments significantly. As a general rule, the size of pigments should not exceed 1/10 of the coating thickness to avoid coating defects.

➤ ***Example 34: Production of a gold-colored coating material*** [232]

250 g of a pearlescent pigment mixture consisting of "Iriodin 323" and "Iriodin 120" (Merck KGaA) in a 1:1 ratio was suspended in 3 l of ethanol and ground in a rotary ball mill at 1500 rpm for 3 h. Subsequently, the grinded pigment mixture was added to the sodium silicate coating sol synthesized in Example 33. The pigment content was 3 % by weight based on the coating sol.

Coating of metallic surfaces: This pigment-containing coating sol was used to coat stainless steel plates ($10 \cdot 10$ cm^2) by a manual spray coating process. After evaporation of the solvent at room temperature, the coating was densified at 475 °C. A shiny champagne-colored coating was obtained, which, in terms of its roughness and its sliding behavior, did not differ from a pigment-free coating produced in the same way.

Pigmented high-temperature coating

In general, sol-gel coatings are applied at thicknesses below 10 µm. Coatings which consist of mostly inorganic components even are applied at coating thicknesses below 2 µm in order to avoid crack formation.

If those coating formulations are to be pigmented, the choice of suitable pigments can become a big problem, because many pigments exhibit dimensions of some micrometers in at least one direction. Those pigments cannot be enclosed completely by the coating material and cause an uneven coating surface.

As a general rule, the diameter of a pigment should not exceed 1/10 of the coating thickness to realize high gloss. In this example, the by far bigger mica pigments are milled to reduce their size. This also causes a significant change of their optical properties, which in this example however still leads to an acceptable result. If possible, especially anisotropic pigments like metal flakes or pearlescent pigments should not be milled to maintain their appearance. The result has to be assessed individually.

The essence of this example however lies in the combination of an inorganic pigment with a flexible inorganic sol-gel resin, which is able to bind the pigments. The alkali silicate has a comparatively low alkali content, so that at the high temperatures, the sensitive pearlescent pigments are not dissolved. Due to its nanoscaled structure, the binder can be densified at significantly lower temperatures than one would expect when considering its chemical composition.

The coating consists of nanoscaled components which have a very large surface area. The high amount of surface energy is released during sintering and contributes to lower the temperature which is necessary to densify the coating (compare Figure 6.12, page 140). Thereby glass- or ceramic-like coatings can be applied at temperatures far below the melting temperature of their corresponding bulk materials. These coatings can serve as host for manifold functions like e.g. color, water repellency, catalysis, etc.

If surfaces which are sensitive to oxidation are to be coated, it is beneficial if the thermal densification above 200 °C is conducted in an inert atmosphere to avoid coloration of the substrate. The organo-alkali silicate which is used in the example, in this point shows a unique advantage compared to acid-catalyzed sol-gel formulations.

The high pH value suppresses rust formation on steel substrates and during thermal densification, the alkali ions catalyze the sintering of the coating so effectively, that the coating becomes gas-tight before oxygen can cause a coloration of the substrate (compare Figure 6.20, page 147) [220].

➤ ***Example 35: Black, high temperature resistant coating for aluminum foil*** [233]

10.89 g 1 % H_2SO_4 is added to 17.80 g (0.1 mol) methyltriethoxysilane under stirring. After 1 h, 24.73 g isopropanol is used to adjust the solid content to 10 % by weight. Subsequently,1 g zirconiumacetate solution (10 % in water) is added as a catalyst. To the so produced sol-gel resin, 2.02 g carbon black "FW200" is added and dispersed by using a homogenizer. The coating is applied by doctor blade onto aluminum foil in thicknesses between 250 nm and 3 μm.

The dark coating material improves the heat transfer through the aluminum foil. The coated foil therefore can be used especially for barbecue. The coating adheres very well in the temperature region of 200 to 400 °C. The foil even can be scrunched without causing delamination or cracks.

Black, high-temperature resistant coating for aluminum foil

The coating of a metal foil with a high temperature resistant coating really is a challenge. Only inorganic compositions have the temperature resistance which is needed, but due to their brittleness, they easily come off and crack if the foil is deformed.

If organic components of the formulation oxidize at higher temperatures, voids are formed and the network loses its mechanical stability (compare Figure 6.22, page 149).

Methylsilsesquioxanes contain the $CH_3SiO_{1.5}$ building block, which is one CH_3 group less than the -$(CH_3)_2SiO$- building block of silicones. They still are flexible, but show higher temperature stability than silicones. Phenyl- modified silicones are also temperature resistant, because the stable phenyl group decomposes at even higher temperatures than the methyl group, but the void which is generated when they decompose is much bigger.

In the example, zirconium acetate is used as a catalyst. Zirconium ions act as condensation catalyst and adhesion promoter. They often are used together with aluminum ions as adhesion primers for coatings and adhesives [234].

Metal colloids

Metal colloids due to their size and their very high absorption coefficient are especially suitable to color sol-gel coatings (compare "Nanoscaled pigments", page 73). Like already discussed in chapter "Metal compounds in the sol-gel process", starting with page 103, salts and/or organometallic precursors can be used as educts to form metal colloids. With the help of suitable solvents and complex formers, a homogeneous solution can be formu-

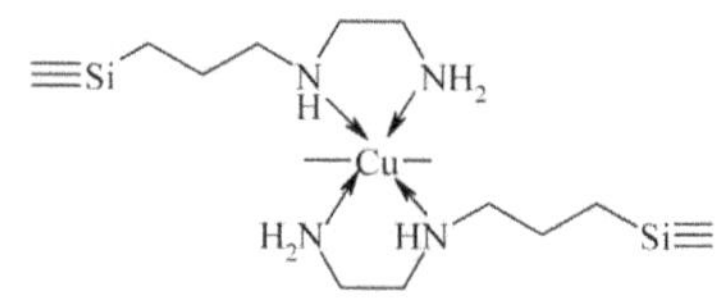

Aminosilane-copper complex

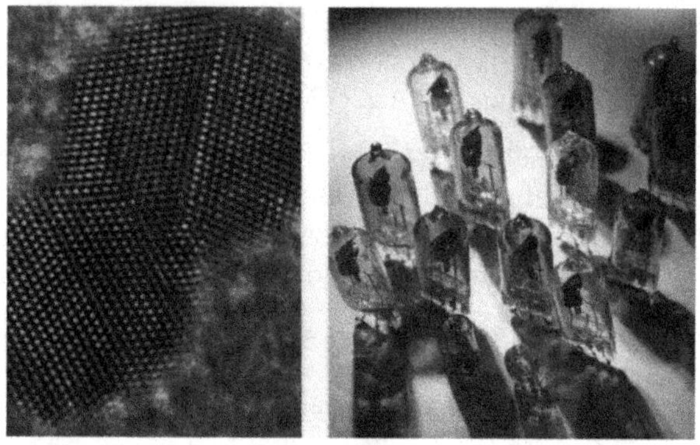

Bildquelle: INM Institut für Neue Materialien gem. GmbH

Gold and silver nanoparticles in a sol-gel matrix

Figure 6.24: Automotive headlight lamps colored with highly temperature resistant sol-gel coatings. Metal nanoparticles of gold, silver and platinum lead to transparent colors with high stability

Source: INM Institute for New Materials GmbH

lated. Metal colloids of noble metals can be generated, if the coating is heated up or irradiated with UV light.

➤ *Example 36: Preparation of colored lamp coatings* [235]

Experiment 1 "Blue": 1.278 g of tetrachloroauric acid ($HAuCl_4 \cdot 3\ H_2O$) are dissolved in 20 g of a colloidal ZrO_2 solution ("ZrO_2-Ac", The PQ-Corporation, Nyacol Products Inc.) containing 20 % by weight of ZrO_2 modified by acetic acid. After stirring for 15 minutes, the sol is ready for coating. In a dip coating procedure, the substrates are drawn from the coating sol at a rate of 2 mm/s. The densification of the layers takes place at 900 °C for 60 minutes. The heating rate is 60 to 100 K/h. The resulting layers show an intense blue color.

Experiment 2 "Red": 2.4 g of N-(2-aminoethyl-3-aminopropyl) trimethoxysilane and 2 g of lead (II) acetate ($Pb(CH_3COO)_2$) are dissolved in 12 g of a colloidal SiO_2 dispersion ("VP-AC 4038", Bayer AG) containing 30 % by weight of ammonia-modified SiO_2.

To this mixture is added a solution of 1.4 g of gold (III) chloride in 3 ml of water to which, additionally, 1 ml of diethylenetriamine has been added. After stirring for 15 minutes, the sol is ready for coating. In a dip coating procedure, the substrates are drawn from the coating sol at a rate of 1 mm/s. The densification of the layers takes place at 600 °C for 60 minutes. The heating rate is 60 to 100 K/h. The resulting layers show an intense red color.

Colored coatings for lamps

Bulk glasses can be colored by introducing the ions of e.g. Fe, Co, Cr or Se. The intensity of such colors however is by far too weak to be seen in a thin coating. Colloids of noble metals in contrast show extremely strong colors and are used since centuries to fabricate gold ruby glass [115, 236].

In sol-gel coatings, metal colloids can be generated by the reaction of soluble metal compounds with the organic components of the sol [237, 238]. These coatings can even simulate a bulk colored glass. The necessary stability and solubility of the metal compounds is usually realized by complexing agents (compare "Metal compounds in the sol-gel process", page 103).

Additionally, the sol-gel process offers the possibility to vary the refractive index of the matrix by adding high and low refractive components (compare "Tailoring of the refractive index of sol-gel coatings", page 186).

The variation of the refractive index leads to a shift of the absorption band of the metal colloids and thereby to a shift of their color. This can be seen in the examples in which gold colloids are generated in both formulations, but a blue color is yielded in titanium dioxide and a red color in the silicon dioxide. In general many metal colloids can be used to generate colors, but only the noble metals can be easily reduced and are stable enough against oxidation to be used for practical applications.

6.3.2.2 Organic pigments and dyes

Concerning their pigment size, the same considerations are true for organic pigments like for inorganic pigments. Organic pigments and dyes show an intrinsically worse temperature and chemical stability than inorganic pigments. This can be improved significantly by the incorporation into a sol-gel matrix.

The famous "Mayan blue" for example is synthesized by the encapsulation of indigo with a layered silicate and withstood centuries in mayan art, whereas "naked" indigo quickly fades [239]. Without doubt this can be regarded as an example of a technical use of nanotechnology, which can be transferred to other pigments and dyes [240].

The encapsulation in a dense, almost inorganic sol-gel matrix can improve the resistance of pigments and dyes against acids or thermal decomposition (compare Figure 6.15, page 143) [241]. Organic dyes can be adsorbed onto exfoliated nanoclays and show an outstanding UV resistivity. Even methylene blue, which in its free form is a very UV-sensitive dye becomes almost UV-stable when adsorbed and encapsulated by layered silicates ("Planomer-concept") [242].

Not always smooth and glossy coatings are desired. If bigger pigments are used, they lead to a matting effect and a roughening of the surface. In combination with high temperature resistant inorganic sol-gel formulations, so the appearance of sandblasted glass surfaces can be imitated. A dispersion of titanium dioxide particles with a tailored size distribution has been used together with the resin of Example 31 (page 140) to simulate the effect of a white flashed glass [243].

A homogeneous matting of the thin sol-gel coatings places high demands on the matting agent and the processing of the coating. Finely grinded aerogels ("Cabot nanogel") [244] have been proposed for this task. But also with conventional matting agents like pyrogenic silica, homogeneous silky surface finishes can be generated, which find application as antiglare coating for picture frames [245].

6.4 Structured sol-gel coatings

Not always coatings should be smooth. In some cases, a defined macroscopic or microscopic structure is desired to realize special requirement profiles.

Whereas their thermoplasticity can be used to structure polymers, sol-gel formulations are duroplastic materials due to their high cross-linking density. A shaping processing step therefore has to take place before cross-linking is complete. After gelation took place, sol-gel formulations still contain many Si-OH groups and condensation is not much advanced.

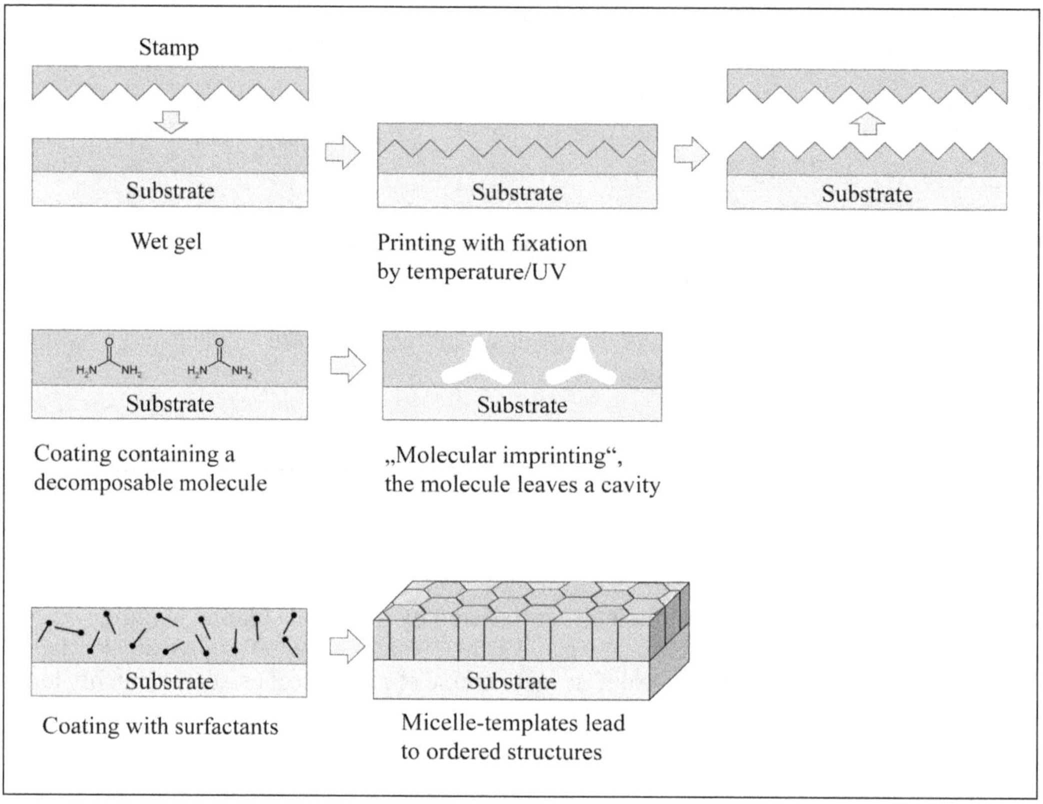

Figure 6.25: Methods to structure sol-gel coatings

Cross-linking in those gels is dominated mainly by hydrogen bridges which can be detached under mechanical load.

The condensation in contrary to hydrolysis is strongly dependent on the temperature. Without strong catalysts and higher temperatures, the Si-OH groups react only slow. A shaping can take place and the resulting structure can be fixed by thermal annealing. Especially formulations with sterically demanding substituents like phenyl- or cyclohexyl- are suitable for this approach. The parallels to the development of sol-gel powder coatings are obvious (compare "Sol-gel powder coatings", page 127).

Multistep curing processes can be realized with organofunctional silane-based coatings. Besides the inorganic cross-linking, also organic cross-linking reactions can be used. A structuring or shaping can be done prior to the final curing step. An example for a deep drawable coating formulation which uses this principle is given on page 156 (Example 37).

It has been described, that sol-gel formulations like Example 31 can be structured by embossing immediately after the dip coating step to yield holograms [246]. Via thermal densification, this hologram can be turned into SiO_2 glass and serves as a security feature. If an embossing step is not possible, the sol-gel coating can also be cured on a structured non-adhesive surface and the resulting film can be transferred to a substrate in a second step [247].

Coatings can be micro-structured also by a deliberate phase separation. Surfactants, especially cationic tetraalkyl ammonium compounds organize themselves into defined phases (lamellar, cubic or hexagonal) as a function of their chain length and addition level. If these phases are fixed by a gelled sol-gel matrix, they lead to micro-structured coatings and bulk materials [248–250].

These materials are also known under the term "mesoporous silica". Depending on the template used, characteristic structures are formed, which have been named according to their production sites MCM (Mobile crystalline materials) or SBA (Santa Barbara amorphous type), followed by a number e.g. MCM-41, SBA-15.

A similar type of structuring is the technique of molecular imprinting. Soluble organic compounds which can be removed in a later step of the processing lead to the formation of an imprint in the coating material during gelling. After their removal, e.g. by a thermal curing step, the imprint is fixed in an inorganic matrix. These structured materials find use as sensors or catalysts [251].

By template molecules, a broad variety of microstructures can be realized in sol-gel materials. Still the prediction of their properties and structures is very difficult, but a growing database of structure-property relationships helps to realize a specific requirement profile.

6.5 Scratch resistant coatings

Sol-gel coatings excel by their outstanding scratch resistance compared against conventional coatings. Their high content of inorganic moieties leads to a high degree of cross-linking and a higher hardness compared to organic polymers.

The inorganic and organic domains exhibit dimensions in the lower nanometer range and therefore do not scatter visible light. So the highly transparent inorganic/organic composites are used as protective and functional coatings for sensitive transparent substrates like polymers (PC, PMMA, ...). The examples which will be discussed in this chapter are also suitable as a base formulation for the integration of further functions like anti-adhesion, easy-to-clean, anti-fogging or anti-corrosion which will be discussed in the following chapters.

In the car industry, the replacement of glass by transparent polymers is pushed forward because the polymers do not form splitters at a crash and contribute much more and longer to the safety and integrity of the passenger cabin in the event of a crash. On top of that, the lower weight and more freedom in the design are further arguments in favor of the use of polymers in car manufacture.

A decisive disadvantage of transparent polymers however is their sensitivity against mechanical abrasion and UV radiation.

One of the first successful sol-gel formulations for the protection of polycarbonate is based on a combination of a methyl-modified silane and colloidal silica sol. It proves, that even without a flexibilizing organic network it is possible to realize a tough scratch resistant coating from almost inorganic components [252, 253, 254]. Example 37 discusses a possible embodiment of this principle.

Figure 6.26: Scratch resistant nanocomposite coating on polycarbonate. The coated company name remains intact, whereas the unprotected polycarbonate sheet is scratched.
Source: Merck KGaA

➤ *Example 37: Preparation of a thermally curable scratch resistant sol-gel coating for polycarbonate* [255]

A colloidal silica filled organopolysiloxane top coat composition containing 37 % solids, 50 % of which are SiO_2 is formulated by adding a commercially available aqueous dispersion of colloidal silica with 13 to 14 nm particle size to methyltrimethoxysilane which has been acidified by the addition of 2.5 weight percent glacial acetic acid. This composition is mixed for four hours and is then adjusted to a pH of 3.9 by addition of more glacial acetic acid.

This acidified composition is then diluted to 18 % solids by the addition of isopropanol and aged for four days to ensure formation of the partial condensate of $CH_3Si(OH)_3$. The coating material can be applied by flooding or spray coating onto primed polycarbonate substrates, followed by a thermal curing step at 125 °C for 1 h.

Thermally curable sol-gel hard coat

This example uses only inorganic cross-linking to realize a tough hard coating. A high content of colloidal silica sol is embedded in a matrix of polymethylsilsesquioxane, which forms tough elastic chain- and band structures (compare Figure 6.23, page 149). The methoxy silane is used, which shows a very high reaction speed and quickly reacts with and stabilizes the colloidal silica.

The reaction is started in a mildly acidic medium with acetic acid and a deionized colloidal silica sol. As already discussed in chapter “Electrostatical stabilization” on page 41, the removal of salts increases the stability of SiO_2 sols, especially when organic solvents are added, which lower the dielectric constant. These SiO_2 sols are stabilized by the “SiO_2 anomaly”, the unusual inertness of Si-OH groups at their IEP (isoelectric point).

Acetic acid catalysis leads to a pH value which promotes hydrolysis, but impedes condensation. Oligomers which are rich of Si-OH groups are generated and slowly undergo condensation reactions to form polymers. The long aging step in the example guarantees an almost complete hydrolysis and the generation of many reactive Si-OH groups. Without a thermal curing step or strong catalysis, these Si-OH groups can remain stable and reactive over a long period of time.

The deionized colloidal silica sol exhibits a low pH and could even be used as the sole catalyst of the reaction. The acetic acid however can speed up the homogenization of the original two phasic system. The catalyst is in part removed by a reaction with the released alcohol as acetic acid methyl ester.

Hard, scratch resistant coatings can hardly be applied without a suitable primer layer on organic polymers. This is necessary on the one hand to ensure a better adhesion and on the other hand to improve the breaking resistance of the substrate. If hard coatings are directly applied on polymers, cracks in the hard coating can propagate even through the polymer, which affects the stone chipping resistance (compare Figure 6.4, page 135).

Another function of the primer can be to host UV-blocking additives in proximity to the substrate surface. The hard coat of Example 37 does not need a UV protection, but the polymer beneath it does. Organic UV absorbers can act as softeners and reduce the mechanical stability of the hard coat, so a primer with a UV filter can be a solution to combine excellent scratch resistance with UV stability. An example of a possible primer formulation is given in the next example.

➤ *Example 38: Primer for hard coatings on polycarbonate* [256]

9.0 g ethylmethacrylate and 1.0 g methacryloxypropyl trimethoxysilane are dissolved in 50 g of a solvent mixture consisting of 20 weight% ethyleneglycol diacetate and 80 weight% butoxyethanol. 0.5 g benzoylperoxide is added and after a nitrogen blanket has been established, the mixture is heated up to 80 °C and stirred for 24 h. When the mixture has reached room temperature again, combinations of UV absorbers and HALS can be added.

The primer is applied by flow-coating onto polycarbonate panels, allowed to drain and heated at 120 °C for 15 to 30 minutes to evaporate the solvent. The primed test panels are then flow-coated with the colloidal silica-filled organopolysiloxane top coat composition prepared in accordance with Example 37.

The primer of this example consists of a silylated polyacrylate polymer (compare Figure 6.35, page 168). The alkoxysilane groups serve as cross-linking points for the sol-gel hard coat and the elasticity of the polyacrylate can stop the propagation of micro cracks which can develop over time in the hard coat on top.

In many cases, a long thermal curing step like in Example 37 is not possible. For those applications, UV-curable formulations were developed.

➤ *Example 39: Scratch-resistant UV-curable nanocomposite coating material for polycarbonate* [257]

A 1 l three-necked flask was charged with 248.4 g (1 mol) of 3-methacryloxypropyl trimethoxysilane. Under intensive stirring, 99.36 g of acetic-acid stabilized AlO(OH) (boehmite, "Sol P_3", Condea) was added and the boehmite was dispersed for 10 minutes.

Subsequently the mixture was heated to 90 °C and stirred for 15 minutes. Then 35.95 g (2 mol) of distilled water was added slowly with stirring and the mixture was heated to 100 °C. After 5 to 10 minutes, the reaction mixture shows severe foam formation due to the release of methanol. After stirring for 2.5 h under reflux at an oil bath temperature of 100 °C, the reaction mixture was allowed to cool to room temperature and subsequently subjected to pressure filtration through a 1 µm membrane filter and kept at -18 °C prior to use.

Scratch resistant UV-curable nanocomposite coating

Acetic acid stabilized boehmite (AlOOH) is one of the few nanoparticle powders, which can be redispersed under mild conditions to primary particle size. Boehmite is a needle shaped aluminum hydroxide/oxide modification, which is commercially available in different particle sizes. The smallest type "Sol P3" can be used to formulate transparent nanocomposites which find application e.g. as coatings for polymer lenses.

Especially if combined with methacryloxypropyl trimethoxysilane, very high loadings of more than 50 % of nanoparticle content relative to the polysiloxane matrix can be realized. At these high nanoparticle loadings, the percolation threshold is exceeded and an inorganic network of connected boehmite particles is formed, which can withstand high mechanical forces.

During hydrolysis and polycondensation of the alkoxysilane, the nanoparticle is covered by the polysiloxane network (Figure 6.13, page 141). If trialkoxysilanes are used, mainly ladder polymers are generated which grow from the particle surface (compare Figure 5.23, page 112) [258].

Since boehmite acts as a catalyst for the polymerization of epoxides, the use of glycidoxypropyl trimethoxysilane as an alternative to methacryloxypropyl trimethoxysilane can be problematic. If the epoxysilane shall be used to formulate a thermally or cationically curable coating system, the hydrolysis has to be done under cooling. The shelf life of such a formulation can be increased by storing at -18 °C.

The acetic acid stabilized boehmite also acts as a catalyst for the hydrolysis and condensation of the silanes. Methanol leaves the formulation at the high reaction temperatures and thereby generates foam, since a viscous resin is generated in parallel. Personal experience showed that the controlling of the reaction is not easy, because the exothermic hydrolysis can lead to a sudden rise in temperature. Appropriate safety measures should be taken.

Residual methanol can be replaced in the resulting formulation by other non-toxic solvents with a more suitable flashpoint. In general, the formulation can be cured with typical UV initiators like 2 to 4 % benzophenone related to the methacryloxypropyl trimethoxysilane content in a UV chamber and thereby forms scratch- and abrasion resistant clear coats. Boehmite is an amphoteric filler, so the resistance of the coating against acids or bases is not excellent.

Example 39 shows, that the formulation is stored at -18 °C. This is one possible measure to increase the shelf life, because the condensation step is temperature dependent. Even if the formulation already has been treated at 90 °C during the production, the residual free Si-OH groups can lead to an increase of viscosity. In chapter "Storage stability of sol-gel formulations" starting at page 126, further measures are discussed which can be used to prolong the shelf life of sol-gel formulations.

Curing by inorganic and organic network formation

Scratch resistant coating formulations which are based on organo alkoxysilanes cure by developing both an inorganic and organic network. These two polymerization reactions do not necessarily have to proceed simultaneously, but can also be performed one after another to realize coatings which can be processed further e.g. by deep drawing, structuring or shaping prior to a final curing step (compare "Structured sol-gel coatings", page 153). This concept is comparable to the dual cure concept of conventional polymer coatings.

➤ ***Example 40: Preparation of a deep-drawable scratch-resistant sol-gel coating*** [259]

To the scratch-resistant UV-curable nanocomposite coating material of Example 39, 37.43 g of tetraethylene glycol dimethacrylate and 134.38 g of a high boiling solvent or solvent mixture are added (e.g. n-butanol or diethylene glycol diethyl ether). For cross-linking, 2 % by weight (2.49 g) of benzophenone, based on 3-methacryloxypropyl trimethoxysilane, are used as photoinitiator to initiate a free-radical polymerization.

Following the application of the coating to thermoplastic substrates (plates or films), the coating is dried in an oven at 90 °C for 2 h. This coated substrate, which can be stored in the absence of light, can then be shaped and subsequently cured by UV exposure. The coating exhibits a high abrasion resistance after the UV curing process. With the exception of PP and PE there is no need to pretreat the thermoplastic substrates (PMMA, PC, ABS, etc.) in order to achieve good adhesion of the coating material. Result: Cross cut/tape test (0/0) and Taber Abraser abrasion resistance (DIN 68861, 1000 cycles; CS 10 F, 5.4 N), 15 % haze.

Deep drawable scratch resistant coating

Organofunctional silanes have two reaction sites which can serve as cross-linking points. On the one hand Si-O-Si bonds and on the other hand organic polymerization reactions can be used to build up networks. Whereas the Si-O-Si cross-linking reaction is mostly initiated thermally, the organic part of the molecule can undergo almost every organic polymerization reaction including UV curing. The very high network density which is reached by this twofold polymerization is one reason for the high scratch resistance of such nanocomposites.

In order to realize deep drawable coatings, it is important to carry out at least one cross-linking step after the shaping process. Because before and during shaping, the substrate is heated up to temperatures above T_g which is typically well above 80 °C, it is obvious, that the organic cross-linking by UV polymerization should be the second and final curing step.

To increase the scratch resistance, in this example boehmite (AlOOH) nanoparticles are used as reactive fillers. Boehmite however is amphoteric and is attacked easily by acids and bases. An alternative could be SiO_2 to realize a good stability against acids or ZrO_2 to realize a good resistance against alkaline media.

The use of high boiling solvents and/or UV reactive thinners in the coating material eases the stress relaxation during the shaping step by hindering the Si-O-Si condensation. A completely cross-linked coating would be too brittle to be deep drawn.

6.5.1 Abrasion resistance

Abrasion resistance often is confused with scratch resistance in general linguistic usage. Both requirement profiles differ significantly and a scratch resistant coating can easily be destroyed by abrasion or abrasion resistant coatings can be scratched.

When scratching a surface, the coating experiences a point shaped deformation and reacts depending on its mechanical properties with a plastic or plastic/elastic deformation. If the deforming force exceeds the mechanical stability of the material, cracks appear.

Abrasion resistance is measured in mg/cycles material abrasion, a value which typically describes the quality of flooring materials like parquet. While scratch resistance can be realized by increasing the network density and the viscoelastic properties of the coating material, abrasion resistance can only be achieved by adding significant amounts of exceptionally hard filler materials in the micrometer range.

➤ ***Example 41: Preparation of an abrasion resistant coating*** *[260]*

75 g of a hexafunctional aliphatic urethane acrylate are mixed with 30 g pentaerytritol triacrylate and 4.4 g 3-aminopropyltriethoxysilane. The mixture is strirred for 15 minutes and then 0.5 g maleic anhydride is added.

Subsequently 20 g trietyleneglycol divinylether, 5 g ethylpolysilicate ("Dynasil 40", Evonik), 1.5 g deionized water, 40 g "Aerosil OX 50" (Evonik) and 40 g corundum ("Placor", by ESK-SIC GmbH) are added under vigorous stirring.

The mixture is sonicated for 30 minutes which raises the temperature to 50 to 60 °C. Coated onto polypropylene panels and cured by electron beam treatment, the coating shows excellent adhesion and abrasion resistance in the Taber Abrasor Test (according to DIN 68861 part 2). The weight loss was determined to be only 4 mg.

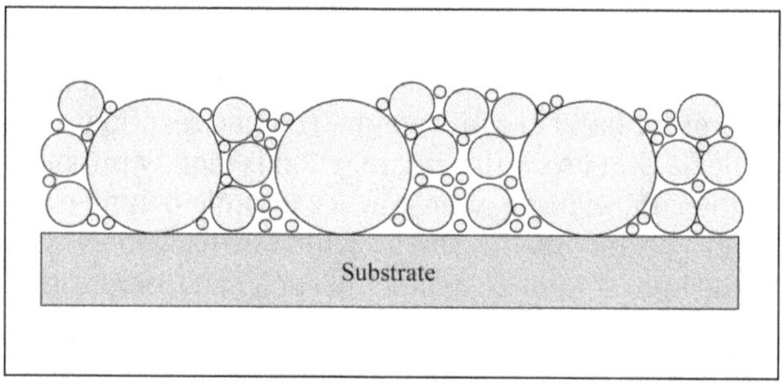

Figure 6.27: Dense packing of particles of different size can improve the abrasion resistance of coatings

Abrasion resistant coating

In a first step, a polyfunctional aminoacrylate silane is synthesized by a Michael addition of the aminosilane to the acrylates (compare Figure 5.31, page 117). Aminofunctional acrylates are known for their high reactivity in UV curing processes and therefore often are used to speed up slower resins [261].

Maleic acid anhydride as polymerizable monomer neutralizes residual amine groups under amide formation and adjusts a mildly acidic pH which promotes Si-OH group formation.

Ethyl polysilicates consist of Si-OR group stabilized SiO_2 oligomers. They are produced by sub-stoichiometric hydrolysis of tetraethoxysilane or tetrachlorosilane. They are used as inorganic binders or as modifiers for organic resins.

The purpose of the formulation is to increase the abrasion resistance of polypropylene which means that high abrasive forces like applied by the Taber Abrasor test have to be withstood. Micrometer sized fillers of a high hardness are the material of choice to realize this requirement profile if the coating can be translucent and does not have to be transparent.

The covalent cross-linking of the filler particles with the matrix is realized by the ethyl polysilicate and the aminoacrylate silane. This close interaction is necessary to increase the mechanical resistance of the composite. When analyzing the formulation, it becomes obvious, that the inorganic domains of the coating are arranged in different sizes. Micrometer-sized corundum, sub-micrometer pyrogenic silica and nanometer-sized domains of polysilicate complement each other to achieve a high packing density with the smaller particles filling the voids generated by the larger particles.

When measuring the abrasion resistance, the value which counts most is the abraded material in mg/cycles which is measured e.g. by the Taber Abrader test. Scratch resistance measurements often focus on the optical appearance and measure the haze increase or the gloss of a coating.

Whereas the scratch resistance can also be increased by making the coating materials softer and by enabling a self-healing characteristic, the abrasion resistance can only be improved by increasing the hardness and/or adding enough hard fillers so that the percolation threshold is exceeded (compare Figure 4.15, page 84).

6.6 Easy-to-clean/anti-adhesive coatings

Driven by the raising awareness about bacteria and fungi in our environment, the demand for hygienic, easy-to-clean surfaces is growing stronger and stronger. In the last years, therefore a variety of low surface energy easy-to-clean coating formulations has been developed. Meanwhile, easy-to-clean coatings are an often found feature of shower cabins or sanitary ware because it has been proven, that these coatings can reduce the time which is necessary for cleaning and the use of surfactants significantly.

Also for the producers of sanitary ware, these coatings are very interesting. Customers readily pay a premium for this function which improves the margin in this commodity market (compare "Historical and actual facts", page 15).

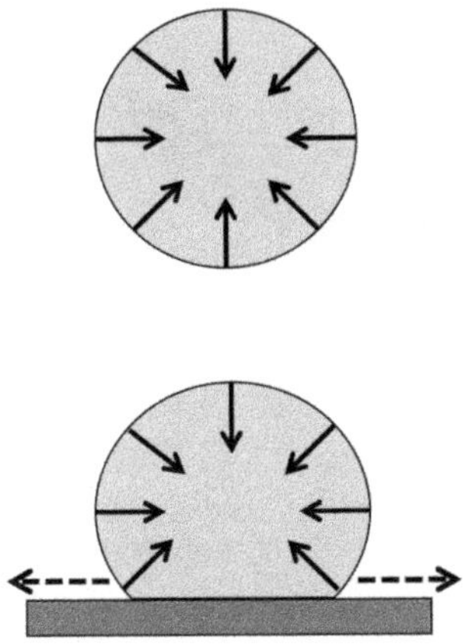

The surface tension of water tries to keep the spherical shape of a water droplet.

When in contact with a surface, energy is released by interaction of the liquid with the surface. The higher the surface energy the larger is the amount of energy released and the tendency of the liquid to wet the surface. The shape of the droplet is determined mainly by the equilibrium between the surface tension of water and the surface energy of the surface.

Figure 6.28: Factors influencing the wetting of surfaces

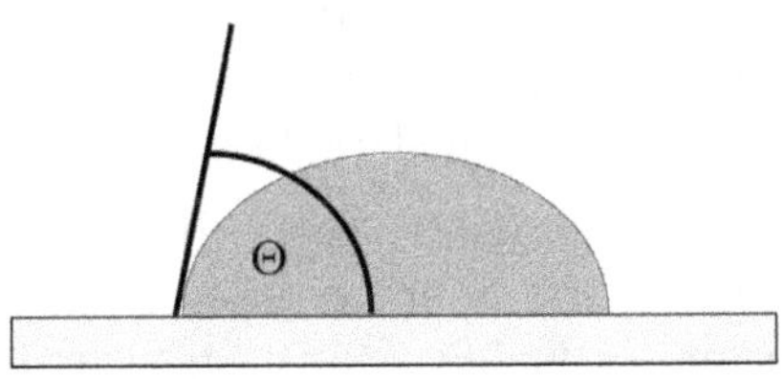

$$\cos\Theta = \frac{\sigma_s - \sigma_{ls}}{\sigma_l}$$ Young's equation

Θ: contact angle
σ_l: surface tension of the liquid
σ_s: free surface energy of the surface
σ_{ls}: interfacial energy between liquid and surface

Figure 6.29: Factors which influence the contact angle of a liquid wetting a surface

If a water droplet wets a surface, an equilibrium of forces competes over the shape of the droplet. The higher the surface energy of the wetted surface, the higher is the energy gain of the interaction between water and surface and the higher is the tendency of the droplet to wet more of this surface. The competing force is the surface tension of the water droplet, which tries to draw the droplet into a spherical shape. From of these considerations, it can be followed, that surfaces with low surface energy have to be synthesized, if the surface tension of water should dominate the wetting and water should easily roll off as droplets.

In an equilibrium state, a characteristic shape of the droplet evolves with a wetting angle which is a measure for the water repellency of the surface. Values smaller than 15° indicate a complete wetting, uncoated glass and ceramic surfaces exhibit values between 40° and 60° and values larger than 80° are characteristic for anti-adhesion surfaces.

When the contact angle is measured and used to access the hydrophobicity of surfaces, the hysteresis between advancing and receding contact angle should be considered. Usually a droplet of water is placed on a surface via a syringe. This can be done manually or automatically. When water flows into the droplet and it becomes bigger, a larger contact angle can be measured as when water is sucked out of the droplet and it becomes smaller.

The same phenomenon can be observed when the surface is tilted and the droplet is deformed by gravity. So an advancing contact angle can be distinguished from a receding

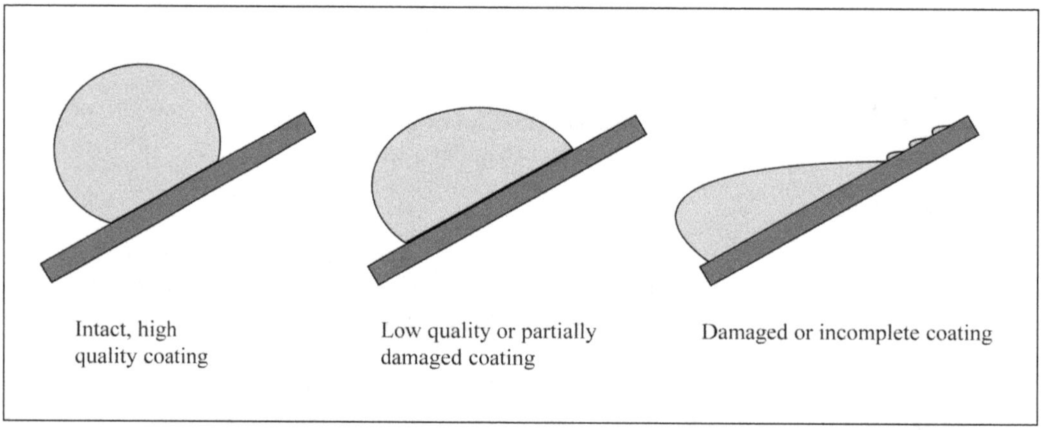

Figure 6.30: Differences between advancing and receding contact angles can be used to evaluate the quality of an anti-adhesive coating

contact angle. For the assessment of the quality of a hydrophobic coating, the receding contact angle and the hysteresis is decisive, because both react much more sensitive to abrasion or coating defects than the advancing contact angle.

Besides measuring the contact angle, a simple tilting test can give a sound feedback about the quality of a hydrophobic coating. If a coated surface with a drop of water on it is tilted, bad coatings are revealed by showing a visible deformation of the droplet. If the droplet additionally shows a “tail” or even small droplets are separated from the tail when the droplet rolls off, a clear hint is given that the coating is damaged or of no good quality.

Besides the hysteresis and the receding angle, the sliding angle, which is the tilting angle of a surface at which a water droplet starts to roll off, is important for the assessment of a hydrophobic coating. In general, a small sliding angle (approximately 2 to 10°) is favorable.

An uncoated clean glass surface is hydrophilic. A droplet of water wets the glass and shows a contact angle below 30°. In order to generate a water repelling coating, the interfacial energy which is a measure for the interaction between the droplet and the surface has to be minimized. This can be done with low surface energy materials like silicones (approx. 20 mN/m), waxes (approx. 25 mN/m) or fluorinated polymers (approx. 15 mN/m) [262, 263].

For the formulation of anti-adhesive sol-gel coatings, a variety of organofunctional silanes are available. Alkyl-, branched alkyl-, fluorinated alkylsilanes and silicones can be used. One of the simplest hydrophobing agents for mineral building materials is sodium methyl siliconate (Figure 6.31).

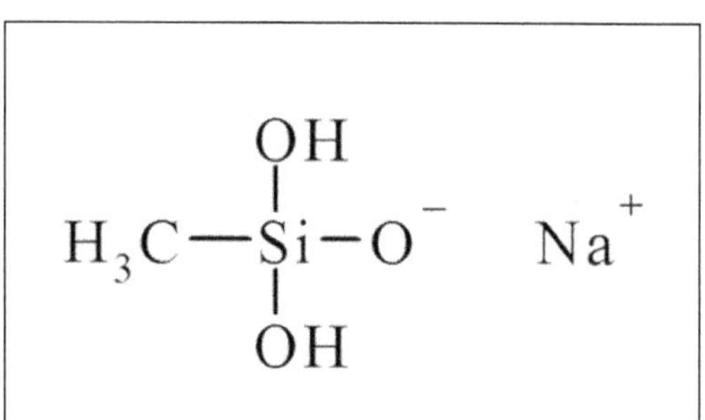

Figure 6.31: Sodium methylsiliconate is a simple but effective hydrophobic treatment e.g. for construction materials

Its aqueous alkaline solution is applied via dipping or spray coating. The methyl siliconate reacts with Si-OH groups of the building material and forms a water repelling coating. Due to its alkalinity, a fast and complete reaction is guaranteed. In a similar way, the alkalinity of concrete fosters the condensation of octyltrialkoxysilanes which are used as anti-efflorescence and hydrophobic impregnation [264].

When flat, nonporous surfaces are coated, their optical appearance should remain

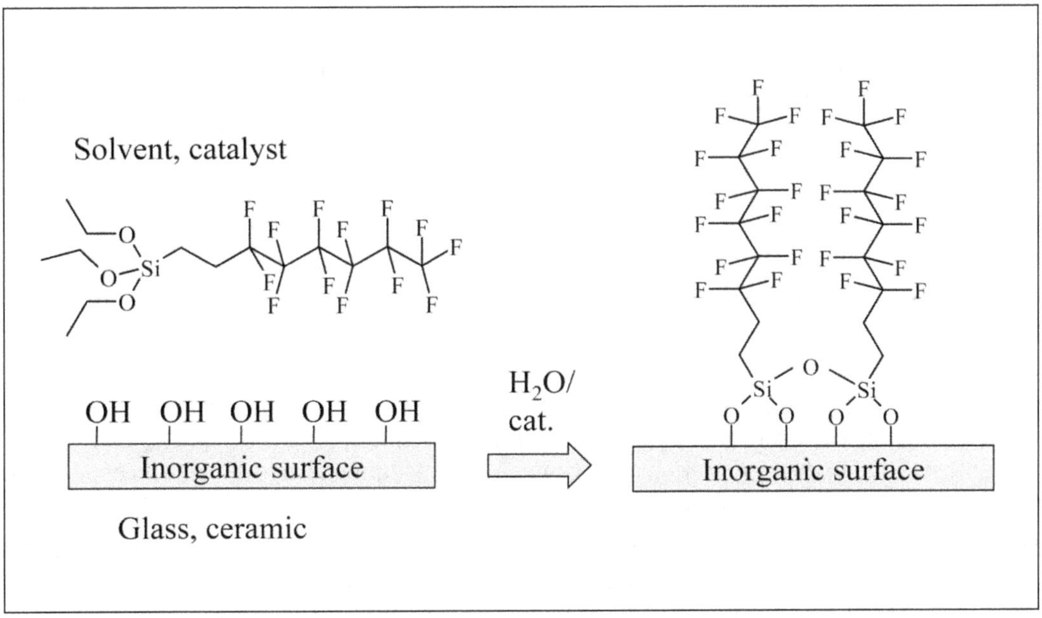

Figure 6.32: Surface modification with water- and oil-repellent partially fluorinated silanes

unchanged. Typical coating thicknesses should not exceed 100 nm to avoid undesired interferences. In order to realize the best possible anti-adhesion action, partially fluorinated materials are necessary.

A typical formulation could contain 1 % tridecafluorooctyltriethoxysilane in a water/isopropanol/acetic acid mixture of a pH of approximately 3 which is applied to cleaned glass (compare: "Pre-treatment of surfaces", page 133).

By hydrolysis of the alkoxysilane, hydrophilic Si-OH groups are generated, which can react with the Si-OH groups of the glass. If the fluoroalkyl chains have arranged themselves, the surface even repels the anti-adhesion coating liquid, which beads up and runs off in droplets.

This behavior, which is called "autophobicity" is astonishing, since the coating liquid contains alcohols as a solvent.

"Autophobicity" makes dipping or spraying as application methods difficult, because the irregularities caused by the droplet formation lead to an uneven coating result with spots and streaks. The application method of choice for such formulations is polishing. The coating liquid is spreaded over the glass and polished until all droplets of the solvent have evaporated. Any excess of active material is not repelled, but deposited on the first layer as oil or wax depending on the mechanical properties of the polycondensate.

Alternative to polishing, "autophobicity" can be avoided by using partially fluorinated solvents like fluoro alkyl ether ($C_4F_9OCH_3$ "Novec HFE", 3 M). The high price of these solvents however in many cases limits their use. They find application in anti-fingerprint coatings for electronic devices like smartphones and tablets (compare Example 49, page 177).

In order to realize a durable coating, covalent bonds between the silanes and the glass have to be generated. Like it is shown in Figure 6.2 on page 134, the first bonds a silane develops to a glass surface are hydrogen bridges, because the reactive sites are blocked by adsorbed water molecules. The building of covalent bonds can be supported by a temperature treatment, strong catalysts or a sufficiently long post curing time.

➤ *Example 42: Preparation of a solvent-based easy-to-clean coating formulation* [265]

64 g isopropanol, 2.78 g isopropoxyethanol and 1 g heptadecafluorodecyl trimethoxysilane (Shin Etsu) were added to a reaction vessel. After stirring the mixture briefly, 0.2 g sulfuric acid was added slowly, followed by stirring for 15 min.

Easy-to-clean coatings

A partially fluorinated alkyl trimethoxysilane as active material is dissolved in alcohol. Via its perfluorinated part of the alkyl group, this silane leads to distinct water and oil repellent properties of coated surfaces.

Remarkably, the addition of water to promote hydrolysis and condensation is avoided. The silane is hydrolyzed by using the adsorbed water on the substrate and the humidity of the surrounding air. The used catalyst is a strong mineral acid. During the application, the acid concentrates due to the evaporation of solvents. As a consequence, hydrolysis and especially condensation is strongly catalyzed.

This concept realizes long storage stability combined with a high reactivity. Since a methoxy silane was used, the reaction speed is even higher than that of an ethoxy silane, however the release of methanol could lead to restrictions concerning safety and handling of this formulation.

When concentrated sulfuric acid is mixed with alcohols, alkylsulfates could be generated, which, being strong alkylating agents, could exhibit a potential mutagenic and cancerogenic action. Appropriate safety measures should be taken when working with comparable formulations.

Via extraction with fluorinated solvents, it can be found that coatings applied by polishing consist of a covalently bonded part which cannot be detached and a weekly bonded part of "free" fluoroalkylpolysiloxanes. The mechanical and physical properties of these fluoroalkylpolysiloxanes are a decisive factor for the durability of the coating.

A polycondensate of tridecafluorooctyl triethoxysilane is an oily liquid which can be rubbed off from a surface quickly. If the longer chain homologues are used, which contain one or more C_2F_4 building blocks more, then a hard, solid wax is generated which can hardly be removed from the surface. Furthermore, investigations showed that long chain perfluorinated moieties arrange themselves in a quasi-crystalline way and thereby increase their hydrophobicity [266]. This effect approximately starts at 8 perfluorinated C-atoms and is one reason why these substances have been used in the past for textile impregnation or coatings.

Since 2010 all substances which contain more than 6 perfluorinated C-atoms are de facto banned, because they can decompose to the corresponding perfluoralkylcarbonic acids, which are persistent, bioaccumulative and potentially toxic [261, 267, 268]. So as a consequence, the use of the silane of Example 42 is no longer possible in industrial applications. The allowed compounds with 6 or less perfluorinated C-atoms do not show the quasi-crystalline self-arrangement and often show a worse durability and hydrophobicity. New developments are alkoxysilylated perfluoropolyethers like the "Dow Corning 2604" coating.

Another important factor which determines the durability and chemical resistivity of a coating is the bonding to the substrate. Surfaces which are to be coated should be accessible for the reactive groups of the coating material. Without the removal of the always present gel layer, technical glass- and glaze surfaces cannot form a durable connection between silane and surface (compare "Pre-treatment of surfaces", page 133). The best pre-treatment are

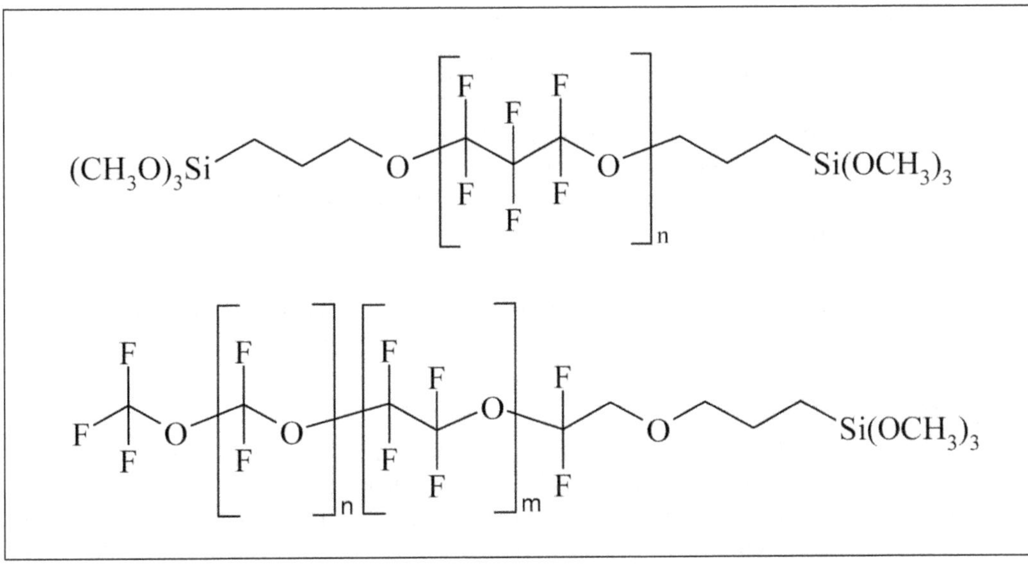

Figure 6.33: Alkoxysilylated perfluoropolyethers can be used for easy-to-clean coatings

abrasive cleaners with a tailored hardness of the abrasive material which remove the gel layer but do not damage the substrate surface.

For the coating of glass there is another important but rarely considered aspect, which affects the durability of sol-gel coatings. Glass plates are most often produced by the float glass process. The glass melt floats on a bath of molten tin and is heated by gas burners on the fire side. After the process, a glass plate exhibits a different surface chemistry on its two sides. With a fluorescent lamp, the tin side can be easily determined. The UV lamp is placed on the glass and the glass edge is observed. If the UV lamp is placed on the tin side, a bluish fluorescence can be seen. Only the tin-side shows fluorescence!

In practice, the durability of coatings on the fire side is much higher than on the tin side. Besides the classical application methods, nanometer thin anti-adhesion coatings can also be applied by cleaners. If a surface gets cleaned regularly with a doped cleaner, with time an anti-adhesive coating can be realized. Example 43 proposes such a solution for an industrial cleaner based on mineral spirit.

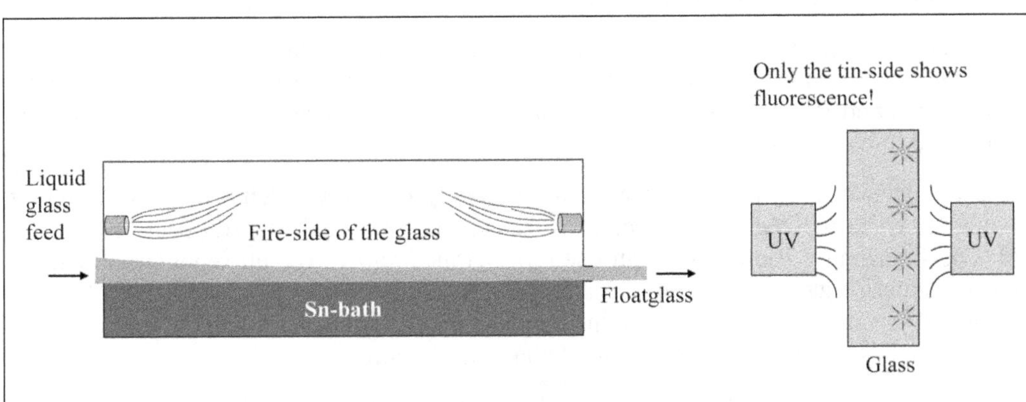

Figure 6.34: Caused by the production process, float glass exhibits a different surface chemistry on its both sides. The side which had been floating on the tin bath shows a characteristic fluorescence under UV light.

➤ *Example 43: Preparation of an easy-to-clean coating material for steel* [269]

25.5 g tridecafluoroctyltriethoxysilane ("Dynasilan F8161", Evonik) are mixed with 13.8 g octyltriethoxysilane. In a separate vessel, 0.026 g HClc is added to 8.83 g of glacial acetic acid. Both mixtures are united under stirring, followed by aging in a closed container for 20 d at ambient conditions. 50 g of the reaction product are added to 950 g of mineral spirit (boiling point 50 to 80 °C).

A soiled stainless steel surface is first precleaned with unmodified mineral spirit and then intensively cleaned with the modified mineral spirit. After the cleaning procedure a water- and dirt-repellent surface results.

Easy-to-clean coating

There is a huge demand for cleaners which ease the following cleaning procedures or which prevent a quick re-soiling of the cleaned surface. For waterborne household cleaning formulations, often quaternary ammonium compounds are used which adsorb onto the cleaned surfaces. In industrial cleaners, mineral spirit still is often used to remove oil, inks or old coatings.

The example uses the non-aqueous condensation of silanes with acetic acid to synthesize a co-condensate of an alkylsilane and a partially fluorinated silane. The solvent ethylacetate is generated during the condensation as a byproduct from the reaction between acetic acid and the ethoxy groups of the silane. The strong mineral acid HCl acts as a catalyst of this reaction. The concept has the advantage, that only reaction products are generated, which are soluble in mineral spirit (compare: "Water-free sol-gel techniques", page 100 and Example 28, page 123).

The co-condensate still contains residual alkoxy groups, because the added amount of acetic acid is not sufficient for reacting with every alkoxy group. Remaining reactive alkoxy groups are advantageous with regard to the targeted use, because they can react with surface groups of the material which is going to be cleaned. With the amount of acetic acid, the degree of polymerization can be tailored.

The difference to Example 28 lies in the hydrophilicity and polarity of the resulting condensate. Example 28 uses the technique of phase separation by an excess of water to separate the reaction product from byproducts and solvents.

A material containing many Si-OH groups is generated, which due to the absence of catalysts is relatively stable against condensation. The oligomer in this example is stabilized by Si-OR groups and much more nonpolar. What reactivity and polarity is the best has to be determined individually.

The cleaned surfaces exhibit water- and oil repellent properties which is an indication that the oligomer has reacted with the metal surface. One could ask the question, why not a 100 % perfluorinated silane had been used to further improve the oil repellent effect. An oligomer consisting only of a perfluorinated polysiloxane however would only be soluble in fluorinated solvents and therefore is of no use. The nonfluorinated octylsilane therefore has the function of solubilizing the fluoroalkyl silane in the mineral spirit.

In many industrial processes, flammable solvents are not possible. So early on, waterborne easy-to-clean coatings were demanded. Sanitary ware for example is coated by a waterborne easy-to-clean coating material which is sprayed into a circulation furnace. At the high temperature, the coating material is activated and bonds to the glazed ceramic surface [298, 270, 271].

➤ *Example 44: Preparation of a waterborne easy-to-clean coating formulation* [272]

Under continuous agitation, 36.4 g of a tetramethylammonium hydroxide solution (25 weight%) are added to a mixture consisting of 51 g tridecafluoroctyl triethoxysilane ("Dynasilan F8161", Evonik) and 4 g isopropanol. 0.7 g perfluoro octanoic acid are added as surfactant. The reaction mixture is emulgated by intense stirring until it becomes clear. At this point in time the temperature of the formulation raises due to the exothermic reaction to approx. 40 °C.

After 1 h of aging, 700 g of water are added to the reaction mixture in small portions under vigorous stirring. The resulting clear solution can be immediately used for coating purposes. The coating formulation can be applied by wiping glass, ceramic or steel surfaces with a soaked cloth, until a homogenous liquid film results. After the water has evaporated, the coated parts are placed in a circulating air furnace and are kept 1 h at a temperature of 260 °C. After the coated parts are cooled down to room temperature, no visible sign of a coating can be detected. Water and oil do not wet the treated surfaces and run off in droplets. The contact angle against water is 92°, against edible oil 65°.

Waterborne easy-to-clean coating

The production of waterborne easy-to-clean coatings is not easy, because the coating material consists of extremely water repellent substances and the alkoxyfunction of the silanes reacts with water (compare: "Waterborne formulation", page 124) to a polycondensate.

One possibility would be the co-condensation of the hydrophobic silanes with hydrophilic silanes which could act as a dispersing aid or solubilizier. These hydrophilic components however would disturb the water-repellent effect in the final product. Another approach could be to emulgate the perfluorinated polycondensate in water with the aid of fluorosurfactants. But also these surfactants would impede the formation of a durable anti-adhesive coating and would disturb the water-repellent effect, because they have to be used in a quite high amount up to 25 %.

Silane surfactant

In Example 44, the first step of the synthesis is the formation of a thermolabile silane surfactant, which during curing decomposes and yields a highly reactive silane.

A surfactant consists of a hydrophobic and a hydrophilic moiety. The hydrophobic part in this example is the fluoroalkyl chain. The hydrophilic part has to be generated with the help of the silane moiety.

The alkoxysilyl group is hydrophobic, the Si-OH group which is generated by hydrolysis however is hydrophilic. This effect can be experienced during the hydrolysis of tridecafluorooctyl triethoxysilane. After the addition of 1M HCl to a 1 % solution of the silane in isopropanol, during a short period of time, foam can be generated by shaking the formulation. After a while, the foam quickly breaks down again, which can most likely be attributed to the beginning condensation of the silane which destroys the hydrophilic Si-OH group and the intermediate surfactant structure.

The object therefore is to find a way to stabilize the Si-OH groups and increase their hydrophilicity. This problem is solved by the addition of tetramethyl ammoniumhydroxide. Quaternary ammonium salts show a high affinity to silicates which is known from

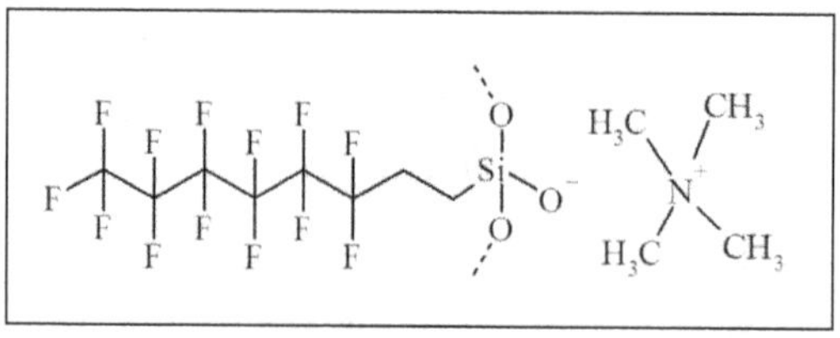

Figure 6.35: Silane-surfactant made of tridecafluoroctyltriethoxysilane and tetramethyl ammoniumhydroxide [272]

the chemistry of cationic surfactants. The tetramethyl ammonium group, together with the silane forms a hydrophilic ammoniumsilanoate group which transforms the hydrophobic fluoroalkylsilane to a stable silane surfactant.

Aqueous solutions of the surfactant foam heavily and wet many surfaces well. The hypothesis is being discussed, whether by diluting the solution of the silane surfactant stable surfactant micelles are formed in analogy to conventional surfactants. This would explain the long storage stability of the coating solution.

The tetramethyl ammonium group is thermolabile and decomposes at temperatures exceeding 150 °C into trimethylamine which, being a gas, readily leaves the coating. The deblocked silanol groups are very active in the alkaline atmosphere and react with the glazed surface of the sanitary ware to a durable easy-to-clean coating.

In the example, perfluoro octanoic acid is added to further improve the wetting behavior of the formulation. This substance must not be used anymore, since it is bioaccumulating, persistent and potentially toxic [267]. If necessary, other suitable fluorosurfactants are commercially available (e.g. by DuPont, 3M or Merck KGaA).

Besides coating of non-porous surfaces like glazed ceramics, the water repelling impregnation of porous building materials plays an important role. Silicone derivatives have been used for this application since a long time. In the course of the trend to graffiti vandalism, fluorinated coating materials have gained more and more interest. Their combined hydro- and oleophobicity repels paints more effectively and eases the cleaning of sprayed surfaces significantly.

➤ *Example 45: Preparation of a waterborne easy-to-clean coating for mineral building materials* [273]

14.2 g of “Dynasylan RTM 1203” and 10.0 g of “Dynasylan F 8161” are introduced into a vessel and 3.1 g of water are added under continuous stirring. During this operation, the temperature rises to about 30 °C. The reaction mixture is heated up to and stirred at 50 °C for 3 h. A mixture of 220.0 g of water and 4.2 g of formic acid (84 % by weight in water) is then metered in over the course of 5 minutes. An ethanol/methanol/water mixture is removed by distillation over the course of about 2 h (pressure: 150 to 133 mbar; temperature: 30 to 48 °C). When the overhead temperature is about 50 °C and the product comprises only water, the distillation is ended and the product is diluted with water to 1000 g.

Brick stone, lime, sandstone and concrete were cut into ashlars having an edge length of about 5 cm and immersed in the aqueous formulation for about 5 minutes. After the stones have been dried at room temperature or in a drying cabinet at about 120 °C, water and also oil applied to the surface of the stones no longer penetrates into the surface. The beading effect of the liquids mentioned is very good. On untreated specimens, the liquids penetrate immediately into the surface. The product is therefore suitable for simultaneous hydro- and oleophobization of mineral building materials.

Waterborne easy-to-clean coating for mineral building materials

Partially fluorinated silanes are well known starting materials for the formulation of water- and oil repellent coatings. If they are used to formulate a waterborne coating material, they have to be combined with a hydrophilic component which acts as an emulsifier. In these general principles, the synthesis shows many parallels to conventional paint and coatings manufacture.

Waterborne paints often use anionic sulfonate or carboxylate groups which are covalently linked to the binder resin. Example 45 uses the cationic ammonium group as hydrophilic moiety (compare Figure 5.39, page 124). This is done on the one hand for practical reasons because aminosilanes are readily available, but on the other hand the cationic ammonium group shows a high affinity to the predominantly negatively charged surfaces of the building materials.

Formiate is used as the smallest possible organic counterion for the ammonium group. According to own practical experience, the choice of the counterion shows a remarkably strong influence on the storage stability of the formulation. Formiate indeed leads to a much higher storage stability compared to e.g. acetate or chloride. An exact explanation for this behavior is still lacking, but the partial charge model of Livage [150, 151] may give some hints for own thoughts. Chloride ions in general are not recommended for the application on construction materials due to their corrosion promoting properties.

During the first hydrolysis step, only a small amount of water is added which corresponds to a R_{OR} of 0.5. This ensures that the silanes react with each other and that a phase separation is mostly avoided. The basic aminogroup already catalyzes a high hydrolysis and condensation speed, so that no further catalyst is necessary. When excess water is added in the second step of the reaction, formic acid neutralizes the amino groups and increases their hydrophilicity.

The example uses 3 mol aminosilane to disperse 1 mol fluorinated silane. In order to maximize the hydrophobicity of the coating however, the hydrophilic part of the co-condensate should be much lower. Optimization of the formulation shows that it is possible to use 2 mol of the fluorinated silane and 1 mol aminosilane for a significantly better hydro- and oleophobic action [274].

The formulation of Example 45 is not suitable for the synthesis of transparent high-quality coatings on glass. For this purpose, preferably solvent-based coatings formulations like Example 42 or Example 46 are used due to their superior optical properties.

Thick anti-adhesive coatings

For some applications, anti-adhesive coatings of a thickness < 100 nm are not suitable or the application method of polishing cannot be applied. If thicker coatings are necessary which can be applied by conventional methods like spray-coating, the anti-adhesive effect can be realized by a deliberate de-mixing and enrichment of components at the top of the coating. These thicker coatings can also deliver some mechanical and chemical protection for the substrate.

The unpolar components orientate themselves towards the air interface to minimize the interfacial energy, whereas the more polar components of the coating material establish a good adhesion towards the substrate. By this strategy of deliberate phase separation, an anti-adhesive coating with excellent adhesion can be realized. Figure 6.36 illustrates this principle.

The strategy of stratification or "leafing" of certain components of a coating is the foundation of the working principle of many surface active additives like the silicone polyether leveling

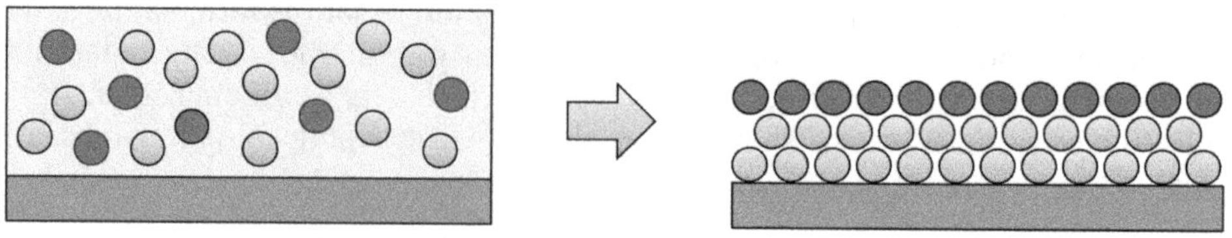

Figure 6.36: Thick anti-adhesive coatings with excellent adhesion to the substrate by a deliberate phase separation process

agents. It is also used to formulate special colors by surface modified pigments, which orientate themselves to the surface during the coating process [71, 194, 275]. A modification with e.g. stearic acid causes a partial incompatibility and the pigments try to minimize the surface which is exposed to the resin by an orientation at the surface of the coating.

Even nanoparticles can be enriched at the surface of scratch resistant (compare "Scratch resistant coatings", page 155) or hydrophilic coatings [276] (page 179).

➤ *Example 46: Preparation of a glass-like anti-adhesive coating* [277]

Methyltriethoxysilane (0.2 mol, 35.7 g) and tetraethoxysilane (0.054 mol, 11.3 g) are stirred in a suitable vessel and 0.1 mol of SiO_2 (in 20.0 g of silica sol 300/30 %) are added thereto. After 5 minutes, 0.4 g of concentrated HCl are added with intensive stirring. The initial two-phase reaction mixture turns white after 2 minutes, heats up and becomes transparent and single-phased again. After 15 minutes of reaction, 3.0 g of "Dowex RTM 50W2" cation exchanger are added and the resulting mixture is stirred for 10 minutes. Subsequently, a pressure filtration through a fiberglass prefilter is carried out.

Immediately thereafter 1.78 g (1 % by mol based on Si) of 1H,1H,2H,2H-perfluorooctyltriethoxysilane are added and the sol is stirred again for 15 minutes. "Amberlyst RTM A-21" (4.0 g) anion exchanger is added and stirring is continued for 30 minutes. 140 g of isopropanol are added for dilution, followed by pressure filtration through membrane filters (pore size 1 μm and 0.2 μm).

The application onto a stainless steel surface is carried out by spray coating (1 to 2 μm coating thickness) followed by curing at 350 °C in air. The resulting coating is a transparent anti-adhesive layer showing contact angles of 110° against water and of 60° against hexadecane respectively.

Glass-like anti-adhesive coating

The example uses the base formulation of Example 31, page 140 and shows how it can be modified to show further functions. The aqueous colloidal silica sol introduces the necessary water for the hydrolysis and the SiO_2 nanoparticles are effectively stabilized by the generated silane derivatives.

The reaction is started without addition of a compatibilizing solvent to homogenize the silanes with the aqueous silica sol. Thereby the sol is not destabilized at a point in time at which no reaction partners are present to build up a sterical stabilization. As ethanol is released by the hydrolysis of the silanes, also in parallel Si-OH group bearing silanes are present to occupy the surface of the SiO_2 nanoparticles.

The low surface energy compound is a partially fluorinated silane, which is added after the reaction of the silanes with the colloidal silica sol. One may ask the question, why the silane is not added at the beginning of the synthesis process.

The objective of this synthesis is to generate a glass-like anti-adhesive material which adheres well to the substrate. If the anti-adhesive component is homogenously incorporated and distributed throughout the coating material, adhesion to the substrate is hard to realize. Furthermore, “autophobicity” would be provoked, because the first adsorbed layers of the coating material would repel the coating liquid.

So it is the deliberate inhomogeniety which is generated by adding the fluorinated compound at the end of the synthesis which allows the movement of the fluorinated compounds to the surface of the coating. To prevent a cross-linking with the other components of the coating, the reaction mixture is ion exchanged prior to the addition of the fluorinated compound. Thereby a pH of about 2 to 3 is adjusted which fosters hydrolysis but not condensation. The fluorinated silane is hydrolyzed, but stays mobile for the movement to the top of the coating.

During the coating step, the fluorinated component orientates itself at the top to minimize the interfacial energy. When the coating gels and densifies, the fluorinated silane cross-links with the other components and is fixed in the upper nanometers of the coating. A “smart” self-organizing coating was synthesized.

The high temperature stability of the anti-adhesion effect is impressive. Even after the thermal treatment at 350 °C, an excellent anti-adhesive effect can be observed in spite of the fact that the pure fluorinated component readily degrades at temperatures > 250 °C.

This is one example for the stabilization of organic molecules by sol-gel matrices, especially in the presence of nanoparticles (compare “High temperature resistant coatings”, page 139).

If steel or stainless steel is treated at temperatures > 300 °C, usually coloration of the steel takes place by the formation of oxidation products. This mostly undesirable coloration can be suppressed by using inert gas ovens or suitable primers (compare Example 33, page 147). The removal of the chloride ions by the ion exchanger process step contributes further to protect coated metals from corrosion.

➤ *Example 47: Preparation of a high-temperature resistant alkaline-silicate anti-adhesive coating* [278]

A suitable vessel is charged with 0.1 mol (17.9 g) of methyltriethoxysilane, 0.027 mol (5.6 g) of tetraethoxysilane and 0.016 mol (8.16 g; 11 % by mol based on Si) of 1H,1H,2H,2H-perfluorooctyltriethoxysilane. Subsequently, 0.018 mol NaOH (0.72 g) is added with stirring and stirring is continued for about 16 hours until all the NaOH has dissolved.

Thereafter, 0.15 mol (2.7 g) H_2O are added dropwise under intense stirring to the solution which meanwhile has turned yellow. After completion of the addition, stirring is continued for about 30 minutes. Subsequently the product is diluted with 35 g of ethanol and filtered through a 1 μm filter.

The coating formulation is applied on stainless steel by spray coating and cured at 350 °C in air. During this process, the steel substrates do not show the typical discoloration which is usually observed at these temperatures. The coating is completely transparent and shows contact angles against water of 95° and against hexadecane of 40° respectively.

Alkali-silicate anti-adhesive coating

Glass-like anti-adhesive coatings which maintain their anti-adhesive properties even at high temperatures are interesting not only for the food processing industry, but also for all industrial applications where decomposition products might adhere to critical parts and cause failures. For example reports have been published which describe the successful coating of automotive motor parts with glass-like coatings to prevent clogging [279].

It is remarkable that with this formulation, thick anti-adhesive coatings can be deposited in spite of the fact that the fluorinated silane was added right from the beginning of the synthesis. This result stands in contrast to Example 46, in which only a coating can be deposited, if the fluorinated active material is added at the end of the synthesis to prevent a premature cross-linking with the other components of the formulation.

The main difference between this example and the example before lies in the presence of a high concentration of sodium ions and the high pH. At this high pH, condensation reactions are very fast and aggregates with few Si-OH groups are generated. The building blocks of the resulting formulation are stabilized with residual Si-OR groups and Na^+ ions.

Sodium ions form strong bonds to Si-O- groups. They can exhibit a binding strength in the same order of magnitude like covalent Si-O-Si bonds [199]. This should be the reason why the "autophobicity" effect in this formulation is not as dominant as in the sodium-free composition of Example 46.

The contact angle measurement shows the disadvantage of the homogenous incorporation of the fluorinated component. At the surface of the coating a smaller amount of the active material is available which expresses itself in a lower contact angle against water. In comparison to Example 46 the contact angle however remains high over a longer period of time and after abrasion because the fluorinated component is also present in the deeper layers of the coating.

When handling fluorinated silanes, the volatility of these components should be considered. Even a monolayer is sufficient to coat lab equipment with a low surface energy coating. Own experience showed that after some months, equipment and devices can exhibit a contact angle > 90 °C. If in the same lab conventional coatings are handled, wetting and adhesion problems are inevitable.

Coatings with "lotus effect"

After easy-to-clean coatings became popular, the next step in the development of coatings which can ease the daily life were self-cleaning coatings, which can keep surfaces clean without involving a manual cleaning step [280, 281].

In nature, such surfaces already are reality. The leaves of the Lotus plant, like many other leaves, show a self-cleaning effect when it rains. The water droplets run off and take with them soil and dirt without leaving any trace on the surface of the leaves. The discovery and investigation of this "lotus effect", as well as the efforts to imitate this effect with technical surfaces caused an immense hype by researchers and companies alike.

On a low surface energy coating, water contracts, because the interaction between the surface and the water is minimized. Now, only the surface tension dominates the shape of the water droplet, so that it becomes more and more spherically shaped. Furthermore, the interaction of dirt and other contaminations with the surface is also reduced, so that it can be removed more easily.

If a surface is structured, the structure amplifies the wetting character of the surface. If it is hydrophilic, the contact area and the interaction between the surface and the water droplets is maximized and spreading is improved until no contact angle can be measured anymore (compare "Hydrophilic coatings", page 179).

If it is hydrophobic, the contact area and interaction between droplets and surface is minimized, until the droplet only touches the tips of the structured surface. Then, a lotus effect can be observed, if the structure shows the right width and height.

The change in the contact angle of a liquid in contact with a structured surface can be calculated with Formula 6.2

Formula 6.2

$$\cos \Theta_{CB} = \phi (\cos \Theta + 1) - 1$$

Θ: contact angle on a flat surface
Θ_{CB}: contact angle on a structured surface
ϕ: area of the surface which is in contact with the liquid

Formula 6.2: Change in the contact angle of a liquid which is in contact with a structured surface according to the model of Cassie & Baxter[282]

Cassie and Baxter found, that the contact angle in this state is dependent on the surface area with which the droplet is in contact. The interaction with the surface is reduced to only a fraction of that of a flat surface and it is mainly the surface tension of the droplet which determines its shape. Therefore the droplet forms almost perfect spheres, if no surface active substances are dissolved in the liquid[283-285].

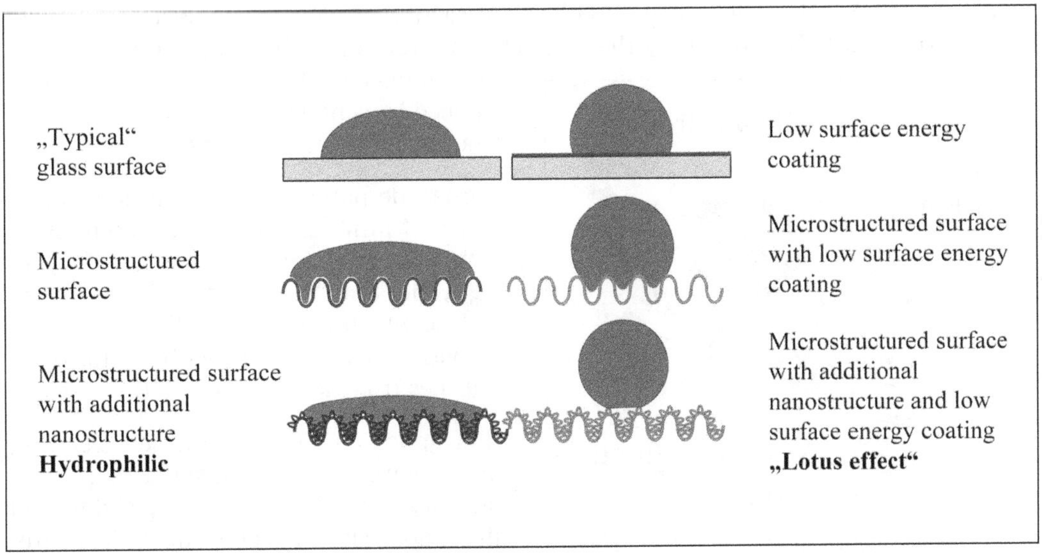

Figure 6.37: Illustration of the influence of structure and surface energy on the wetting behavior of surfaces

Microstructured „Lotus“ surface

Figure 6.38: Pot saucer made of clay with a superhydrophobic “Lotus“-coating Source: G. Jonschker

Three factors for a lotus effect:

For a lotus effect, three prerequisites have to be fulfilled at the same time

- intact structure with the right width and height
- low surface energy anti-adhesive coating
- water droplet with a high surface tension, which means no surfactants or contaminations are present

The structure has to be designed in a way that a minimization of the contact area between water and the surface is reached. This is realized by the lotus plant by a combination of a coarse and a fine structure. For technical surfaces, the width and the height of the structure have to be determined individually for each application. A coarse structure of 40 μm and a fine structure of 250 nm can serve as a starting point for own investigations.

Dirt particles show a stronger interaction with the water droplet than with the low energy surface. The dirt particles, which lie on the tips of the structured surface attach to the water droplet when it rains. If they are hydrophilic, they are taken up by the water droplets, if they are hydrophobic, they remain attached to its outside.

If the structure gets damaged or contaminated by fingerprints, oil or fat, water droplets adhere again to the surface. The difference in the wetting behavior between damaged and intact surfaces is eye-catching and optically disturbing. This overemphasized perception of defects by customers can lead to problems in marketing technical applications.

As already discussed, a micro-structured easy-to-clean coating can only be a prerequisite for the lotus effect. The driver for the characteristic roll-off of the water droplets is the surface tension of the water droplet. If it is reduced by contaminations, lotus- and self-cleaning effect cannot work.

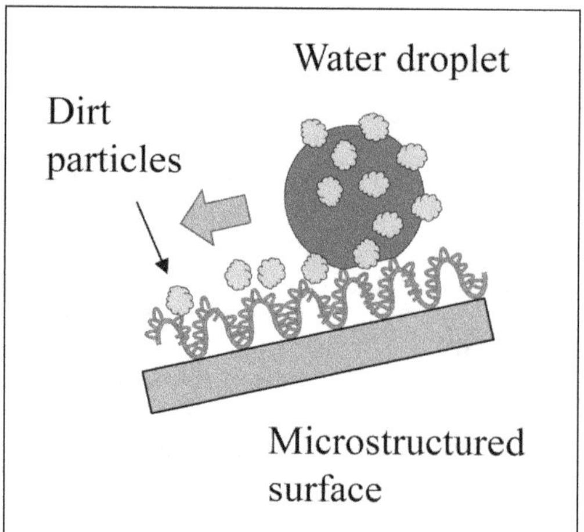

Figure 6.39: Working principle of self-cleaning “Lotus“-effect coatings.

The facade paint “Lotusan” uses a combination of mineral fillers in different sizes with silicone resins to realize a self-cleaning wall paint [286]. The part of the silicone surface which is destroyed by weathering, is continuously replenished by diffusion of silicones from deeper layers of the coating.

This approach to realize a self-cleaning façade however is not indisputable. Investigations showed that the take-up of dust and dirt is not so much dependent on the hydrophobicity or hydrophilicity of a surface, but that the decisive factor is the adhesiveness of the surface. This adhesiveness is determined mainly by the T_g of the used binder polymer and how the coating swells when it comes into contact with water. Hydrophilic inorganic façade paints based on alkali silicates also show an excellent long-term anti-dirt behavior, compared to conventional organic resin-based paints [287, 288].

Structuring impedes gloss

Since lotus effect surfaces have to be structured, they cannot be glossy since the width and the height of the structures are larger than the wavelength of visible light which causes scattering.

➤ *Example 48: Preparation of a transparent "lotus effect" coating on glass* [289]

Step 1: Preparation of a silica-polystyrene sol-gel formulation

Two solutions (named A and B, respectively) were prepared. For solution A, 500 nm PS (polystyrene) spheres with a number average molecular weight of $9{\cdot}10^4$ Daltons were prepared first via emulsion polymerization without using emulsifiers. A certain amount of PS spheres were then dispersed into 20 ml ethanol. 3 ml ammonia solution was used to adjust the pH value, and the suspension was stirred at 45 °C for 1 h.

For solution B, 3 ml tetraethyl orthosilicate (TEOS) was dispersed into 25 ml ethanol. To prepare a silica-PS sol-gel formulation, solution B was mixed with solution A, and the mixed solution was stirred at 45 °C for 1 h.

Step 2: Coating of glass substrates with the silica-PS sol-gel formulation

A transparent flat-glass substrate was coated via dip-coating at 2.65 cm/min. The substrate was immersed into the sol for 5 minutes before the first dip-coating, and 5 s before the subsequent run. After each coating, the substrate was dried at room temperature (25 °C) for 5 minutes and this step was repeated 5 times. With increasing coating thickness, some of the glass substrates became semi-transparent.

The coated glass substrates were heated at 500 °C for 10 minutes to remove the PS nano-spheres. The substrates were cleaned with an H_2SO_4/H_2O_2 (50/50 wt.% "piranha") solution for 1 h, sonicated in acetone for 10 minutes, and then rinsed with distilled water before the CVD coating step.

Step 3: Chemical vapor deposition (CVD) of perfluoroalkysilane:

First, a solution containing 1 % 1H,1H,2H,2H-perfluorooctyl triethoxysilane and 1 % H_2O in methanol was prepared. The coated glass substrates were put into a sealed vessel containing 0.3 ml of the perfluorooctyl triethoxysilane solution. The distance between the glass substrates and the solution was 55 mm. Then the vessel was kept at 150 °C for 3 h to cover the coatings by a monomolecular layer of perfluorooctyl triethoxysilane.

The contact angle against water was 160° and the sliding angle almost 0°. If a coated and an uncoated glass plate were placed on a piece of paper with writings on it, almost no difference could be observed.

Transparent lotus effect coating

In order to realize a coating with lotus effect, a hydrophobic structured surface is needed. Basically this can be realized via two approaches. Either the structure is realized by the

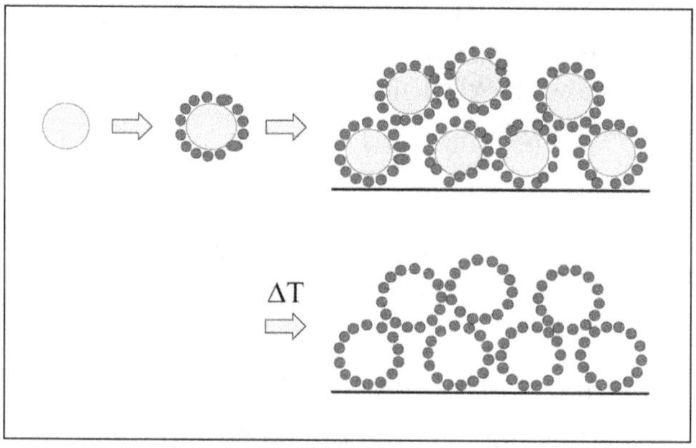

Figure 6.40: Transparent "Lotus effect"-coatings can be realized by thermal treatment of PS/SiO_2 "blackberry" core/shell particles (Example 48)

addition of bigger particles or by removing parts of the coating by a special treatment like a thermal decomposition or selectively dissolving parts of the material.

Polystyrene nanoparticles are readily accessible via emulsion polymerization. If introduced into inorganic coating materials, they serve as a template and later can be removed by heating up the composite above their decomposition temperature. This approach is also used for the synthesis of inverse opals [290].

A lotus effect coating however does not only need a coarse but also an additional fine structure on top. This is the reason, why the inorganic component tetraethoxysilane is not processed with an acidic pH but with ammonia at a high pH and high temperatures under "Stöber" conditions. The result is a dispersion of 50 nm sized SiO_2 spheres, which arrange themselves during the application around the larger polystyrene spheres and thereby generate a finely structured surface. The identical composition, processed at an acidic pH would result in a smooth unstructured SiO_2 shell around the polystyrene which would not show such a pronounced lotus effect.

The SiO_2 coating is activated in an acid bath before the hydrophobic coating is applied. This process step is recommended for thermally treated glasses to increase the number of Si-OH groups which are necessary for the adhesion of the fluorinated silane. H_2O_2 serves to remove last traces of organic contaminations. The mixture of sulfuric acid and H_2O_2 is called "piranha" in lab jargon and very suitable for the pre-treatment of glass substrates prior to a coating step. Its applicability in industrial applications however is limited.

The structured surface is hydrophobized by coating with vapors of a fluorinated alkoxysilane. A wet coating by with solvents would be an alternative, but often leads to inhomogeneity on structured surfaces during drying. The nanometer scale fine structure must not be impaired by the coating material if the lotus effect shall remain intact.

The result which is described in Example 48 is impressive. "Transparent" however must not be confused with "clear". The view through the glass will be impaired by the coating, therefore it is not suitable e.g. as window pane. The targeted application is that of an antiglare coated picture frame in which the glass is in direct contact with e.g. a photography and avoids a directed reflection of light.

Another possibility to realize superhydrophobic structures by a combination of bigger and smaller particles has been described in literature [291]. A raspberry-like structure forms, which then is coated with a low energy coating.

6.7 Anti-fingerprint coatings

During the day, the skin segregates fat, water, diverse metabolic products and salt which are carried over when we touch a surface. Besides the unaesthetic appearance, the salt content in fingerprints can even cause corrosion on metals.

Many attempts were undertaken to prevent the carry-over of fingerprints. A fingerprint is only carried over, if the adhesion of fat and sweat on the surface is higher than the cohesion of the fat and sweat on the finger.

The known principles to realize anti-fingerprint coatings are:

- prevention of the carryover
- easy-to-clean surfaces
- decreasing the visibility of fingerprints via structured surfaces.

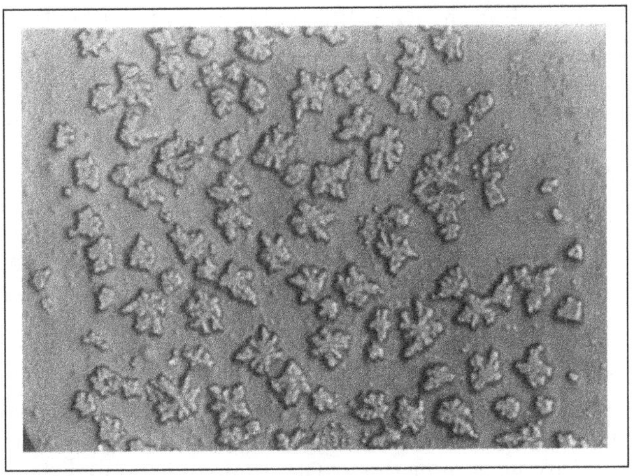

Figure 6.41: Microscopic image of a fingerprint on a glass surface Source: M. Opsölder (www.hobby-photo.de)

The easiest protection against a carry-over of fingerprints is a thin layer of silicone or mineral oil. This approach is often used for brushed stainless steel surfaces e.g. in hotel halls. The fingerprint is not carried over, because the cohesion of the finger fat is higher than that of the oil. So when such a surface is touched, a thin layer of oil is carried over to the finger. This however also is the reason why this solution is not popular and cannot be used for everything.

Since the first partially fluorinated anti-adhesive coatings were realized, they were also tested as anti-fingerprint coatings. It quickly became obvious, that even the most hydro- and oleophobic coatings could not prevent the carry-over of the fingerprints but only ease their cleanability.

A proposed solution is a coating of plasma cross-linked silicones. It is claimed that such coatings can prevent the carry-over of fingerprints [292]. Silicone components are also one of the active components in the following example of an anti-fingerprint coating for frosted/ mat glass.

➤ *Example 49: Preparation of an anti-fingerprint coating formulation for mat glass [293]*

20 g sulfuric acid was added to 3,500 g isopropanol and 850 g "Dowanol PM". After a brief period of stirring, 130 g 3,3,4,4,5,5,6,6,7,7,8,8,8-tridecafluorooctyl triethoxysilane and 12 g tetramethoxysilane were added. As a last step, 15 g "Silopren C10" (Si-OH terminated polydimethylsiloxane) was added after it was dissolved in 1,400 g isohexane. The mixture was stirred for 30 min after the last addition.

The coating formulation was polished onto an etched glass substrate by using a microfiber sponge in two quick passes, each time using 2 ml of the coating solution. The treated glass surfaces were allowed to dry overnight at room temperature.

Anti-fingerprint coating for frosted/mat glass

Fingerprints on mat glasses are perceived as particularly disturbing, because they create glossy areas which attract more attention than on unstructured glass surfaces. The removal of the fingerprints also is more difficult in comparison to unstructured glass, because the fat sits in the structure. Therefore, the need for anti-fingerprint coatings on mat glass is very high.

The first thing which stands out in Example 49 is that no water is added to the formulation to hydrolyze the silanes. Therefore the necessary water has to come from the humidity of the air or from adsorbed water. Without the addition of water to the formulation, no reaction can proceed, so this approach is beneficial for realizing long storage stability (compare Example 43, page 166).

To use the adsorbed layers of water for hydrolysis can even be useful with regard to another aspect. Adsorbed layers of water block the access to the reactive Si-OH groups of the glass. If they are consumed by the hydrolysis reaction, the coating material generates reactive Si-OH groups in the proximity of the surface -OH groups of the glass which increases the possibility of a covalent bond (compare "Pre-treatment of surfaces", page 133).

It is known that silicones can decrease the carry-over of finger fat to the surface [292]. In this example, silicones are combined with oleophobic, partially fluorinated silanes to enhance this effect and ease the cleanability. The Si-OH functionalized silicones either can react with the alkoxysilane or the Si-OH groups of the glass substrate and thereby are cross-linked with the coating.

Besides addressing the carry-over or the removal of fingerprints, also the optical visibility of fingerprints can be reduced with a suitably structured coating. An example for such a coating with a quite coarse structure can be found here [294]. It is obvious that these coatings are not suitable for a glossy or transparent substrate surface.

➤ *Example 50: Preparation of an anti-fingerprint coating on an anti-reflex coated PET substrate [295]*

A PET film (100 µm thick) was coated with a sol-gel hard-coating formulation consisting of 135 parts by weight of silica sol, 129 parts by weight of gamma-glycidoxypropyl triethoxysilane and 70 parts by weight of gamma-chloropropyl trimethoxysilane. The thickness of the coating layer was 3 µm.

On the hard-coat layer, SiO_2 and TiO_2 were alternately deposited each to a lambda/4 optical film thickness by sputtering whereby an antireflection layer consisting of five layers was formed. Next, a coating solution was prepared by dissolving 0.2 g of a perfluoropolyether- modified silane $CF_3O(CF_2CF_2O)_{30}CF_2CH_2OCH_2CH_2CH_2Si(OCH_3)_3$ (compare Figure 6.33, page 165) in 99.8 g of perfluoro(2-butyltetrahydrofuran). The coating solution was applied onto the antireflection layer by spin coating. The coating was allowed to stand for 24 hours in an atmosphere of 25 °C and 70 % humidity, whereby the coating cured into an antifouling layer.

The cured coating exhibits a contact angle against water of 110° and a sliding angle against oil of 2°. Fingerprints can be easily wiped off with a dry fabric.

Anti-fingerprint and anti-reflex coating on PET

Anti-fingerprint coatings on electronic devices like tablet PC's have to be extremely thin, especially when they are applied on anti-reflective surfaces. The coating thickness must not exceed a few nanometers, if the effect of the anti-reflective coating shall not be disturbed.

For devices which are operated with the fingertips, besides the easy cleanability a low coefficient of friction is decisive if a smooth and pleasant user experience is desired.

Coatings based on short-chain silanes cannot fulfill such a requirement profile, they create rather blunt surfaces. Therefore in this example, an alkoxysilane-terminated perfluoro polyether is used as hydro- and oleophobic coating material. These polymers exhibit an exceptionally low coefficient of friction and lead to a smooth feeling.

Like already discussed (compare Figure 6.32, page 163), fluorinated coating materials cannot be deposited in a homogeneous way from conventional solvents, because the first layer of the adsorbed fluorinated molecules repels the coating liquid ("autophobicity").

Fluorinated solvents are an expensive, but doable way to overcome this repellency and to deposit thin, homogeneous coatings of perfluorinated coating materials.

The typical coating thickness of anti-fingerprint coatings on electronic devices is only 10 to 20 nm. Coatings of such a "thickness" can easily be hydrolyzed by the adsorbed water layers on the glass substrate and do not need additional water. Even catalysts are usually not necessary because the acidity of the Si-OH groups of the glass is sufficient to catalyze the hydrolysis and condensation. As an alternative, more reactive groups like acyloxy- or chlorosilanes can be used.

The authors of the example do not disclose the synthesis parameters for the hard coating formulation, but in this book many examples can be found how such a formulation could be processed. One possibility could be to stir the deionized colloidal silica sol with the silanes and isopropanol for 1 to 2 days until hydrolysis is complete. Then curing can be initiated by the addition of 1 % methylimidazol prior to the application as a catalyst. The role of the chloropropylsilane is unclear. Typically it is used to promote adhesion to polymers and to adjust a higher index of refraction to match that of the substrate.

6.8 Hydrophilic coatings

Hydrophilic coatings are needed in many applications. They lead to a spreading of water droplets, thereby a faster drying of wetted surfaces and serve as anti-fogging coatings for e.g. mirrors or traffic signs.

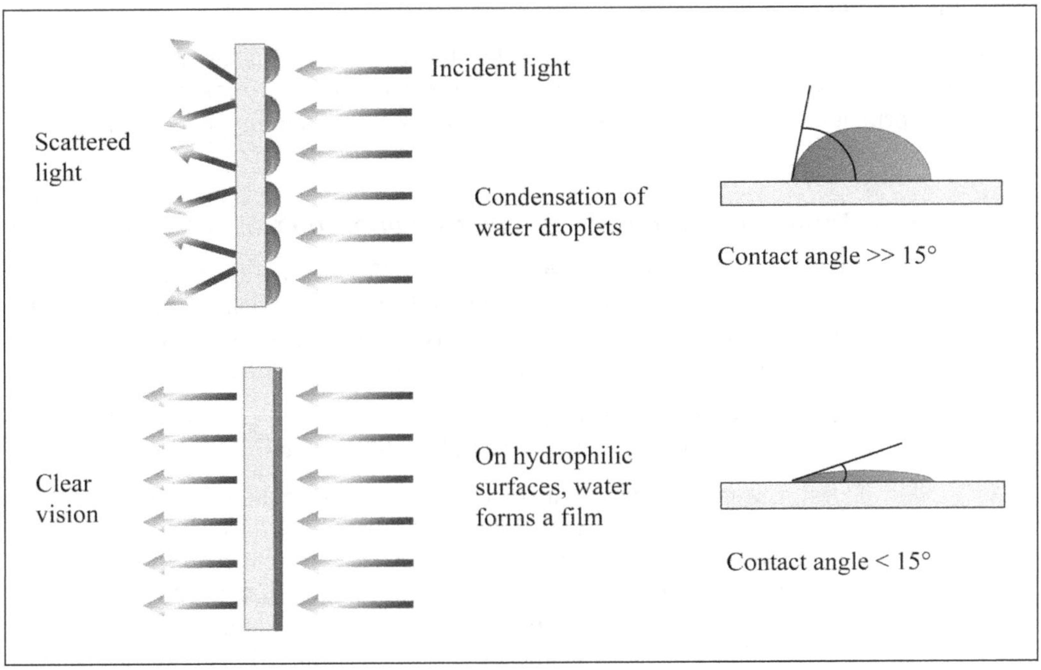

Figure 6.42: Working principle of anti-fogging coatings

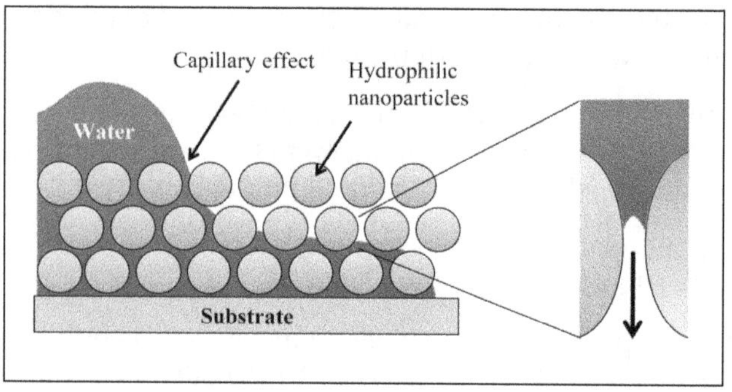

Figure 6.43: On nanoporous hydrophilic surfaces water spreads due to the capillary effect

In general, on hydrophilic coatings water droplets exhibit a contact angle of smaller than 30°, on anti-fogging coatings < 15°.

Fogging happens, if water droplets condense on surfaces which exhibit a lower temperature than the surrounding air. The anti-fogging action bases on the spreading of the condensed water droplets to a film, which does not disturb the transmission of the light.

The sol-gel process offers a variety of possibilities to realize temporary and permanently active hydrophilic coatings. Four principles are used to generate the hydrophilic effect:

- nano-porous coatings
- surfactant-depot coatings
- coatings with polar groups
- photocatalytically active coatings

If a liquid comes into contact with a pore (i.e. capillary), the liquid is sucked into the void. The interaction between the liquid and the pore walls and the surface tension of the liquid are the driving forces for this action (Figure 6.43). The smaller the diameter of the capillary the deeper the liquid is drawn into the pore.

A nanoparticulate coating consists of a system of interconnected nano-capillaries which quickly leads to a spreading of water droplets which come into contact with such a surface (Figure 6.43). Due to this effect, a thin coating of colloidal silica is hydrophilic and leads to a spreading of water.

Hydrophilic coatings are also used to improve the cleanability of surfaces. They are wetted easily and dirt is easily detached from their surface [296]. Example 51 explains the application of this principle for a window cleaner which leaves a nanoporous SiO_2 coating on the cleaned surface.

➤ *Example 51: Formulation of a glass cleaner with protection against re-soiling* [297]

9 g colloidal silica sol ("Levasil 300/30", 30 % SiO_2 12 nm particle size) is diluted with 85.9 g deionized water and 5 g isopropanol. 0.1 g sodium dodecylsulfate is added to the solution which then is filtered through a 1 µm filter and filled into a conventional pump trigger bottle.

The glass cleaner is used to clean window glass. As a reference, the identical cleaning solution without colloidal silica is used.

First, it was observed, that the cleaning action of the solution which contains nanoparticles is rated significantly better than that of the reference. After the cleaning procedure, a wetting test was performed with tap water. Both windows showed a homogenous spreading of the water film and a quick drying. After one week, the test was repeated. Only the window glass which had been cleaned with the nanoparticle-containing cleaner still shows excellent wetting, quick drying and no formation of dirt spots.

Glass cleaner with anti-soiling effect

Hydrophilic surfaces dry quicker, because the water spreads to a film, thereby exposing a larger surface area for evaporation. Especially on window panes this is advantageous, because the non-volatile components of the raindrops like salts or dust are spreaded homogeneously and are not deposited as clearly visible spots.

After the cleaning fluid of Example 51 has dried, a thin layer of nanoparticles remains on the glass and forms a nanoporous coating. When it is raining, the dirt is washed off because the water spreads homogenously, even under the dirt which is collected at the bottom of the glass pane. The window appears cleaner for a longer period of time [298, 299].

Another possible field of application for nanoporous coatings is fabric softeners. Nanoparticles with cationic surface modification adsorb during the rinsing cycle to the predominantly negatively charged textiles and improve the wettability and water transport capability of e.g. textiles made of microfiber [300].

A disadvantage of nanoporous, hydrophilic coatings is their tendency to adsorb hydrophobic compounds which disturb their capability of spreading water. Their effect therefore in most cases is not durable.

If the porosity of a nanoporous network is used to incorporate surfactants, a depot-coating with a significantly longer lifetime is generated. Figure 6.44 depicts a mirror which has been coated with an anti-fogging coating in a room saturated with water vapor. The uncoated glasses fog up while the surfactants of the depot-coating lead to a spreading of the water droplets on the mirror.

Figure 6.44: Sol-gel anti-fogging coating on a mirror. A surfactant depot ensures the homogenous wetting of condensed water in contrast to the uncoated glasses

Source: INM Institute for New Materials GmbH

The mechanical durability of depot coatings generally is not very good, because the surfactants act as softeners and lead to a swelling in contact with water. Figure 6.45 illustrates the structure of a surfactant depot coating [301].

If the depot is depleted, the coating loses its hydrophilic properties. Replenishment could be realized via the application of a concentrated surfactant solution, but in practice this is hardly manageable. This is one reason, why depot coatings were so far not accepted by the market on a broad basis. If the hydrophilic action shall be realized without a

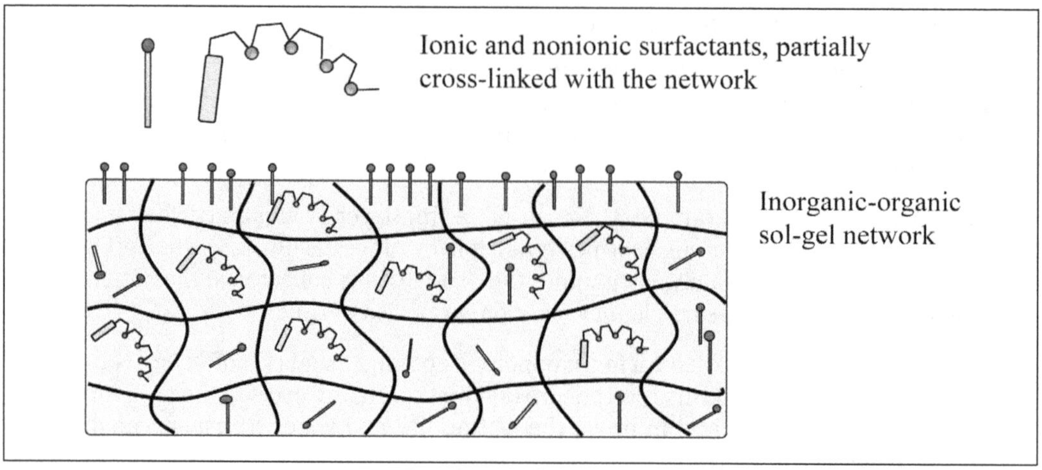

Figure 6.45: Illustration of a sol-gel anti-fogging coating containing immobilized surfactants and a surfactant depot

surfactant depot and without porous coatings, then ionic or hydrophilic groups have to be a part of the coating network itself.

As non-ionic groups, free -OH or polyether groups are possible. Ionic groups could be chosen from e.g. -COOH, $-SO_3H$ or $-PO(OH)_2$. The degree of hydrophilicity or hydrophobicity respectively can be estimated via their HLB increment.

HLB concept to assess surfactants

The HLB concept (hydrophilic lipophilic balance) describes the hydrophilicity and hydrophobicity of surfactants with the help of a number. In order to determine the HLB value, the surfactants are formally splitted into fragments whose HLB increments are known (e.g. $-CH_3$, $-CH_2-$, $-SO_3$, $-CF_2-$) and the values are added up to yield the HLB number. This number is a first hint, whether a surfactant is rather hydrophilic or hydrophobic [302, 303].

According to this concept, the most hydrophilic groups are sulfonate groups $-SO^{3-}$ [304]. So it seems to be a logical step to incorporate as many sulfonate groups as possible into a hydrophilic coating. Example 52 und Figure 6.46 show a possible realization of this concept.

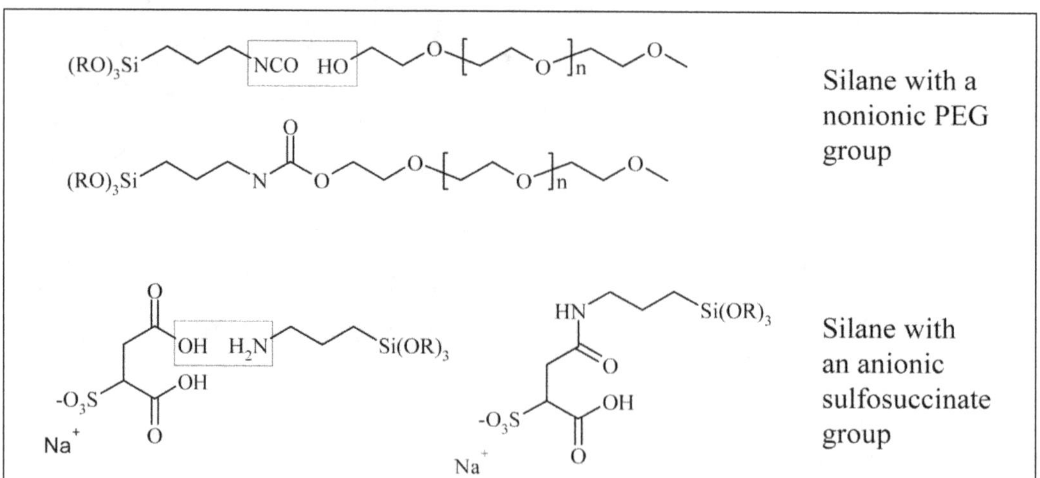

Figure 6.46: Two examples how silanes with hydrophilic groups can be synthesized

➤ *Example 52: Preparation of a durable anti-fogging coating on glass* [305]

221.29 g (1 mol) 3-aminopropyl triethoxysilane are added to 444.57 g sulfosuccinic acid while stirring. The mixture is then heated to 120 °C in a silicone bath for 5 h. After the reaction mixture has cooled down to room temperature, 20 g of the viscous fluid are mixed with 80 g (0.38 mol) tetraethoxysilane and is dissolved in 100 g ethylalcohol. The solution is then mixed with 13.68 g (0.76 mol) of a 0.1 N HCl solution and tempered in a water bath overnight at 40 °C.

The resulting coating solution is diluted with a mixture consisting of 1/3 water and 2/3 N-methyl pyrrolidone (NMP) to a solid content of 5 % and applied to glass plates by spray coating in a wet film thickness of 10 to 20 μm. Subsequently, the substrates are cured in a circulating air drying cabinet for 3 h at 150 °C. The resulting layer demonstrates a contact angle to water of approximately 10 ° and a very good anti-fogging effect which is stable over the long term.

Durable anti-fogging coating

Anti-fogging action can be realized by surface coatings with very polar groups. In this example, a sulfonate group is used to generate a permanently hydrophilic effect. In a first step, a sulfo-succinic acid functionalized silane is synthesized, starting from the corresponding aminosilane and sulfosuccinic acid.

Ionic groups in a coating always carry with them the problem of building up osmotic pressure in contact with water. This pressure can cause delamination, blisters and cracks. Via co-condensation with tetraethoxysilane, a high degree of inorganic cross-linking is realized which is insensitive to swelling and can stabilize the hydrophilic network.

To complete the hydrolysis, the formulation is stirred a long time at low temperatures. Thereby many Si-OH groups are generated, which are quite stable under these conditions. The curing step at higher temperatures activates the condensation reaction which, in contrast to hydrolysis, is temperature sensitive (compare “Hydrolysis and condensation”, page 94).

The use of NMP (N-methylpyrrolidon) as a solvent today is no longer possible due to its toxicity. NEP (N-ethylpyrrolidon) or N,N-dimethyl lactic acid amide can serve as substitutes.

A proposal for an anti-fogging coating based on a non-ionic, covalently bonded silane with PEO can be found here [306] (compare Figure 6.46, page 182). PEO silane modified coatings show a significantly reduced adhesion of lime and organically derived compounds like proteins, a property profile which on the first glimpse one would have rather attributed to low surface energy anti-adhesion coatings [307].

As a general rule, anti-fogging coatings are more prone to scratches and abrasion than hydrophobic coatings. The hydrophilic components lead to a decrease of mechanical durability when the coating comes into contact with water. This is one reason, why anti-fogging coatings rather are applied on surfaces where they do not have to face high abrasive forces, like on the inside of automotive headlights.

6.8.1 Superhydrophilic coatings

The phenomenon of superhydrophilicity was discovered in 1967 by Akira Fujishima. A titanium dioxide coating showed a perfect wetting of water with a contact angle of 0° after having been irradiated with UV light. Approaches to explain this observation are related to

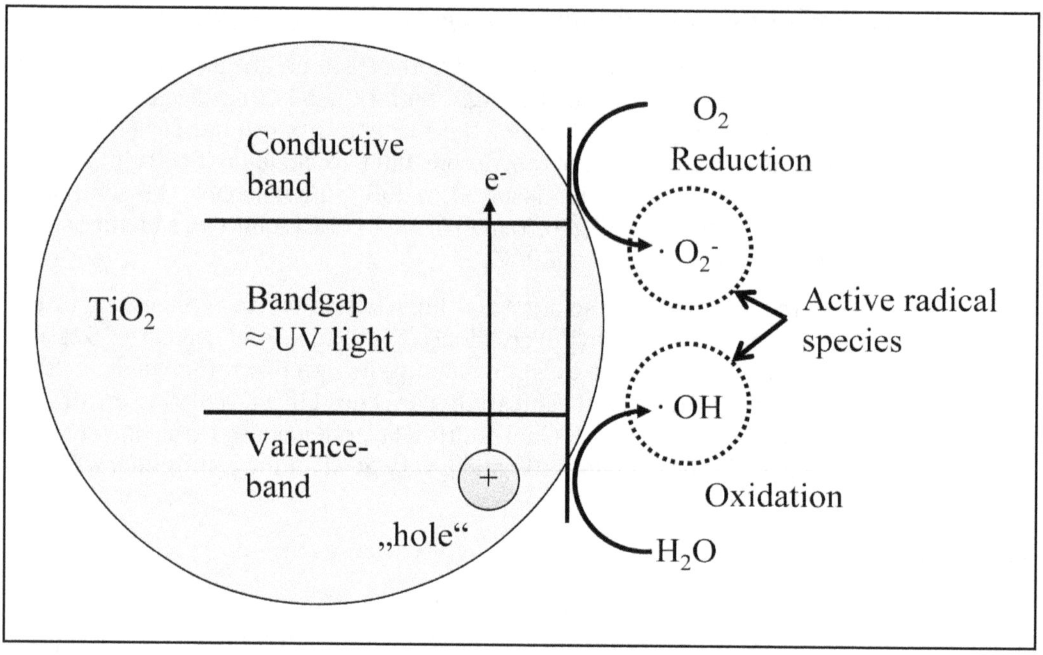

Figure 6.47: Illustration of the photocatalytic effect of TiO_2

UV-induced changes to the crystal lattice which generate a pattern of hydroxylated, high surface energy bonding sites for water [308-310].

Together with its superhydrophilic behavior, titanium dioxide, especially in its anatase form, exhibits a pronounced photocatalytic action (compare "UV absorption", page 78). Via absorption of light with a high enough energy content, electrons are lifted from the valence- to the conductive band. Thus electron/hole pairs are generated, which oxidize/reduce water and oxygen to highly reactive radical species.

Without an appropriate doping of the TiO_2 lattice, only UV light possesses the necessary energy to create electron/hole pairs. By doping, the photocatalytic effect can either be increased (Pt^0 nanoparticles, carbon) or decreased (Al^{3+} doping).

Organic compounds are decomposed or hydrophilized by the aggressive radicals. This is the reason, why titanium dioxide coatings can show a self-cleaning effect and have found many applications like on window panes, traffic signs or tiles [311].

For the deposition of a titanium dioxide coating, titanium dioxide precursors can be brought into contact with the still hot glass surface immediately after the float process [312, 313]. Coated float glass is available from all big glass manufacturers, like e.g. "Pilkington Active".

Hydrophilic titanium dioxide coatings also show disadvantages. Due to their high refractive index, a coated glass reflects more light and is also more sensitive to fingerprints. Like the lotus effect, the superhydrophilicity is a phenomenon which optically emphasizes the areas in which the effect is not working. In dark areas, the available light might not be sufficient to activate photocatalysis and hydrophilicity. Silicones, which might migrate from a window-sealing can be decomposed only very slow, which leads to unaesthetic stains.

If titanium dioxide shall be used in coatings, the coating matrix must be protected against the photocatalytic action. A high content of inorganic components like SiO_2 in the resin can help to keep the coating intact over a longer period of time. As an alternative, a SiO_2 coating on the titanium dioxide can help to separate the radicals from the organic binder.

➤ *Example 53: Preparation of SiO_2 encapsulated TiO_2 nanoparticles* [314]

Under agitation, 1.5 mol 2-(2-methoxyethoxy) acetic acid are added drop-wise to a 2 l round bottom flask containing 0.6 mol titanium tetraethylate. The solution is stirred for 1.5 h and subsequently hydrolyzed by the addition of 10 mol water. After the removal of all volatile constituents in a rotary evaporator at a water bath temperature of about 85 °C, a resin-like product is obtained.

17.5 g of the resin-like product is dissolved after adding 42.5 g ethanol while stirring. When a clear solution has been obtained, it is transferred to a Teflon vessel, which is closed with a sealing ring and lid and placed inside a closed autoclave. The autoclave is placed in an oven, which is heated up to 160 °C and held at this temperature for 4 h. A white-colored, gelatinous mass is obtained which contains titanium dioxide nanoparticles with a mean crystallite size of 9 nm.

21.75 g of the gelatinous mass are mixed with 400 g ethanol, 1.99 g tetramethoxysilane, 725.23 g water and are dispersed after the addition of 16.17 g of 25 % NH_4OH solution (pH approx. 9 to 10). The colloidal solution obtained is reduced to 50 % of its volume by evaporation and 6 g of 3-glycidyloxipropyl trimethoxysilane are added to it.

After overnight agitation, it is concentrated further until a 6 % dispersion (TiO_2) is obtained, which can be absorbed in organic solvents for further processing or can be incorporated free of agglomerates into organic and hybrid polymer formulations (for example, varnishes, resins, polymers).

SiO_2 encapsulated titanium dioxide nanoparticles

In Example 53 the already discussed method of a surface modifier moderated reaction control is used for the nanoparticles synthesis (compare Figure 3.18, page 49). The surface modifier trioxadecanic acid (2-(2-methoxyethoxy) acetic acid) shows a strong affinity towards titanium dioxide and can bind simultaneously to the surface via a chelate with its carboxy group and via ether bridges by dipole interactions (compare Example 2, page 21).

The surface modifier is a carbonic acid and also acts as a hidden source of water (compare "Water-free sol-gel techniques", page 100). In the first reaction step, a part of the alkoxy groups is already hydrolyzed, before with the addition of water the rest of the alkoxy groups is hydrolyzed. The stabilized TiOx clusters aggregate according to their titanium dioxide/surface modifier ratio to yield titanium dioxide nanoparticles of different sizes. The higher the amount of surface modifier, the smaller the nanoparticles are.

The controlled precipitation yields amorphous and partly crystalline particles. If their refractive index and photocatalytic activity is to be maximized, a controlled thermal processing has to follow. Particles should remain dispersed during this process step, so an autoclave is necessary to avoid boiling and evaporation of the solvent.

After the crystallization step, a SiO_2 shell is precipitated on the particles by using an ammonia catalysed sol-gel process. Under these reaction conditions of very high condensation speed, a patchwork-like, maybe porous layer of SiO_2 seeds is formed. If the SiO_2 shell would have been formed under acidic conditions, a rather smooth polymeric SiO_2 shell can be expected. It is rather difficult to anticipate which structure would be better with regard to the application profile; this has to be found out empirically.

The SiO_2 coated titanium dioxide now can be used as UV absorber and the binder resin will not come into contact with the photocatalytically active surface.

6.9 *Tailoring of the refractive index of sol-gel coatings*

The refractive index is decisive for the optical properties of a coating. Coatings with high refractive index reflect more light and show high gloss, whereas low refractive index coatings rather act as anti-reflective coatings.

Combinations of layers with alternating high and low refractive indices are used as highly effective anti-reflex coatings e.g. in optical systems (compare Example 50, page 178).

The reflective index of a material is directly correlated to its polarizability and crystallinity as well as in many cases its density. Amorphous and porous materials show lower refractive indices than crystalline and dense materials.

The sol-gel process enables the formulator to tailor the refractive index of a coating by variation of the raw materials like organofunctional silanes, nanoparticles and metal-organic compounds. Figure 6.48 shows some typical examples of high and low index raw materials.

The refractive index of a coating can be estimated by adding up the volume ratio of its components. The refractive index of a SiO_2 coating (nD = 1.45) can be increased by 10 vol.-% zirconium dioxide (nD = 1.85) to 1.49. The density of zirconium dioxide (approximately 5.5 g/cm^3) is very high, compared to SiO_2 (approximately 2.2 g/cm^3), therefore 10 vol.-% translates into 22 weight% zirconium dioxide which is necessary to realize this volume ratio.

By the sol-gel process, materials are synthesized "from the cold end", which means that loosely packed aggregates and agglomerates in a first step form a porous network, which is compacted further and further during the thermal densification. Due to the high content of

Low refractive index:
- Partially fluorinated alkyl silanes
- Methyl/dimethylsilanes
- Nanoparticles of MgF_YO_X
- Porous coatings

High refractive index:
- Thiosilanes
- Phenylsilanes
- Nanoparticles of TiO_2, ZrO_2 or Sb_2O_3
- Metalorganic complexes

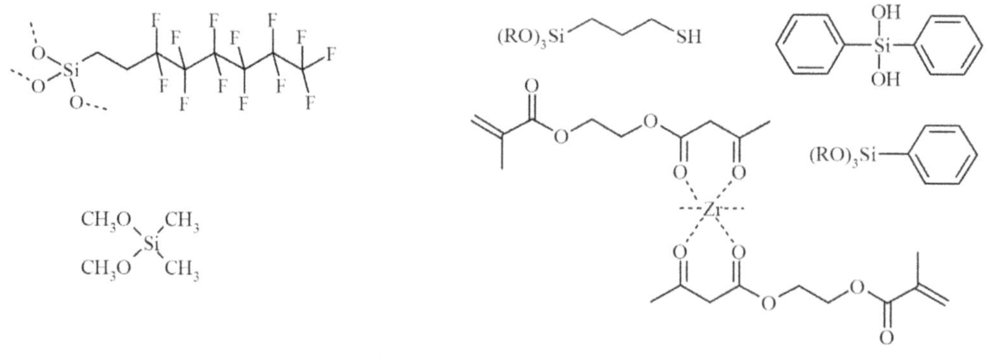

Figure 6.48: Examples for suitable sol-gel precursors to influence the refractive index of the resulting material

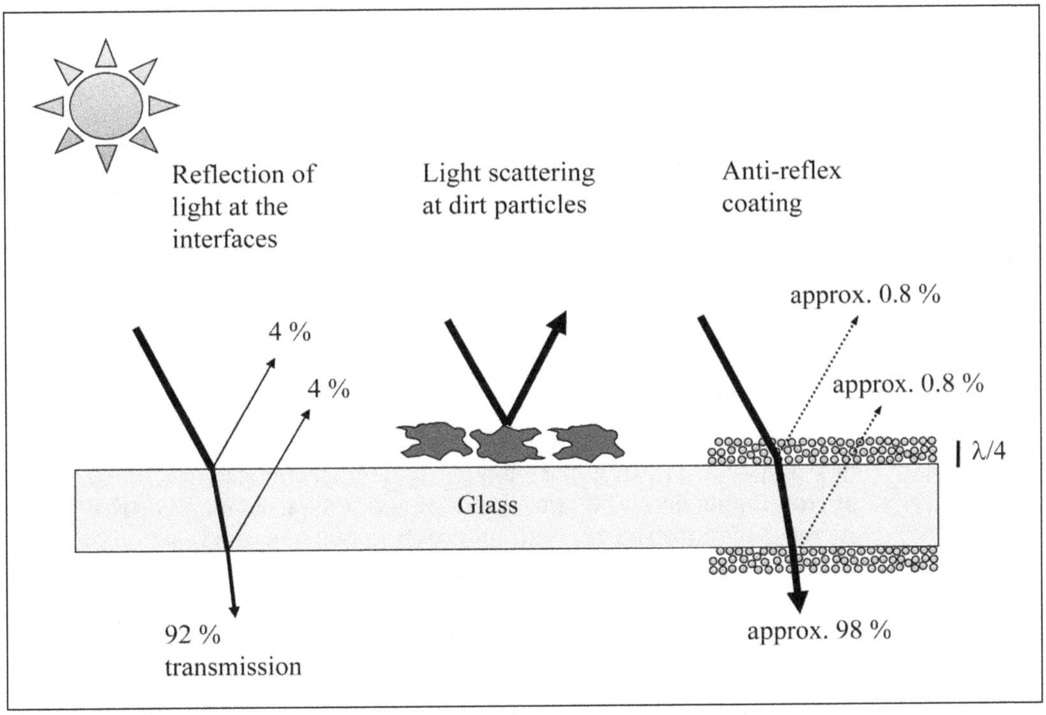

Figure 6.49: Nanoporous antireflex coating on glass

air (n_D = 1.0) in the pores, the refractive index of sol-gel materials first is very low and can be increased as a function of the densification temperature.

The porosity which is generated by the chemical synthesis decreases the refractive index of the material and is reduced gradually during sintering. Porosity can also be used deliberately to realize coatings of low refractive index like it is demonstrated in the example of a broadband anti-reflex coating for photovoltaic devices.

The production process involves a dip coating step in which the porous SiO_2 layer is generated from colloidal silica "Stöber"-particles, followed by thermal densification during the thermal toughening process of the glass substrate.

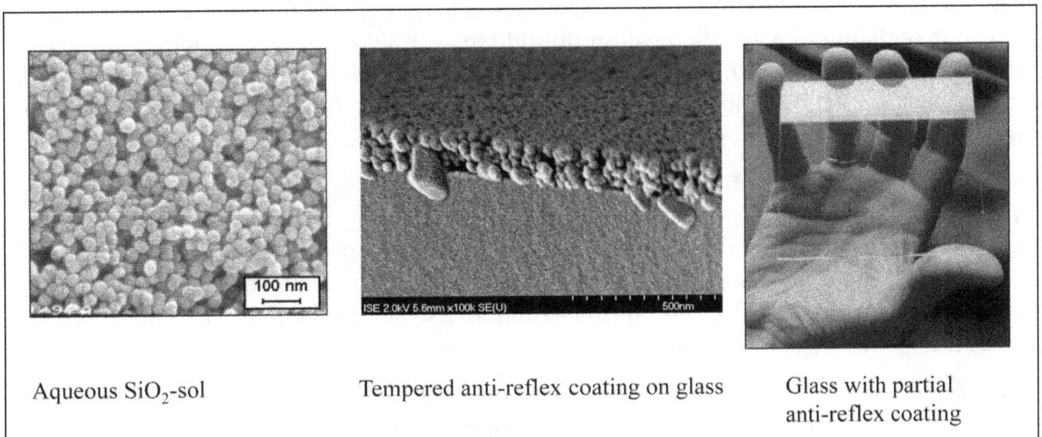

Figure 6.50: Sol-gel anti-reflex coating for glass *Source: Merck KGaA*

By reflection at the interfaces air/glass/air, light loses roughly 8 % of its intensity when it passes through a conventional glass pane before it hits the photovoltaic cells. A porous coating with a thickness of $\lambda/4$ acts as a broadband anti-reflective layer which decreases the losses to only about 2 %.

In contrast to anti-reflex coatings, antiglare coatings improve the visibility of objects which are placed directly behind the glass by a colorless, transparent and structured coating. The irregular structure which avoids direct reflection of light is being used for e.g. picture frames (compare Example 48, page 175) [315]).

Another possibility to synthesize sol-gel coatings with low refractive index is described in Example 54.

➤ ***Example 54: Preparation of a coating with low refractive index on glass*** *[316]*

MgF_2 sol: 25.396 g (0.22 mol) of magnesium ethoxide are added to 522.81 g of 2-propanol. 51.016 g (0.35 mol) of trifluoroacetic acid are added to the stirred dispersion. The start of the reaction is accompanied by a slight increase of temperature. With progressing reaction, the reaction mixture clarifies. Any insoluble constituents are removed by means of a syringe filter (1.2 μm) resulting in a yellow solution. The coating composition is storage-stable for at least 4 weeks at room temperature.

ZrO_2 sol: 24 g (51.3 mmol) of zirconium tetra-n-propoxide (70 % by weight in 1-propanol) are dissolved in 240 g of 2-propanol. 2.553 g (25.5 mmol) of acetylacetone are added while stirring and the mixture is stirred further for 10 min. Subsequently, 1.8 g of concentrated hydrochloric acid are added and the mixture is stirred at room temperature for 1 h. Filtration through a 5 μm syringe filter results in a yellow, clear sol.

MgF_2 composite sols are prepared by mixing the MgF_2 sol with the appropriate amounts of SiO_2 sol, Al_2O_3 sol or ZrO_2 sol in order to adjust the refractive index and mechanical stability of the final coating. Soda-lime silicate glass panes are cleaned by wiping with ethanol and coated with the mixed nanoparticle coating formulations by dip coating (3.5 mm/s withdrawal speed). The coating is cured at 450 °C for 30 min.

Glass coating with low refractive index

There are only few coating materials available which are suitable for realizing a low refractive index without porosity. Magnesium fluoride (n_D = 1.38) is an exceptionally interesting optical material because it exhibits a far lower refractive index than the already low index material SiO_2 (n_D = 1.46) and shows a very high transmission for UV radiation over a broad spectrum.

The synthesis and stabilization of magnesium fluoride via the sol-gel process is challenging. This is the reason why in Example 54 the magnesium fluoride is synthesized in the very last step of the process, the thermal densification. The trifluoromethyl group of trifluoroacetic acid serves as the source for the inorganic fluoride when the molecule decomposes at high temperatures. Trifluoroacetic acid in this example also serves as additive to speed up the dissolution of the magnesium methylate in 2-propanol.

Magnesium fluoride coatings which are prepared according to this method are very sensitive towards abrasion and scratches. Via the combination with zirconium dioxide, not only the scratch resistance can be improved, but the zirconium dioxide also opens up the possibility to adjust the refractive index of the coating in a wide range.

Besides the possibility to synthesize magnesium fluoride by thermal decomposition of organofluorine compounds, magnesium fluoride can also be synthesized by anhydrous fluorolysis of organomagnesium compounds with HF [317-319].

In this process, HF takes over the role of H_2O in the hydrolysis reaction.

M-OR + HF	→ MF + R-OH	Fluorolysis
n MF	→ $(MF)_n$	Aggregation

An anhydrous solution of HF in an alcohol like methanol serves as a source for HF. The process yields 3 to 5 nm sized magnesium fluoride particles dispersed in alcohol, which can be used for low refractive index coatings in optical quality.

By mixing HF and H_2O in a desired ratio, oxyfluorides in a tailorable stoichiometry are accessible, which find application as heterogeneous catalysts.

➤ ***Example 55: Preparation of a sol-gel coating with a high refractive index on polycarbonate*** *[320]*

A partially hydrolyzed organoalkoxysilane sol is prepared by mixing 100 g of γ-glycidoxypropyl trimethoxysilane in 100 g ethanol with 8 g of water and 0.2 g of nitric acid and stirring for 10 minutes at ambient temperature. To this partially hydrolyzed silane formulation are added 20 g of zirconium n-propoxide ($Zr(OC_3H_7)_4$) and 10 g of titanium ethoxide ($Ti(OC_2H_5)_4$). The composition is stirred at ambient temperature for 20 minutes to allow copolymerization of the metal alkoxides with the partially hydrolyzed organoalkoxysilane. Finally, 20 g of water and an additional 60 g of ethanol are added to hydrolyze the composition and dilute it for coating application.

A polycarbonate substrate is cleaned and primed by dipping in aminosilane (A1120 from Union Carbide) for 7 minutes, rinsing with 2-propanol followed by water and drying for 30 minutes at 80 °C. Then the primed polycarbonate is dip-coated with the coating composition, dried at ambient conditions and cured at 130 °C for 2 h. The index of refraction of the coating is 1.6, compared to 1.54 for a comparable coating without the titanium- and zirconium dioxide.

High reflective index sol-gel formulation

Primers are used on difficult to coat surfaces to promote adhesion of the following overcoat. Further functions could be corrosion protection, anti-tarnish action or depot for UV absorbers (compare "Figure 6.4, page 135).

In principle all known adhesion promoters can be used like aluminates/zirconates, organofunctional silanes or polymers. Amino- and chloropropyl-functionalized alkoxysilanes often are chosen to improve the adhesion on polymers like polycarbonate. The amino groups for example can form a salt with free carboxylate groups of the polymer. Experience however has shown that pure aminosilanes are too hydrophilic and lead to adhesion failure under humid conditions.

Hydrophobically modified aminosilanes (e.g. butylaminopropyl-) or co-condensates with more hydrophobic silanes often improve adhesion and show a better flexibility. Another aspect is yellowing. Aminosilanes tend to turn yellow under UV irradiation and this is a reason why primers should be kept as thin as possible and substituted amines are preferred over primary amines.

In the example, the prehydrolysis technique is used to homogeneously combine the faster reacting zirconium and titanium alkoxides with the slower reacting silanes (compare "Metal compounds in the sol-gel process", page 103). In principle, prehydrolysis means to react the slower component with a sub-stoichiometric amount of water (R_{OR} of < 0.5) before adding the faster components and the rest of the water.

When a sub-stoichiometric amount is used, it leads to an almost complete consumption of water and the formation of many Si-OH groups, which can react in step two with the metal alkoxides. Thereby precipitation is suppressed, which otherwise would occur if the metal alkoxides came directly into contact with the water. After the reaction took place, the residual amount of water can be added without causing gelation or precipitation. During prehydrolysis, the pH should be chosen wisely to promote hydrolysis, but to avoid condensation. The Si-OH groups are necessary for the reaction with the metal alkoxides and their amount should be as high as possible.

This is the reason, why prehydrolysis almost always is carried out under acidic conditions. Basic catalysts would foster condensation and reduce the number of Si-OH groups. The catalyst should also be chosen according to its anion. Nitrate or e.g. citrate can help to stabilize the metal oxide which is generated in step two.

During curing, the epoxy-functional silane polymerises to a polyether. Especially zirconium dioxide catalyzes this reaction and contributes to the alkaline resistance of the nanocomposite network. Pure silane condensates often show instability when exposed to alkaline media, which is caused by the cleavage of the Si-O-Si bond. Resistance against alkaline media can also be improved by the dimethylsiloxane group ($-Si(CH_3)_2O-$), but in this case in a trade-off against mechanical stability.

A comparably simple method to synthesize thin (<< 1 μm) metal oxide high refractive index coatings is to use coating liquids which only contain metal alkoxides, dissolved in suitable solvents like alcohols or ether alcohols. When applied in a thin layer, the humidity of the air contributes enough water to complete hydrolysis and condensation directly on the substrate and so metal oxide coatings form without the need of a catalyst.

Metal resinates are complex formulations consisting of natural or artificial materials (like e.g. colophony, uric acid, ricinus oil) and metal salts. They are used since many years to realize iridescent, highly refractive coatings (lusters) on decorative glass and ceramic articles. During the thermal treatment, the organic part of the formulation decomposes and a metal oxide network remains [321, 322]. The process shows parallels to the high-temperature solvent synthesis method for nanoparticles from chapter "High temperature liquid phase synthesis of nanoparticle" page 32.

6.10 Anti-corrosion coatings

Every year, corrosion causes damages which can be valued in billions of euros [323]. If most metals come into contact with water, oxygen and an electrolyte (e.g. salt), oxidation takes place. The type of reaction products which are generated determines whether the metal corrodes or not. If the oxidation products form a passivating, dense protective layer, the progression of the reaction is stopped like it is the case for chromium. Iron, aluminum or magnesium however do not form dense layers which leads to their progressing destruction.

A variety of processes and treatments exist to passivate metals against the undesired reaction with oxygen and water. In many cases still chromates or other toxic substances are used. Here the sol-gel process offers the possibility to protect metals by an environmental benign technology:

- stabilization/passivation of the surface
- coating materials with high network density as a barrier against water and oxygen
- stable coating matrix for cathodic corrosion protection (e.g. by zinc powder)

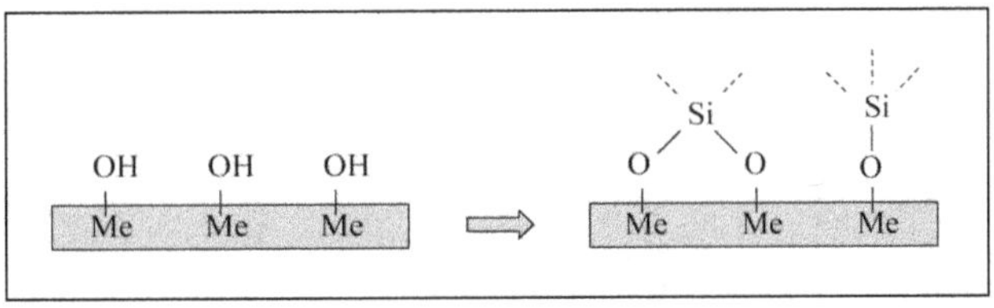

Figure 6.51: Protection of metal surfaces against corrosion by the formation of Si-O-Me bonds

Si-OH groups can react with metal hydroxide groups and form Me-O-Si bonds. By this reaction, the metal surface is stabilized against corrosion. This chemistry is widely known from the application of silanes as adhesion promoters [324-326].

This principle of interface inertisation is not only interesting for sol-gel anti-corrosion coatings, but can also be used for nanometer-thin primers under conventional organic coatings.

The requirement profile for thin silane-based primers is complex and partially contradictionary. The silanes should be stable as formulation in water and exhibit a long shelf life, but on the other hand they should be highly reactive and hydrophobic on the metal surface to be able to act as an efficient barrier against water.

One technical objective is the replacement of chromate and phosphate-based conversion coatings especially for coil coating processes. Example 56 explains the synthesis and application of such a silane-based primer for metals.

➤ *Example 56: Alcohol-based anti-corrosion primer for aluminum* [327]

A 3 % solution of hydrolyzed 1,2-bis-(triethoxysilyl) ethan (BTSE) was prepared as follows:

3 ml of BTSE was added to 0.3 ml of acetic acid with magnetic stirring. 3 ml of water was then added to this solution, followed by 94 ml of methanol.

Samples of 2024-T3 (aluminum; 2.5 cm · 10 cm) were then ultrasonically cleaned for 15 minutes in hexane, 15 minutes in acetone and 15 minutes in methanol. Thc samples were then dipped (at room temperature) into the 3 % BTSE solution for approximately 100 seconds, and then blown dry with air. After drying, the coated samples were then cured at 100 °C for 15 minutes. The dipping/drying/curing steps were then repeated two additional times, such that a total of three layers of BTSE were deposited on the samples.

The coated samples, untreated samples, and samples treated by a conventional chromating process were then partially immersed in a 3 % NaCl solution, and the water line was monitored for evidence of corrosion. Corrosion began to appear on the untreated samples in less than 12 hours, and in 100 to 150 hours for the chromated samples. The BTSE-coated samples exhibited no corrosion after 175 hours.

Anti-corrosion primer for aluminum

Bipodal silanes with two $Si(OR)_3$ groups are used as adhesion promoters with exceptionally good adhesion even at high humidity or as silicate modifier for tires [324, 328].

1,2-bis-(triethoxysilyl) ethane is known as effective anti-corrosion treatment [325]. The silane is hydrolyzed using diluted acetic acid as a catalyst. The mildly acidic pH fosters the genera-

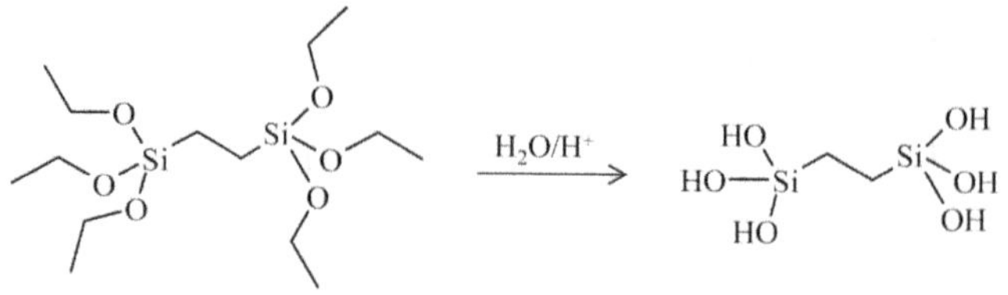

Aqueous primer for Al surfaces based on 1,2-bis-(triethoxysilyl)ethane

Figure 6.52: Aqueous primers for metals based on hydrolysed silanes are currently being used for corrosion protection. The example shows 1,2-bis-(triethoxysilyl)ethane.

tion of Si-OH groups and prevents rapid condensation (compare "Hydrolysis and condensation", page 94). This large amount of Si-OH groups is a prerequisite for the formation of Si-O-Me bonds which passivate the metal surface. The condensation reaction is activated via the thermal treatment. As already discussed, condensation reactions are very sensitive to the temperature. This can be used to increase shelf life (cooling) or to speed up the reactivity of silane coatings by heating.

The simple corrosion test with salt water shows that the silane pre-treatment can yield equally good results compared to a treatment with chromates. Since however in industrial processes the usage of flammable solvents in open baths is difficult, waterborne primer solutions based on silanes were developed [329].

The example shows, that silane based primers are able to suppress corrosion, so it can be assumed that sol-gel coatings, which mainly consist of silanes can be used as anti-corrosion coatings as well.

Based on this assumption, sol-gel anti-corrosion coatings have been developed, which combine the interface stabilization of a primer with a high network density and barrier properties [330-333].

➤ ***Example 57: Preparation of a transparent anti-corrosion coating for aluminum*** [334]

23.64 g of glycidoxypropyl trimethoxysilane were charged to a flask and a solution of 7.8 mg of sodium benzoate in 7.7 g of silica sol ("Levasil 200S", 30 % SiO_2 by weight in water; particle size: 15 to 20 nm) was added with stirring. The initially cloudy two-phase mixture was stirred at room temperature for 5 hours, after which a mixture of 9.13 g of bisphenol-A in 9.2 g of ethanol was added and the combined mixture was stirred for a further 18 hours. Subsequently, 0.41 g of methylimidazole was added and stirring was continued for 40 minutes. The resulting transparent sol had a pot life of more than 6 days.

Aluminium panels (Al 99.5) were degreased with an alkaline cleaner ("P3 Almeco 18", Henkel KGaA) at 50 °C, pre-treated additionally in 20 % strength nitric acid for 10 seconds, rinsed with deionized water and dried at 70 °C in a circulating air furnace.

The coating was applied to the metal panels by dip coating, using a drawing speed of 5 mm/s. Curing took place at 130 °C for a period of 60 minutes. The thickness of the coating was approximately 9 µm. After 240 hours of exposure in the salt spray mist climate (CASS – copper accelerated salt spray) there was no surface damage; the scribe creep was less than 0.4 mm. Coatings produced on steel (ST 37) withstood exposure in the CASS-test for 120 hours without occurrence of corrosion.

Transparent anti-corrosion coatings for aluminum

The example describes an inorganic/organic polyether which is synthesized by a base catalyzed polymerization from an aromatic diol and an epoxy silane. The epoxysilane is hydrolyzed by using the weakly acidic colloidal silica dispersion to prevent a premature ring opening of the epoxy group (compare "Waterborne formulation", page 124). The colloidal silica dispersion "200S" was modified with polymeric aluminum chloride and exhibits a pH value of approximately 3.5.

The mild reaction conditions foster hydrolysis, but not condensation. A storage stable formulation results, which is rich of Si-OH groups and can react with the Al-OH groups on the surface of the aluminum metal.

Bisphenol-A is an aromatic, hydrophobic building block which reduces the polarity of the polyether. An unmodified polyether would swell in a humid environment and lead to delamination of the coating. Methylimidazole is an effective basic catalyst of the polyether polymerization and should be added immediately before the coating step to ensure a long shelf life of the formulation.

Besides by methylimidazole, the reaction of the epoxygroup is catalyzed also by metal-, sulfate-, or phosphate-ions. In the presence of water, quickly the diol is formed which cannot cross-link further. Therefore the aluminum-containing colloidal silica sol may not be the best choice for this formulation. A deionized silica sol at its IEP also shows a pH of approximately 3 and does not contain metal ions, which could affect the epoxy silane. In general, silanes with epoxy groups should not be catalyzed by sulfuric acid or phosphoric acid because these acids convert a significant fraction of the epoxide to the corresponding diol.

Sodium benzoate is added to the formulation as an additional corrosion inhibitor and buffer. The sodium content can catalyze the densification of the SiO_2 network significantly even at low densification temperatures (Figure 6.20, page 147).

Corrosion not only takes place outsides, but can also be caused by high temperatures and aggressive media. If steel is heated up to some hundred degrees, a dark tarnish layer forms which consists of oxidation products (compare Figure 6.24, page 152). Sol-gel coatings with a high content of inorganic building blocks are suitable to suppress this high temperature corrosion. In order to realize a coating thickness of some micrometers without cracks or delamination, they must contain a sufficiently high amount of flexibilizing organic components (compare "High temperature resistant coatings", page 139ff).

➤ *Example 58: High-temperature resistant and protective coatings for metals* [335]

Cleaned substrates (stainless steel 1.4301, brass and aluminum) were dip-coated at a drawing speed of 4 to 6 mm/s with the coating composition described in Example 33, dried for a short period of time (10 minutes) at room temperature and subsequently dried at 80 °C for 1 h. To densify the coating, the dried substrates were heated to 500 °C at a heating rate of 1 K/minute, respectively 400 °C in the case of aluminum and kept at that temperature for 1 hour.

The layer thicknesses after the densification were 2 to 4 μm. HREM studies on the coatings on steel and aluminum showed a layer free of defects. The blushing/tarnishing of the stainless steel during heating was prevented almost completely by the coating. Furthermore the layer showed no damage even after 1000 hours of salt spray test. The coated brass and aluminum showed no corrosion after 3000 hours in the salt spray chamber (according to DIN 50021), whereas the uncoated part was completely corroded.

High-temperature corrosion protection

The organo-alkalisilicate formulation from Example 33 protects the steel substrate by a dual action (compare Figure 6.21, page 148). Its alkalinity protects the steel from acidic gases and the almost dense silica network protects the metal surface from oxidation. Silicate coatings which contain alkaline ions sinter to a dense film already at low temperatures.

In contrast to alkaline ion free coatings, which contain pores even after densification at high temperatures, alkaline ion containing coatings are almost dense after a treatment at 450 °C.

The network of the coating densifies, before a temperature is reached at which the steel begins to oxidize. Even aluminum, which is sensitive to alkaline media, is protected. The content of alkaline ions is low, compared to conventional alkaline silicates ("water glass"). The protective action of the coatings against salt spray is excellent, something which could be expected from an almost glasslike coating, if no cracks are present.

6.11 Coatings with antistatic action

Electrically insulating surfaces are prone to building up an electrostatic charge via the triboelectric effect [336]. Dust particles are attracted which on the one hand give the surface an unattractive look and on the other hand easily lead to scratches. By applying an antistatic coating, the surface charge can be dissipated and the attraction of dust can be avoided.

Inorganic/organic coatings are an attractive base material for antistatic coatings due to their high mechanical and chemical resistance. Their high electric resistivity of 10^{13} to 10^{16} Ω cm has to be lowered by additives to values of about $< 10^{8}$ Ω cm to ensure that a charge carrier transport takes place [337-339].

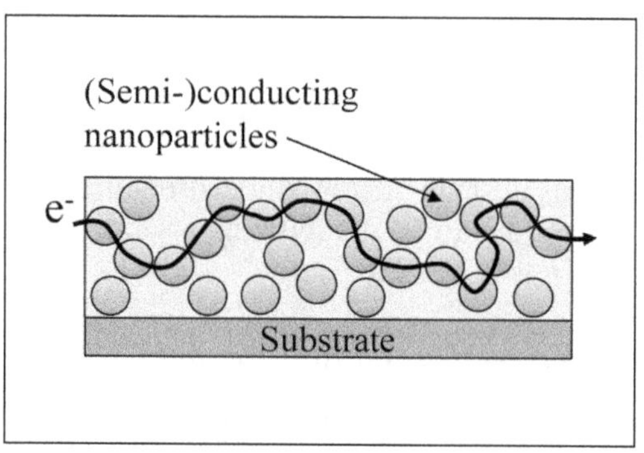

Figure 6.53: Charge carrier transport through a coating containing ITO-nanoparticles

One possibility is the integration of (semi) conducting nanoparticles like ITO, ATO or carbon nanotubes. Using this approach, it is important to reach the percolation threshold to ensure that the particles build up a conducting network within the coating.

Another possibility to dissipate an electrostatic charge is to incorporate charge carrying or hydrophilic organic components like octadecyldimethyl trimethylsilylammoniumchlorid (Figure 6.54) or the polymer from Example 59.

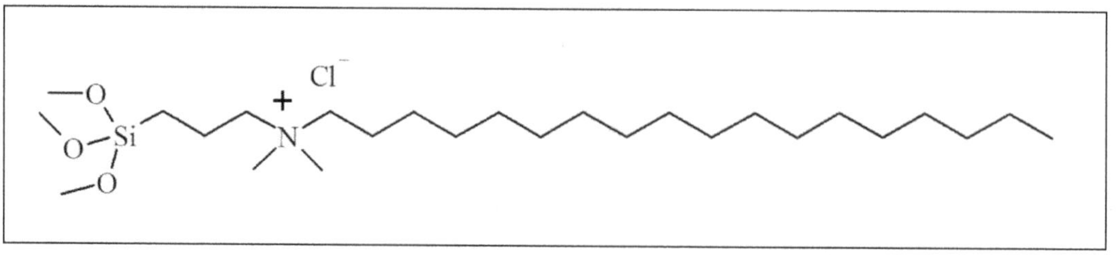

Figure 6.54: Alkoxysilane bearing a quarternary N-atom as a basis for antistatic coatings

➤ *Example 59: Preparation of a scratch-resistant antistatic coating* [340, 341]

Antistatic hard coating formulations were prepared by mixing the antistatic ingredients into an abrasion resistant coating formulation.

Hard-coating: 0.1 N HCI was added dropwise to glycidoxypropyl trimethoxysilane. During mixing, the temperature should be controlled to stay between 25 to 40 °C. Then dimethyldiethoxysilane was added dropwise to the mixture. After completion, the mixture was kept stirring for 16 to 20 h at room temperature to complete hydrolysis. A SiO_2 colloid (MA-ST from Nissan Chemical America Inc, 30 % solid content in methanol) was introduced into the mixture, followed by stirring for about 10 min. Aluminum acetylacetonate, FC4430 (fluorosurfactant, 3M) and methylethylketone were added and stirring was continued for 10 min.

Antistatic ingredient: 6.1 g silylated polyethyleneoxide (Bis-[triethoxysilylpropyl-] polyethylene oxide, 25-30 EO, Gelest Inc) was first hydrolyzed with 1.8 g 0,1 N HCl (added dropwise) at 25 °C for 24 hours then 1.1 g lithiumtrifluoromethanesulfonate was added. The mixture was stirred for 2 hours.

The antistatic abrasion resistant coating formulation was prepared by mixing the antistatic ingredient with the abrasion resistant coating formulation in a designated ratio. The solution was filtered (3 µm filter) and kept at 4 °C for storage. The coating material was applied to CR39 (polydiethyleneglycol-bis-(allylcarbonate) lenses in 1-2 µm thickness and cured at 75 °C for 30 min. The abrasion resistant coating formulation without the antistatic additive was applied on top of the first layer in a thickness of 2.5 to 3.5 µm and cured by heating at 75 °C for 12 min, followed by 105 °C for 3 h.

Scratch-resistant antistatic coating

The example describes a double layer coating on plastic lenses, in which a hard coating is applied on top of the antistatic coating. The base material for both coatings is identical, the antistatic additive for the lower layer is a PEGylated silane and a lithium salt.

Polyethylene oxide groups are hydrophilic and excellent lithium ion carriers. So in this example two principles how to generate an antistatic effect are combined. The basis for the coating material is an epoxysilane hydrolysate, which is cured by the metal-organic catalyst $Al(acac)_3$ to yield a polyether. The hydrophilicity of a pure polyether however would be too high for a scratch resistant coating.

Contact with humidity would cause swelling of the polyether network, accompanied by a loss of its scratch resistance. So in this formulation, dimethyldiethoxysilane, a silicone building block, is used to counteract the hydrophilicity of the polyether and colloidal silica is added as inorganic filler and cross-linker to prevent swelling.

The protection against humidity is the reason, why the antistatic layer was placed under the scratch resistant layer. The hydrophilic antistatic additives weaken the mechanical properties of the coating to such a great extent, that a top coating is necessary to protect not only the plastic lens but also the antistatic coating.

It remains unclear, why the faster reacting methoxide is prehydrolyzed before the slower reacting dimethyldiethoxysilane is added. The SiO_2 nanoparticle dispersion exhibits a low pH of approximately 3 and could be used as a catalyst for hydrolysis instead of the HCl to reduce the salt content in the topcoat.

After all, the performance of the final coating material decides the synthesis strategy and not a theoretical "should-be" know how. It is not always the most logical synthesis strategy which yields the best results. One thing however remains true: First you have to know the rules before you should consider breaking them.

6.12 Antibacterial coatings

Given a suitable environment, bacteria and fungi grow on almost all materials. This is normal and no reason for concerns, since our body is adapted for the ongoing challenge by these pathogens and even needs them to build up and maintain a healthy immune system.

There are however places and situations, where we have to reduce the number of pathogens like in hospitals or in the kitchen. Usually surfaces which are nonporous and easy-to-clean are used. But besides these passive measures there is a need for an active antibacterial action on surfaces. This is possible and can basically be realized by three principles:

- materials showing intrinsic antibacterial activity like silver
- materials with a depot of an active material in a non-active carrier material
- materials which act as catalysts like TiO_2

There are only a few surfaces with catalytic or intrinsic antibacterial action. Silver and titanium dioxide with a photocatalytic action are two examples (compare "Hydrophilic coatings", page 179). The depot principle however can be used for a variety of antibacterial active materials.

Depots of materials with antibacterial action

A series of antibacterial compounds exist, which can be integrated into a sol-gel coating.

Sol-gel coatings are predestined as depot materials, because they can be applied in micrometer thin coating thicknesses and due to their high content of inorganic components, they exhibit the necessary mechanical stability to host the amount of active materials which is needed for a long lasting effect. Organic antibacterial compounds can act as softeners and weaken the mechanical properties of coatings to such an extent, that they become sensitive to scratches and abrasion.

Sol-gel coatings with their high network density compensate this effect and decrease the diffusion speed of the active material to the surface, thus preventing a fast leaching.

Among the materials which can be considered are well known compounds like triclosan, benzoic acid, quarternary ammonium compounds, chlorhexidine, antibiotics .... or silver nanoparticles. Diffusion of the active materials to the surface ensures that always a sufficient amount of active material is present.

Optimized efficacy combined with tailored release

Like already mentioned before, the tailoring of the speed with which the active material is released is decisive for the performance of the coating. If it is too slow, there might not be enough active material present at the surface after cleaning. If it is too high, an unnecessarily high amount of active material is present at the surface and the depot is leached out quickly.

Within this context, an often underestimated problem is to determine, whether the coating still is active or not. Since the coating does not indicate, whether its depot is full or empty, the design of an antibacterial coating has to cover even a worst-case scenario in

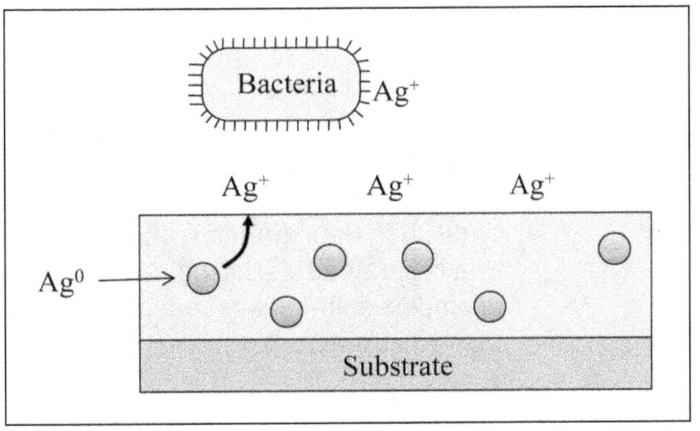

Figure 6.55: Sol-gel coating with antibacterial action. Nanoparticles of silver or silver compounds act as a depot, from which continuously silver ions are released and diffuse to the surface.

order to guarantee a sufficiently high activity over the whole lifetime of the coating. For single use articles this might be feasible and easy, but for the coating of e.g. furniture or tools this might become a difficult task.

Users of contact lenses are familiar with the problem. At the latest after three months of use, the storage and carrying boxes have to be exchanged, even if they look like at the first day of their use. After this time span, the producers cannot guarantee that there are still enough silver ions available in the polymer to prevent the growing of dangerous bacteria. The upper layers of the antibacterial polymer are leached and it takes longer and longer until the active silver ions diffuse to the surface.

Activity spectrum of silver

The antibacterial action of silver is based on its interaction with anionic and thiol-containing building blocks of the bacteria. Thereby the membrane, enzymes, proteins and the DNA is damaged simultaneously. Its action is so broad, that the microorganisms have almost no chance to develop a resistance against it.

The other side of the medal is that silver has become something like a fashion biocide, which is used in a variety of everyday products without the necessity to do so. Even laundry machines are on the market, which release silver ions during the rinse cycle to kill bacteria [342].

The application of silver as a biocide is highly controversial. Whereas biocides which are based on organic molecules can be decomposed in the environment, silver is persistent and keeps its antibacterial action forever. Critics expect difficulties in sewage plants because the increasing amount of silver in the wastewater stream harms the bacteria which clean the water [343]. The long-term consequences of the usage of silver for the environment are not clear and are subject of speculations.

In coatings, silver can be incorporated as metallic silver nanoparticles or as ionic silver compounds. Silver nanoparticles exhibit an intensively yellow color, when a certain size of a few nanometers is exceeded; therefore often silver compounds like silver-oxide are used.

If the antibacterial action of silver is combined with easy-to-clean surfaces which impede adhesion of bacteria, surfaces with a high value at use for customers can be generated [344].

➤ *Example 60: Synthesis of a bactericidal sol-gel coating* [345]

Synthesis of the silver complex solution: 0.28 g of silver nitrate is dissolved in 30 g of ethanol (96 %). After 30 minutes of stirring, 13 g of isopropanol and 4 g of acetone are added and stirring is continued for 15 minutes. To form the complex, 1.7 g of N-(2-aminoethyl)-3-aminopropyl trimethoxysilane are added slowly dropwise with vigorous stirring.

0.03 mol of 3-glycidyloxypropyl trimethoxysilane (GPTMS) (7.09 g) is prehydrolysed with 4.05 g of 0.01 N nitric acid at room temperature for two hours.

0.07 mol of dihydro-3-[3-(triethoxysilyl) propyl]-2,5-furandione (GF20) is introduced as an initial charge and 0.07 mol of 3-aminopropyl trimethoxysilane (APTMS) is added slowly dropwise with ice cooling and vigorous stirring. The mixture then is diluted with 33.9 g of isopropoxyethanol (IPE) and stirred at room temperature for 30 minutes. 18.90 g of 0.01 N nitric acid are added and the mixture is stirred at room temperature for 15 minutes. It is then diluted with 43.3 g of IPE and the GPTMS prehydrolysate is incorporated by stirring. After 15 minutes of stirring, 28.38 g of the silver complex solution are added and stirring is continued at room temperature for 15 minutes.

Application to the substrate may be done by dipping, flooding or spinning. The coatings are cured at 130 °C for 1 h on glass and at 80 °C for 6 h on PE.

Coating with biocidal action

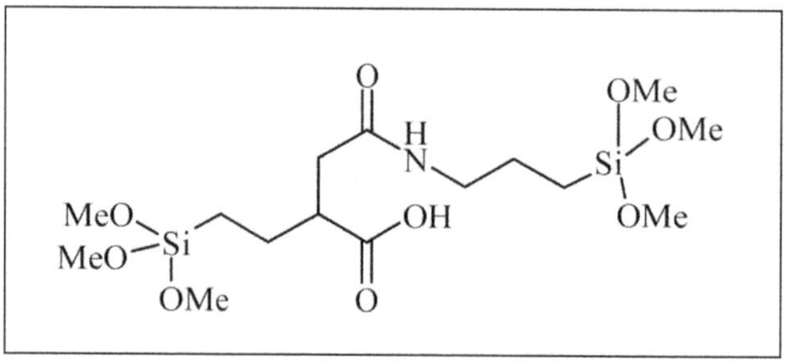

Figure 6.56: The reaction product of a carboxyanhydride- and an aminosilane can be used for the formulation of coatings, which adhere well to a variety of surfaces.

The main component of the coating is a carboxy-amide-disilane precursor, which after hydrolysis is cured with an epoxy silane.

The usage of silver nitrate as antibacterial agent demands a catalyst with an anion, which does not lead to a precipitation of a silver salt (e.g. HNO_3). In this example, silver ions are stabilized by complex formation with a diaminosilane (compare Figure 6.24, page 152).

Alcohols or other organic components of the coating formulation can reduce part or all of the silver nitrate to elementary silver which causes a yellowing if the particles grow too large. This reduction is also easily induced by UV light.

A naturally occurring alternative to silver and other biocides can be chitosan, which is produced from crab shells and which can also be integrated in sol-gel coatings [346].

6.13 Barrier coatings

Barrier coatings are needed to protect a material against the influence of an aggressive medium. Oxygen, water or alkali ions are examples for gaseous, liquid or solid media which can cause deterioration of materials. Corrosion as an example is a consequence of the presence of oxygen, water and ions at the metal surface and paper or sensitive pigments might be attacked by acid-contaminated air e.g. in industrial areas.

A barrier coating seals the substrate and protects it from the aggressive environment. To be able to do so, the coating itself must be resistant and must not change e.g. under the influence of water or solvents. An aggressive medium can use several pathways to penetrate a protective coating. For example

- cracks and defects
- ion conductivity
- diffusion
- interaction with the coating like swelling.

If the network density is low, small molecules can easily pass through the voids of the polymer network. A low network density also eases swelling of the polymer when liquid media come into contact with the coating. If the interaction between medium and polymer is strong enough, the medium can easily diffuse through the coating. In lithium ion batteries for example, the high mobility of lithium ions in polyethylene oxide is used.

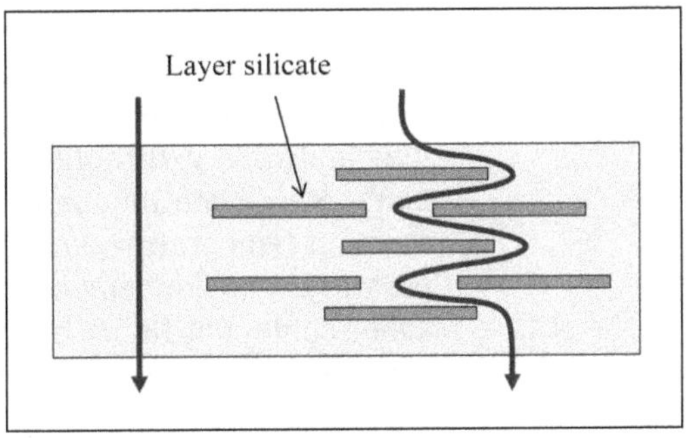

Figure 6.57: By the introduction of anisotropic fillers like layer silicates, the pathway for diffusion is elongated ("tortuous path"). Those coatings can serve as barriers against water, gases

Inorganic/organic hybrids show exceptionally good barrier properties [347]. The inorganic network stabilizes the coating and inhibits swelling. The network density is significantly higher compared to traditional polymer coatings. Furthermore, in proximity to an inorganic particle, organic polymers are forced into a denser packing due to the interaction with the particle at the interface. A denser packing can contribute to better barrier properties (compare Figure 6.14, page 142) [24, 220].

Anisotropic nanofillers like platelet shaped silicates are suitable to improve the barrier properties of coatings. They increase the effective length of the diffusion through the coating ("tortuous path") and thereby lead to a smaller permeation rate compared to unmodified coatings. This principle is used e.g. for packaging foils.

Layer silicates can also increase the mechanical properties of coatings if they are completely exfoliated and dispersed throughout the polymer.

➤ *Example 61: Preparation of a polymer-clay nanocomposite dispersion* [348]

A 2 l four-necked flask was charged at 20 to 25 °C (room temperature) under a nitrogen atmosphere and with stirring with 489 g of deionized water, 20 g of "Laponite RDS" phyllosilicate in powder form (trademark of Rockwood Holdings, Inc.; average diameter in the delaminated, dispersed state: 20 to 50 nm) over the course of 5 minutes.

In order to delaminate the phyllosilicate completely, the initial charge was stirred further for 15 minutes and then heated to 82 °C. Subsequently 1.2 g of (3-methacryloxypropyl) trimethoxysilane was metered in continuously over the course of one hour. Then a solution consisting of 40 g of deionized water, 2.1 g of 10 % strength by weight aqueous sodium hydroxide solution and 0.6 g of sodium peroxodisulfate was added over the course of 2 minutes via a separate feed line and 5 minutes were allowed to pass.

Subsequently the reaction mixture was heated to 85 °C. In parallel with this a homogenous emulsion was prepared as feed 1, consisting of 401 g of deionized water, 8.9 g of 45 % strength by weight aqueous solution of "Dowfax 2A1", 18.4 g of a 10 % strength by weight aqueous sodium hydroxide solution, 4 g of methacrylic acid, 118 g of n-butyl acrylate, 60 g of methyl methacrylate, 16 g of ethyl acrylate and 1.7 g of (3-methacryloxypropyl) trimethoxysilane and, as feed 2, a mixture of 161 g of deionized water, 8.5 g of 10 % strength by weight aqueous sodium hydroxide solution and 2.4 g of sodium peroxodisulfate.

After heating, the two feeds were metered in, beginning at the same time, via separate feed lines. Feed 1 was metered in continuously over the course of 2 hours, the metering rate being 3.14 g/minute in the first 40 minutes and 6.28 g/minute thereafter. In parallel, feed 2 was metered in 2 hours with a continuous volume flow. The reaction mixture was stirred at reaction temperature for 30 minutes and finally was cooled to room temperature. The aqueous composite-particle dispersion thus obtained had a solids content of 18.5 % by weight, based on the total weight of the aqueous composite-particle dispersion.

Polymer/clay nanocomposite dispersion

Layer silicates are anisotropic nanofillers of natural or synthetic origin. For applications in which emphasis is put on transparency and color neutrality, often synthetic nanoclays are used like e.g. “Laponite RDS” in this example.

With a thickness of 1 nm and a length/width of 25 nm, nanoclays should be the ideal nanofiller, if they

a) could be dispersed to primary particle size in polymer systems and

b) if a covalent bond could be established between the polymer and the filler.

Nanoclays show some Si-OH groups which are capable of a chemical reaction only at the edges, whereas in plane adsorbed ions stabilize the nanoparticle. If they are dispersed in water, viscosity increases significantly. A house of cards structure is built because the different charges at the edges and the plane attract each other (compare Figure 3.6, page 39). If the dielectric constant of the dispersion medium is lowered or the ion concentration is increased, the electrostatic stabilization breaks down and the nanoclay precipitates.

In this example, a phosphate-stabilized nanoclay is used. The polyphosphate neutralizes the positive charge of the edges of the nanoclay and thereby inhibits/delays the formation of the house of cards structure. So, for a limited time, high concentrations of exfoliated nanoclay can be prepared which show only a modest increase of viscosity.

Unfortunately the dispersion of nanoclays in organic polymers is not trivial, because the hydrophilic silicates have to be compatibilized with the unpolar polymer. For this purpose, in most cases high amounts (more than 20 % by weight) of cationic alkylammonium compounds are used, which adsorb to the nanoclay and can disturb further processing steps. In this example, methacryloxypropyl trimethoxysilane is used to integrate the nanoclay in the running polymerization process.

The silane can only react chemically with the few Si-OH groups at the edges of the clay or adsorb physically to the plane. The edges however in this case are occupied by the phosphate ions, so it can be assumed that adsorption here is the dominant mechanism to incorporate the nanoclay. The dispersion of the nanoclay exhibits a high pH, which fosters a quick condensation of the hydrolyzed silanes.

During the following radical polymerization, the nanoclay can be incorporated into the growing polymer by different mechanisms.

The methacrylate silane can take part in the polymerization reaction of the monomers which are present in the system. The anionic radical initiator peroxodisulfate can replace the phosphate and occupy the positively charged edges of the nanoclay. The polymerization process then would start from there, which would lead to an embedding of the nanoclay in the growing polymer droplet.

Free polymer chains can adsorb to the large (> 300 m^2/g) nanoclay surface and thereby lead to an incorporation of the clay in the polymer. Given the fact that the phosphate ions block the reactive edges of the nanoclay, most probably the adsorption of polymer chains is the predominant mechanism.

In this example, all nanoclay particles have been homogeneously incorporated into the resin droplets of the dispersion and we can expect that the mechanical and barrier properties of the resulting coating will be substantially enhanced.

6.13.1 Pigments

Even pigments can fall victim to the attack of aggressive media in their environment. Aluminum flakes which serve as silver colored pigments can only be dispersed in waterborne coatings, if they have been protected before by inorganic SiO_2 barrier layers. Otherwise, they discolor, decompose and release dangerous hydrogen gas.

If interference pigments are synthesized by a combination of aluminum flakes and iron oxide layers, a nanometer-thin passivation layer prevents that a thermite reaction can take place. And even in conventional interference pigments, the upper layer of titanium dioxide has to be coated to protect the polymer resin from its photocatalytic activity [349].

Further examples for barrier coatings on pigments are the encapsulation of sensitive organic pigments and dyes by inorganic sol-gel layers as described in chapter "Nanoscaled pigments", starting at page 73.

7 Health, safety and environmental aspects of nanoparticles

There is an emotional discussion going on about the possibility that nanoparticles might be harmful for the environment and the human body. Some NGOs demand a moratorium, which would mean a complete stop of research, development and production of nanoparticulate materials. On the other side groups argument, that the current regulations for chemicals are sufficient and suitable for also handling nanoparticles and nanostructured materials.

The central part of the discussion is the question, if a separate and specific registration process is necessary for a nanoparticulate form of a material, given the fact that the chemical and toxicological properties might change when the particle size is reduced further and further.

But before the discussion starts, it would be wise to come to an agreement about the criteria to identify a nano-product. It certainly is problematic, that todays sophisticated high-resolution microscopes are able to detect nanoparticles and nanostructures in almost any given material. So a threshold for a minimum nanoparticle content has to be defined which turns an "ordinary" material into a nano-material. Otherwise, the sand on the beach and mineral water might be defined as nano-materials.

Considering that even < 1 % by weight of nanoparticle content translates to a very high number of nanoparticles, the involved groups are discussing whether the weight percent or the number of nanoparticles per volume should be used as a threshold.

The EU commission published a recommendation for a definition of nanomaterials in October 2011:

- natural, incidental or manufactured material,
- containing particles, in an unbound state or as an aggregate or as an agglomerate
- and where, for 50 % or more of the particles in the number size distribution, one or more external dimensions is in the size range 1 nm to 100 nm.
- In specific cases and where warranted by concerns for the environment, health, safety or competitiveness the number size distribution threshold of 50 % may be replaced by a threshold between 1 and 50 %

As a consequence it would be sufficient, if more than 50 % of the primary particles of the material are in the nanoscale to define the whole material as a nanomaterial. Even 1 % nanoparticle content could to be enough for this declaration in specific cases. However, it is planned to revise the definition in December 2014.

One important point of this recommendation would be that now almost all pigments, even those which have been used since centuries, are now nanomaterials and will have to be reassessed by the authorities.

As a quite absurd consequence, even the sand on the beach now might fall under the definition of a nanomaterial, because it contains many calcium carbonate nanoplatelets and

Gerhard Jonschker: Praxis der Sol-Gel-Technologie

ISBN: 978-3-86630-875-6

particles stemming from seashells and diatoms. A further discussion seems to be necessary to find a best practice mode and to establish laws and rules which are beneficial for the environment and can be established not only in the EU, but also worldwide [350].

Most of the discussions are organized within a framework of funded projects like the nano-dialogue [351, 352]. Since the discussion about a definition of nano-materials and the regulatory consequences might take some time, every user and every company should take all preventive measures to protect people and environment from potential hazards [353, 354].

Most nanoparticles however are processed anyway as dispersions to prevent agglomerates. After further processing steps like formulation, coating and curing they are embedded and cross-linked within the coating.

Often the question is being discussed, whether the nanoparticles might be released from a coating during its lifetime. In order to clarify this open question, the association of the German paint and printing industry conducted a study to assess the release of nanoparticles from a coating material e.g. by polishing. The results were published in 2010.

"There are no significant differences observable between conventional coatings and coatings which contained nanoparticles. The number of released particles is … low, e.g. in comparison to a road with heavy traffic. The nanoparticles are embedded into the resin; the released particles are fractions of the resin. There is no evidence for an additional risk coming from polishing coatings which contain nanoparticles".

Besides the ongoing discussions about further regulatory hurdles for nanoparticles, the laws and rules which already are in place are simple and clear:

Companies are responsible for the safety of their employees and their products and are fully liable for all consequences, whether the product is a nanomaterial or not. Therefore it is important for all companies to make sure, that their products and raw materials are safe and that all precautionary measures are taken to prevent any harm for people and environment.

8 Resume

Within this book, only a fraction of the possibilities and applications of sol-gel and nanotechnology could be touched. Of course it is incomplete and one or another aspect or example is missing.

I tried to cover some principles of synthesis and application of nanomaterials in paints and coatings by discussing concrete examples. This book should give the reader an easy start to practical lab work in the field of nanotechnology, but it cannot replace deep studies with the help of conventional textbooks.

I hope it became clear, that sol-gel and nanotechnology formulations are fractal systems. Even small changes of the reaction parameters can have a significant influence on the end product.

On the one hand this increases the complexity, but on the other hand it offers a big chance to develop an own, individual problem solution, often by a minimum variation of given formulations.

By using the discussed principles, techniques and hints, it should be possible to realize your own sol-gel innovation. Even if nanoparticles often are stubborn fellows, it is fascinating to work with them. Now it is your turn!

Gerhard Jonschker: Sol-Gel-Technology in Praxis

ISBN: 978-3-86630-886-2

9 List of Examples

Gerhard Jonschker: Sol-Gel-Technology in Praxis

ISBN: 978-3-86630-886-2

10 Literature

[1] Drew, Myers, "Surfaces, Interfaces and Colloids", 2nd Edition, Wiley-VCH, 1999
[2] D. Everett, "Basic principles of colloid science", The royal society of Chemistry, 1988
[3] Bundesministerium für Bildung und Forschung. "Nano-Initiative - Action plan 2010", www.bmbf.de
[4] W. Fetzer, "J. Kunckel. Leben u. Werk eines großen dt. Glasmachers des 17. Jh.", Berlin 1977
[5] T. Sugimoto; "Monodispersed Particles", Elsevier 2001, ISBN: 0 444 89569 8
[6] Taniguchi, "On the Basic Concept of Nano-Technology" Proc. Intl. Conf. Prod. Eng. Tokyo, Part II, Japan Society of Precision Engineering, 1974
[7] W. Ostwald, "Die Welt der vernachlässigten Dimensionen", Verlag von Theodor Steinkopf, 1937
[8] R. Iordanova et al. "History of Sol-Gel Science and Technology (Review)", Journal of the University of Chemical Technology and Metallurgy, 43, 2, 2008, 181-192
[9] Michael Chrichton - "Prey", Harpercollins UK ISBN-13: 978-0007153794
[10] G. M. Whitesides, (2001): "Lernen von der ältesten Nanomaschine", in: Spektrum der Wissenschaft Spezial 2 (Nanotechnologie)
[11] Nano.DE-Report 2011 (www.bmbf.de/pub/nanoDE-Report_2011.pdf)
[12] "Growth Market Nanotechnology", N. Malanowski et al. (Eds.), 2006 Wiley-VCH Verlag GmbH &Co KGaA Weinheim
[13] BCC 2010: "Nanotechnology: A Realistic Market Assessment", BCC Research Inc.
[14] R. Anselmann, private communication
[15] G. Jonschker, "Smaller is better. Easy cleanable surfaces via Nanotechnology", 50th SEPAWA Congress 08 - 10 October 2003, Bad Dürkheim
[16] Patent DE 10304849
[17] Patent WO 00/14017
[18] Patent US 7169834
[19] P. G. McCormick, "Nanopowders Synthesized by Mechanochemical Processing", Advanced Materials, 13 (2001), 1008-1010
[20] Patent US 6,001,326
[21] Reiss, H.; La Mer, V.K.; J. Chem. Phys.; 18; 1; 1950
[22] La Mer, V. K.; Oinegar, R. H.; J. Am. Chern. Soc.; 72; 4847; 1950
[23] Everett, D. H.; "Grundzüge der Kolloidwissenschaft"; Darmstadt; Steinkopff Verlag; 1992
[24] Unpublished results
[25] Y. Xia et. al. "Monodispersed colloidal spheres", Advanced Materials 12: (10) 693-713
[26] "The Colloidal Domain"; 2nd Edition; Wiley-VCH;1999
[27] Jander, Blasius, "Lehrbuch der analytischen und präparativen anorganischen Chemie" S. Hirzel Verlag Stuttgart 1985
[28] Jun Ji et al., 2011, Advanced Materials Research, 233-235, 1184
[29] K. Toyokura et.al. "Controlled Double-Jet Precipitation of Sparingly Soluble Salts. A Method for the Preparation of High Added Value Materials", Chem. Mater. 1992, 4, 545-555
[30] Patent US 2010/0144518
[31] Patent US 2010/0124532
[32] E. Matijevic, Langmuir 10, 8 (1994)
[33] E. Mattijevic, Prog. Colloid Polym. Sci., 101, 38 (1996)
[34] Patent US 4775520
[35] Patent US 2010/0144918
[36] Patent US 2574902
[37] Patent EP 1366112
[38] Yoshida, A.; et.al. Chem. Soc. Jp.; 64; 1682; 1991

Gerhard Jonschker: Sol-Gel-Technology in Praxis

ISBN: 978-3-86630-886-2

[39] Patent US 5221497
[40] G. Engelhardt et al. "Zur Abhängigkeit der Struktur der Silikatanionen in wässrigen Natriumsilikatlösungen vom Na:Si-Verhältnis", Z. anorg. allg. Chemie Bd. 418 S. 17-28
[41] P. Christophliemk, "Herstellung, Struktur und Chemie technisch wichtiger Alkalisilikate", Glastech. Ber. 58 (1985) Nr. 11 S. 308-314
[42] C. Feldmann et al. "Polyol-vermittelte Präparation nanoskaliger Oxidpartikel", Angewandte Chemie 2001, 113, No. 2, p. 372-374
[43] C. Feldmann, "Preparation of nanoscale pigment particles" in: Advanced Materials 2001, 13, No. 17, September 3, p. 1301-1303
[44] Patent EP 1 386886
[45] Patent WO 2006/048030
[46] Anselmann, R.: Journal of Nanoparticle Research 3, (2001), p. 329-336
[47] Anselmann, R.: In: "Die Schlüsseltechnologie des 21. Jahrhunderts in der industriellen Anwendung". Drittes Fachforum Nanotechnologie, OTTI-Kolleg, 22./23.10.02, Regensburg
[48] R. Hempelmann, "Elektrochemische Herstellung amorpher und kristalliner Metalloxide mit Teilchengrößen im Nanometerbereich", Patent DE 198 40 842
[49] Patent WO 00/14302
[50] K. M. Stephen et al. "Unintentional silver intoxication following self-medication: an unusual case of corticobasal degeneration". L. Ann Clin Biochem. 2009 Sep 3
[51] Kwon H. B. et al., "A case of argyria following colloidal silver ingestion", Ann. Dermatol. 2009 Aug; 21(3): 308-10. Epub 2009 Aug 31
[52] Monnoyer et al.; Colloid Surf. A 100 (1995) 233-243
[53] Winkler, J., Proft, B. "Vor UV schützen und selektiv blaues Licht streuen - Eigenschaften und Anwendungen von nanokristallinem Titandioxid". Farbe & Lack 2 (2001) 28-36
[54] Iler, R. K.; "The Chemistry of Silica"; J. Wiley and Sons; New York; 1979
[55] Overbeek, J. Th. G.; Verwey, E. J. W.; "The Theory of the Stability of Lyophobic Colloids"; Elsevier; New York; 1948
[56] Derjaguin, B. V.; Landau, L.; Acta Physicochim., USSR; 14; 633; 1941
[57] Dumont, F.; "The Colloid Chemistry of Silica"; Bergna; H.; Advances in Chemistry Series 234; American Chemical Society; Washington OC; 1994
[58] Zeta-Meter Inc. "Everything you want to know about coagulation and flocculation", 1993, www.zeta-meter.com
[59] Parks, G. A.; Chem. Rev.; 65; 177; 965
[60] G. Lagaly, "Dispersionen und Emulsionen", Steinkopff-Verlag Darmstadt
[61] Patent DE 103 53 507
[62] Patent DE 19840527
[63] H. Schmidt, "Organically modified silicates and ceramics as two-phasic systems: Synthesis and processing", Journal of Sol-Gel Science and Technology 8, 557-565 (1997)
[64] R. Iden, "Nanotechnology - Vision becomes reality", BASF presentation at: Chemistry. Elements of Life, Tokio, Nagoya, Osaka, 7-11.11.2005, Japan
[65] Matijević, E. Journal of Colloid and Interface Science, Volume 288, Issue 2, 15 August 2005, p. 489-495
[66] Sugimoto, T. (Ed.), "Fine Particles, Synthesis, Characterization, and Mechanisms of Growth", New York: Marcel Dekker, 2000
[67] E. Matijevic, "Nanosize Precursors as Building Blocks for Monodispersed Colloids"; Colloid Journal, 2007, Vol. 69, No. 1, pp. 29-38.
[68] Patent US 2801185
[69] Patent EP 0 699 626
[70] Patent EP 1544257
[71] G. Meichsner, T. Mezger, J. Schröder, "Lackeigenschaften messen und steuern", Vincentz Verlag 2003
[72] Advincula, R.C; European Coatings Journal Issue 3, 2011, p. 96-101
[73] Patent EP 2102292
[74] G. Jonschker et al. "Matched to the medium", European Coatings Journal (ECJ) 10 (2008) 46
[75] Offenlegungsschrift DE 10 2008 024 868
[76] Patent WO 2008/101581
[77] Patent US 2801185
[78] Patent WO 2007/134712
[79] Patent US 2692863
[80] A. Frantzen, Dissertation Universität des Saarlandes, 2001
[81] Patent WO 2009/015724
[82] Patent WO 2007/048570
[83] Patent WO 2007/134712

[84] Patent US 2801185
[85] Patent WO 2005/082514
[86] Gächter/Müller, Taschenbuch der Kunststoff-Additive, Carl Hanser Verlag München/Wien, 1979
[87] H. Schmidt et.al "The sol-Gel Process as a basic technology for Nanoparticle-Dispersed Inorganic-organic nanocomposites", Journal of Sol-Gel Science and Technology 19, 39-51, 2000
[88] I. Cabrera et al. "Nano keeps water and dirt at bay", ECJ European Coating Journal 03/2011, p. 108ff
[89] A. di Gianni et al. "Hyperbranched Polymer/TiO_2 Hybrid Nanoparticles Synthesized via an In Situ Sol-Gel Process", Macromol. Chem. Phys. 2007, 208, 76-86
[90] M. Mennig et.al "Colored coatings on glass based on noble metal colloids" in: Sol-gel technologies for glass producers and users, Kluwer, 2004, S. 161-168
[91] Patent DE 102007030285
[92] Patent WO 2005/119359
[93] Patent EP 1366112
[94] Patent EP 1008631
[95] D. Distier, "Wässrige Polymerdispersionen, Synthese, Eigenschaften, Anwendungen", Wiley-VCH Weinheim, 1999
[96] Pickering, S. U.: "Emulsions". J. Chem. Soc. 91 (1907) 2001-2021
[97] Lagaly, G., Reese, M., Abend, S.: "Smectites as colloidal stabilizers of emulsions", Applied Clay Science 14 (1999) 83-103
[98] Patent WO 2008/071285
[99] Patent WO 2006/008120
[100] Patent DE 10004499
[101] Patent US 7488769
[102] Patent EP 2072554
[103] M. Bock, "Polyurethane für Lacke und Beschichtungen" Hrsg. Ulrich Zorll, Vincentz Verlag 1999
[104] Patent EP 1 773 919/ WO 2006/008120
[105] Patent EP 2072554
[106] Patent EP 1754685
[107] Patent WO 2008/017364
[108] U. Nolte, European Coatings Journal, 06/2010, p. 20 ff
[109] R. A. Slepetys, W. F. Sullivan, "Light Scattering by Pigmentary Rutile in Polymeric Films", Ind. Eng. Chem. Prod. Res. Dev., 1970, 9 (3), pp 266-271
[110] Patent EP 0270472
[111] Matijevic, E.; "Ultrastructure Proeessing of Advanced Ceramics"; Wiley; New York; 1987
[112] Matijevic, E.; Langmuir; 10; 8; 1994
[113] Matijevic, E.; J. Coll. Int. Sci; 165; 1; 141; 1994
[114] Patent WO 03/076521
[115] Hayat, M. A. (ed.) "Colloidal Gold, Principles, Methods and Application" (Academic Press Inc., NewYork, USA, 1989)
[116] Patent US 7,211,135
[117] D. I. Gittins and F. Caruso, "Spontaneous Phase Transfer of Nanoparticulate Metals from Organic to Aqueous Media", Angew. Chem. Int. Ed. 2001, 40, No. 16, p. 3002-3004
[118] R. Maisch et. al., "Perlglanzpigmente", Vincentz Verlag, 1996
[119] R. Maisch, M. Weigand, "Perlglanzpigmente", Verlag moderne Chemie, 1991
[120] Patent US7255736
[121] Patent WO 2006030001
[122] Patent WO 031050193
[123] Patent WO 2010/046030
[124] Buchmeiser, M. "Zinc Oxide Based Coatings for the UV-Protection of Wood for Outdoor Applications." Macromolecular Symposia, (2011), 301: 23-30
[125] Patent EP 1 832 624
[126] Patent WO2003002651
[127] Patent US 2011/0130273
[128] Patent EP 1204701
[129] Patent DE 102005006870
[130] B. Ramezanzadeha et al., "Studying the role of polysiloxane additives and nano-SiO_2 on the mechanical properties of a typical acrylic/melamine clearcoat", Progress in Organic Coatings 72 (2011) 621- 631
[131] Crockmeter - Test EN ISO 105 - X 12
[132] Prüfung der Kratzfestigkeit von Beschichtungen mit dem AATCC Crockmeter. Daimler Prüfanweisung PBODC390
[133] Patent DE 19846660

[134] T. Lan, Chem. Mater. 6 (1994) 2216
[135] W. Helbert, J. Y. Cavafile, A. Dufresne, Polym. Composites 17 (1996) 604
[136] G. Jonschker, "Nanocomposites - The new Generation of scratch resistant coatings", IIR Fachkonferenz "Kratzfeste Beschichtungen", 13.-14.01.01 Stuttgart
[137] C. J. Brinker, G.W. Scherer, "Sol-Gel Science - The physics and chemistry of Sol-Gel processing", Academic Press Inc. 1990
[138] S. Sakka, A. Ceram. Bull. 64, 1985, 1463
[139] S. Sakka (ed), Handbook of Sol-Gel Science and Technology, Processing, Characterization and Application, Kluwer Acad. Publish., 2005
[140] Journal of Non-Crystalline Solids. Elsevier Science B.V.
[141] Journal of Sol-Gel Science and Technology. Editor-in-chief: S. Sakka; Kluwer Academic Publishers
[142] H. Dislich, J. Non-Cryst. Solids, 73, 1985, 599
[143] V. Gottardi (ed.), J. Non-Cryst. Solids, 48, 1982
[144] Patent WO 021064664
[145] S. Sepeur, "Nanotechnologie", Vincentz Network Gmbh & Co KG, 2008
[146] A. Vioux, "Nonhydrolytic Sol-Gel Routes to Oxides", Chem. Mater. 1997, 9, 2292-2299
[147] M. Niederberger, "Nonaqueous Sol-Gel Routes to Metal Oxide Nanoparticles", Acc. Chem. Res. 2007, 40, 793-800
[148] Niederberger, M.; Garnweitner, G.; "Organic reaction pathways in the nonaqueous synthesis of metal oxide nanoparticles". Chem. Eur. J. 2006, 12, 7282-7302
[149] Patent US 4754012
[150] J. Livage et al, in: "Chemical Processing of Advanced Materials", ed.: L. L. Hench and J. K. West, John Wiley and Sons, New York, 1992, p. 223
[151] J. Livage, M. Henry, C. Sanchez, Progr. Solid St. Chem, 1989, 18, 259-341
[152] R. C. Mehrotra et. al. "Alkoxo and Aryloxo Derivatives of Metals", Academic Press; 1st edition (2001)
[153] Patent DE 10 2004 029 303
[154] Spin-on gläser, Product information from Honeywell Inc. (www.honeywell.com)
[155] H. Schmidt et. al. "Principles of hydrolysis and condensation reaction of alkoxysilanes". J. Non-Cryst. Sol. 63 (1984) 1-11
[156] Hartmann-Thompson, C. (Ed.), "Applications of Polyhedral Oligomeric Silsesquioxanes", Advances in Silicon Science, Vol. 3, 1st Edition, 2011
[157] Patent PCT/EP03/04586
[158] W. Noll, "Chemie und Technologie der Silikone", Wiley-VCH 1984
[159] J. Wen and G. L. Wilkes, "Organic/Inorganic Hybrid Network Materials by the Sol-Gel Approach", Chem. Mater. 1996, 8, 1667-1681
[160] Patent DE 198 01 004
[151] H. Zou, S. Wu, and J. Shen, "Polymer/Silica Nanocomposites: Preparation, Characterization, Properties, and Applications" Chem. Rev. 2008, 108, 3893-3957
[162] P. Uhlmann et al. Progress in Organic Coatings 58 (2007) 122-126
[163] B. Arkles, "Silane coupling agents", Gelest Inc, 2003, www.abcr.com
[164] Wacker Chemie AG, "Geniosil" broschure, www.wacker.com
[165] A. Behr et al. in Chemie Ingenieur Technik (1999), 71 (5), 490-493
[166] Patent EP 0 693 492
[167] M.G. Voronkov et.al., Herald of the Russian Academy of Sciences, 2010, Vol. 80, No. 6, pp. 514-521
[168] Patent US 5468789
[169] Patent DE 102007061854
[170] Patent U.S.4499150
[171] Patent US-A-4,499,151,
[172] Patent EP-A-0 571 073
[173] Patent DE 10353507
[174] Loy, D. A.; Shea, K. J. Chem. Rev. 1995, 95, 1431
[175] Patent PCT/DE2007/001602
[176] Patent DE 102004050748
[177] Patent DE 10106342
[178] Patent WO 02150191
[179] Patent DE 10063519
[180] H. Schmidt et al. Thin Solid Films 351 (1999) 209-211
[181] H. Schmidt et al., "Aqueous sol-gel derived nanocomposite coating materials", Materials Research Society symposia proceedings 1998; 519, S. 297-308
[182] N. Joubert et al, Macromol. Rapid Commun., 2005, 602-607
[183] Akzo Nobel. Technisches Datenblatt "Bindzil CC"

[184] Patent WO 99/52964
[185] Patent DE 10326538
[186] Patent WO 2009/129932
[187] Patent EP 1088036
[188] Patent WO/1999/054412
[189] Patent EP 1088036
[190] Patent US 2011/0130273
[191] Patent GB2041956
[192] Patent EP 1728834
[193] S. Hofacker, M. Mechtel, M. Mager, H. Kraus; Progress in organic coatings, 45 (2002) 159-164
[194] Brock, Groteklaes, Mischke, "Lehrbuch der Lacktechnologie", Vincentz-Verlag, 1998
[195] Patent WO 2008/110229
[196] Patent WO 01/09231
[197] H. Scholze, "Glas. Natur, Struktur und Eigenschaften". Springer, Berlin; Oktober 1988
[198] "Physikalische Chemie der Glasoberfläche"; ed. Prof. H. Dunken. VEB Deutscher Verlag für Grundstoffindustrie, Leipzig 1981
[199] A. Michalske, K. D. Keefer., "Adhesion of hydrated silica films", Mat. Res. Soc. Symp. Proc. Vol. 121 (1988) 187-197
[200] D. R. Uhlman et al., "Strengthening of silica glass by gel-derived coatings"., J. Non-Cryst. Sol. 82 (1986) 349 ff
[201] C. J. Brinker, W. D. Drotning und G. W. Scherer., "A Comparison between the densification kinetics of colloidal and polymeric silica gels"., Mat. Res. Soc. Symp. Proc. Vol. 32 (1984)
[202] www.sura-instruments.de, "Pyrosil" product information
[203] DIN 55996-1
[204] A. Goldschmidt, H.-J. Streitberger, " BASF Handbuch Lackiertechnik", Vincentz-Verlag 2002
[205] Seth, et al. "Novel Water-Based High-Performance Primers That Can Replace Metal Pretreatments and Chromate-Containing Primers", Journal of Materials Engineering and Performance, Aug. 2004, pp. 468-474, vol. 13 (4), ASM International
[206] Patent US 7732016
[207] Patent GB 2041956
[208] G. W. Scherer, "Drying gels, I. General Theory", J. Non-Cryst. Sol. 87 (1986) 199-225
[209] G. W. Scherer, "Stress and fracture during drying of gels", J. Non-Cryst. Sol. 121 (1990) 104-109
[210] M. Yamane et al., "The structural evolution of alkoxy-derived gel during drying". J. Non-Cryst. Sol. 100 (1988) 215-219
[211] S. S. Kistler, Nature. 127, 1931, S. 741ff
[212] Fricke, J. and Emmerling, A. (1992), "Aerogels". Journal of the American Ceramic Society, 75: 2027-2035
[213] Patent WO 99/026880
[214] Patent DE 19801 004
[215] S. S. Prakash, C.J. Brinker, A.J. Hurd, and S.M. Rao, "Silica Aerogel Films at Ambient Pressure via Surface Derivatization and Reversible Drying Shrinkage", Nature, 374, (1995), 439-443
[216] F. Lange;"Liquid precursors for ceramic particles, fibers and thin films", Proceedings of the International Symposium on Molecular Level Designing of Ceramics, p. 14-28, March 16th, 1991, Nagoya, Japan.
[217] E. M. Rabinovich et al. "Shrinkage and swelling of alkoxide silica gels on heating", J. Non-Cryst. Sol. 82 (1986) 78-85
[218] L. L. Hench et al., "Structural evolution during sintering of optical sol-gel silica", J Non-Cryst. Sol. 121 (1990) 124-127
[219] W. L. Vasconselos, L. L. Hench., "Mechanical properties evolution during sintering of optical sol-gel slica", J. Non-Cryst. Sol. 121 (1990) 132-5
[220] G. Jonschker, "Synerese und Spannungsrelaxation von Sol-Gel Systemen", PhD thesis, Universität des Saarlandes, 1998
[221] M. Mennig et.al. "Sol-gel derived thick coatings and their thermomechanical and optical properties". Sol-gel optics II: 20-22 July 1992, San Diego, California ; proceedings/Bellingham, Wash.: SPIE, 1992 (SPIE proceedings series; 1758), p. 125-134
[222] Patent EP 0973958
[223] Patent EP 0642475
[224] Patent EP 1 284307
[225] Patent PCT WO9711035
[226] C. J. Brinker, W. D. Drotning und G. W. Scherer., "A Comparison between the densification kinetics of colloidal and polymeric silica gels"., Mat. Res. Soc. Symp. Proc. Vol. 32 (1984)
[227] W. L. Vasconcelos, R. T. DeHoff, L. L. Hench; "Structural evolution during sintering of optical sol-gel silica" J. Non-Cryst. Sol. 121 (1990) 124-127

[228] J. C. Ro, I. J. Chung; "Structures and properties of silica gels prepared by the sol-gel method", J. Non-Cryst. Sol. 130 (1991) 8-17
[229] Patent WO 98/45502
[230] Patent WO 2005/066388
[231] M. Mennig et.al. "Flame retardent coatings on glass", Glas-Ingenieur. 4. 1994, 3, p. 54-60
[232] Patent US20100028631
[233] Patent EP 1484372
[234] G. Habenicht, "Kleben", Springer-Verlag Berlin 2009
[235] Patent EP 093845781
[236] W. Vogel, "Glaschemie", 3rd Edition, Springer-Verlag, 1992
[237] Patent DE 4338360
[238] S. Sakka et al., Proc. SPIE-Int. Soc. Opt. Eng. (1994), 2288 (Sol-Gel Optics 111), 108-119
[239] P. Gomez-Romero, C. Sanchez, New J. Chem., 2005, 1, 57
[240] U. Schubert, N. Hüsing, and A. Lorenz, "Hybrid Inorganic-Organic Materials by Sol-Gel Processing of Organofunctional Metal Alkoxides" Chem. Mater. 1995, 7, 2010-2027
[241] Patent WO 2006/030001
[242] www.tno.nl, "Planocolor" concept
[243] M. Mennig et al. "Sol-gel coatings for the substitution of fluoride or lead containing white decorations on glass", Journal of Non-Crystalline Solids 218, 1997, p. 395-398
[244] www.nanogel.com, "ENOVA"
[245] H. Schmidt et al. "One step antiglare sol-gel coating for screens by sol-gel techniques", Journal of Non-Crystalline Solids 218, 1997, p. 163-168
[246] Mennig, M. et al., "Preparation of micropatterns with profile heights up to 30 microns from silica sols", Proceedings of SPIE - The International Society for Optical Engineering, Volume 3136, 1997, p. 480-485
[247] Patent DE 102007008073
[248] C. J. Brinker et al. "Self-assembly and characterization of mesostructured silica films with a 3D arrangement of isolated spherical mesopores", Adv. Funct. Mater., 13, 1, 47-52, 2003
[249] J. Köhler, "Funktionelle mesostrukturierte Beschichtungen auf Basis von SiO_2 und TiO_2", Dissertation Universität Ulm 2009
[250] C. Sanchez et al., "Applications of hybrid organic-inorganic nanocomposites", J. Mater. Chem., 2005, 15, p. 3559-3592
[251] M. E. Diaz-Garcia, "Molecular Imprinting in Sol-Gel Materials": Recent Developments and Applications, Microchim. Acta 149, 19-36 (2005)
[252] Patent US 3708225
[253] Patent US 3986997
[254] Patent US 3976497
[255] Patent WO 80/01006
[256] Patent WO 80/01006
[257] Patent WO 03/093383
[258] Bauer, F., Ernst, H., Hirsch, D., Naumov, S., Pelzing, M., Sauerland, V. and Mehnert, R. (2004), "Preparation of Scratch and Abrasion Resistant Polymeric Nanocomposites by Monomer Grafting onto Nanoparticles", 5. Macromolecular Chemistry and Physics, 205: 1587-1593
[259] Patent Patent WO 200014149
[260] Patent EP 1153090
[261] "Lackharze: Chemie, Eigenschaften und Anwendungen", Hrsg. D. Stoye, W. Freitag, Carl Hanser Verlag 1996
[262] W. M. Haynes: CRC Handbook of Chemistry and Physics. 91. Auflage. Taylor & Francis, 2010
[263] Ullmann's Encyclopedia of Industrial Chemistry
[264] Technical data sheet "Wacker BS 1701" octylsilane-based concrete protection
[265] Patent WO/2004/037936
[266] Honda et al in: Macromolecules, 2005, 38, 5699 ff
[267] H. Fromme, "Perfluorinated compounds - Exposure assessment for the general population in western countries". Int. J. Hyg. Environ. Health 2009; 212: 239-70
[268] PFOA Stewardship Program. http://www.epa.gov/oppt/pfoa/pubs/stewardship/index.html
[269] Patent PCT/EP02/01424; WO2002064664A3
[270] Patent WO 2004/026793
[271] Patent WO 01/23101
[272] Patent W0 01/18135/ PCT/DEOO/02984
[273] Patent U.S.6,054,601
[274] Unpublished results
[275] Patent EP 1655349

[276] Patent WO 03/106164
[277] Patent EP 0973958
[278] EP 0973958
[279] Patent WO 02123033
[280] I. P. Parkin, R. G. Palgrave, "Self-cleaning coatings", J. Mater. Chem., 2005, 15, 1689-1695
[281] V. A. Ganesh et al ,"A review on self-cleaning coatings", J. Mater. Chem., 2011, 21, 16304
[282] Cassie, A. B. D., Baxter, S. (1944), "Wettability of Porous Surfaces", Trans. Faraday Soc. 40: 546-551
[283] Fürstner, R., Barthlott, W., Neinhuis ,C. & Walzel, P. (2005): "Wettting and self-cleaning properties of artificial superhydrophobic surfaces", Langmuir 21(3), 956-961
[284] Neinhuis, C., Barthlott, W. (1997): "Characterization and distribution of water-repellent self-cleaning plant surfaces". Annals of Botany 79 (6), 667-677
[285] Ma, M., & Hill, R. M. (2006), "Superhydrophobic surfaces", Current Opinions in Colloid and Interface Science, 11(4), 193-202
[286] www.lotusan.de
[287] E. Bagda. "What does the customer get from the depearling effect?" Farbe und Lack Volume 112, Issue 3, 2006, Pages 36-42
[288] E. Bagda. "Sticky facades become grey more rapidly". Farbe und Lack Volume 108, Issue 2, 2002, p. 98-104+8
[289] Patent WO2009118552
[290] M. Lanata et al. "Titania inverse opals for infrared optical applications", Optical Materials Volume 17, Issues 1-2, June-July 2001, p. 11-14
[291] Ming W. et al. Nano Letters 2005, 5 (11), 2298
[292] Project description "Antifingerprint", www.ifam.fraunhofer.de
[293] Patent EP 1555249
[294] Patent DE 10159288
[295] Patent EP 1 300 433
[296] Broschüre: "Bindzil colloidal silica for hard surface cleaners". Akzo Nobel Chemicals
[297] Patent DE 10201596
[298] G. Jonschker, "Nanotechnology is disrupting the future", Proceedings of the 5th World Conference on Detergents, A. Cahn (ed.) 13-17.10.2002, Montreux
[299] Patent US 2010/0092765
[300] Patent DE 10248583
[301] Patent DE 2068494
[302] H.-D. Dörfler, "Grenzflächen- und Kolloidchemie", VCH, Weinheim, 1994, p. 198
[303] Brezesinski, G.; Mögel, H.-J.: "Grenzflächen und Kolloide"; Spektrum Akademischer Verlag; Heidelberg, Berlin, Oxford; 1993
[304] H. Mollet. "Formulierungstechnik - Emulsionen, Suspensionen, Feste Formen", Wiley-VCH 2000, p. 72
[305] Patent US 2004/0237833
[306] Patent US 2004/0237833
[307] Siegmann, K. et al., "Sol-gel coatings with lime repellent properties" Journal of Sol-Gel Science and Technology, Volume 59, No. 3, 2011, pp. 574-579(6)
[308] A. Fujishima, K. Hashimoto, T. Watanabe, TiO_2 Photocatalysis, 1. Aufl., Bkc Inc., Tokyo, 1999
[309] Wang et.al, Nature 388, S.431
[310] Bahnemann "Selbstreinigende Oberflächen: Nanotechnologie für Jedermann" Nanoday 2005, 06.10.2005 Hannover
[311] Ball, P. "Window, clean thyself!", Nature Volume 377, Issue 6547, 1995, p. 290
[312] K. D. Sanderson et al. "The Use of Titanium Dioxide Coatings Deposited by APCVD on Glass Substrates to Provide Dual Action and Self Cleaning", Proceedings, Annual Technical Conference - Society of Vacuum Coaters 2003, p. 203-208
[313] Mills, A., Elliott, N., Parkin, I., ONeill, S., & Clark, R. (2002). "Novel TiO2 CVD films for semiconductor photocatalysis", Journal of Photochemistry and Photobiology, 151(1-3), 171-179
[314] Patent US 20110259244
[315] H. Schmidt et al. Journal of Non-Crystalline solids, 218, 1997, p. 163-168
[316] Patent WO 2005/120154
[317] E. Kemnitz. Angew. Chem. 115 (2003) 4383
[318] St. Rüdiger, E. Kemnitz, Review in Dalton Trans., 9 (2008) 1117-1127
[319] E. Kemnitz et al., Thin Solid Films 2008, 516, 4175; Phys. Stat. Sol. 2008, 205, 821
[320] Patent US 4754012
[321] Patent EP 514073
[322] Patent EP 1559693
[323] T. N. Nguyen, J. B. Hubbard, G. B. McFadden, J. Coatings Technol. 63 (1991) 43

[324] B. Arkles, "Silane coupling agents: Connecting Across Boundaries", 2003 Gelest Inc. www.abcr.de
[325] Van Ooij, Wim J., "Replacement of Chromates by Silanes-An Overview", Presented at ICEPAM, Jun. 15-18, 2004, pp. 1-12, Oslo, Norway
[326] Van Ooij, et al., "A Novel Self-Priming Coating for Corrosion Protection!, 3rd Int. Surface Engineering Congress, Aug. 2-4, 2004, pp. 8-13, Orlando, FL, USA
[327] Patent US 6261638
[328] P. Pape et al, in: "Silanes and other coupling agents", ed. K. Mittal, 1992, VSP, S. 105
[329] Patent WO 2010/100187
[330] M. Guglielmi, "Sol-Gel Coatings on Metals", Journal of Sol-Gel Science and Technology 8, 443-449 (1997)
[331] Langenfeld, S., Jonschker, G. and Schmidt, H. (1998), "Neue Sol-Gel-Beschichtungen als Korrosions- und Verschleißschutz für NE-Metalle", Materialwissenschaft und Werkstofftechnik, 29: 23-29
[332] Feil, W. Fürbeth, "Nanopartikel-basierte Korrosionsschutzschichten für Leichtmetalle" in: "Jahrbuch Oberflächentechnik 2010" (Ed.) Leutze-Verlag, 176-187
[333] F. Feil, W. Fürbeth, "Thin, Nanoparticulate Coatings for the Improvement of the Corrosion and Passivation Behavior of AZ Magnesium Alloys", Advanced Materials Research, Light Weight Metal Corrosion and Modeling 2010, 138, 47-53
[334] Patent EP1068372
[335] Patent EP 1284307
[336] E. Németh, "Triboelektrische Aufladung von Kunststoffen" Dissertation TU Freiberg, 2003
[337] T. Textor, B. Mahltig, "A sol-gel based surface treatment for preparation of water repellent antistatic textiles". Applied Surface Science, Volume 256, Issue 6, 1 January 2010, p. 1668-1674
[338] K. Rose, S. Amberg-Schwab and M. Heinrich In: N. Auner and J. Weis, Editors, "Organosilicon Chemistry IV", VCH Weinheim, Germany (1999), p. 613
[339] M. Wouters et al., "Transparent UV curable antistatic hybrid coatings on polycarbonate prepared by the sol-gel method". Progress in Organic Coatings Volume 51, Issue 4, 20 December 2004, p. 312-319
[340] Patent WO/2010/128148
[341] Patent EP2010/056277
[342] www.samsung.de, "Silver active"-series
[343] M. Burkhardt et al. "Verhalten von Nanosilber in Kläranlagen und dessen Einfluss auf die Nitrifikationsleistung in Belebtschlamm", Umweltwissenschaften und Schadstoff-Forschung, Volume 22, No. 5, 529-540
[344] Patent US 7799432
[345] Patent EP 1330501
[346] Patent EP 1992230
[347] L. C. Klein et al, "Development of a Hermetic Barrier Using Vinyl Triethoxysilane (VTEOS) and Sol-Gel Processing", Material Matters 2006, 1.3, 11
[348] Patent EP 2 147 020
[349] M. Jäger, U. Schmidt, "Die Barriere macht den Unterschied", Farbe und Lack 8, 2007, 20-25
[350] Press release of The European Chemical Industry Council "Practical nanomaterials definition needed to push forward next great innovation breakthroughs" Brussels, October 18, 2011 www.cefic.org.
[351] David B. Warheit, "Debunking Some Misconceptions about Nanotoxicology". Nano Lett. 2010, 10, 4777-4782
[352] S. M. Hussain et.al. "Toxicity Evaluation for Safe Use of Nanomaterials: Recent Achievements and Technical Challenges". Adv. Mater. 2009, 21, 1549-1559
[353] VdL-Guidance for the Handling of Nano-Objects at the Workplace German Paint and Printing Ink Association. June 2010. www.lackindustrie.de
[354] "Product stewardship on nanomaterials" and "Responsible Care", Download under www.vci.de

Acknowledgments

I would like to thank all the people who contributed by one way or another to this book. The first place belongs for sure to my family. They suffered a lot and contributed by their support and understanding.

Many thanks also go to my colleagues, which contributed literature search, lab work, pictures, ideas, analytical measurements and inspiration. Only to mention a few, I thank Christian Depner, Gorica Dimitrowska, Reiner Friedrich, Yvonne Geubert, Wolfgang Hechler, Thomas Hölzer, Matthias Koch, Sebastian Müller, Jörg Pahnke, Jörg Peschke, Jens Pradella, Sabine Renker and Johanna Schütz-Widoniak.

A very warm "thank you" also goes to my employer Merck KGaA especially to Monika Kursawe and Manfred Weigand for their support and the freedom to use photos, graphs or results from Merck.

No one is an island – without you, all this would not have been possible. Thank you!

Gerhard Jonschker

Heppenheim, Germany, January 2014

Author

Dr. Gerhard Jonschker, born 1966, studied chemistry and made its PhD in materials science at the University of the Saarland/Germany in the area of sol-gel chemistry. From 1995 on, he headed the department material- and process development and the application Center NMO at the Institute for New Materials in Saarbrücken/Germany. As one of the founders of the company Nanogate, he was responsible for research and development of functional coatings, holding a position as Director Nanocomposites within the management team.

Since 2004 he is with Merck KGaA, Darmstadt/Germany, working as senior manager in the field of performance materials with over 100 patent applications and a great number of developed products he contributed to the commercial application of sol-gel and nanotechnology.

Index

D

E

F

G

H

I

L

R

S

T

U

V

W

X

Y

Z

www.ingramcontent.com/pod-product-compliance
Lightning Source LLC
LaVergne TN
LVHW081321300926
849161LV00008B/1844